ELEMENTS OF
COMPILER DESIGN

By
Adesh Kumar Pandey
Department of Computer Science,
(Krishna Institute of Engg. and Tech.)
Ghaziabad (U.P.)

S.K. KATARIA & SONS®
Publisher of Engineering Books
4885/109, Prakash Mahal, Dr. Subhash Bhargav Lane,
Opposite Delhi Medical Association, Daryaganj, New Delhi-110002
Phone: +91-11-43551243; Mobile: +91-9871775858, 7678425209
e-mail: katariabook@gmail.com, katariabooks@yahoo.com
Website : www.skkatariaandsons.com
(e-books, subscriptions and bulk buying also available on our website)

Published by :
S.K. Kataria & Sons®
4885/109, Prakash Mahal, Dr. Subhash Bhargav Lane,
Opposite Delhi Medical Association, Daryaganj, New Delhi-110002
Phone: +91-11-43551243; Mobile: +91-9871775858, 7678425209
e-mail: katariabook@gmail.com, katariabooks@yahoo.com

First Edition: 2011, Second Edition: 2012, Reprint: 2026

Price: ₹ 395/-

Laser Typesetting by: Aisa Enterprises, Delhi

Printed at : A.S. Offset

Preface

The book addresses a full spectrum of compiler design. The emphasis is on solving the problems universally encountered in designing a compiler, regardless of the source language or target machine. Concepts of designing and practical implementation are the main focus of book.

The book is written as a text, with problems and exercises, with particular emphasis on the examples. In every chapter of this book, due consideration has been given to the pattern and standard of various universities.

To facilitate easy understanding, the text and problems have been put in simpler form. The emphasis is on simpler set of examples relating to every concept in each chapter. This will certainly help the readers to understand the subject in better way.

Suggestions are always welcome from the readers.

Author

Acknowledgement

The book itself is an acknowledgement to the intensity, drive and technical competence of many individuals, who have contributed to it.

I am thankful to Almighty God, without whose grace nothing is possible for any one. I am very thankful to my friends and students who have inspired me to write this book.

I extend my gratitude to my wife, Gunjan Pandey, my daughter Shreya, my parents and my in-laws for their strength and support, failing which, this book wouldn't have been complete.

I convey my thanks to publisher of this book, for their constructive critism and suggestions for improvement in the manuscript and bringing out this book.

Adesh K. Pandey

Contents

SECTION-A .... **1—93**

1. INTRODUCTION TO COMPILERS .... **3—36**

1.1. Introduction to Programming Languages 3

1.1.1. The History of Computer Programming Languages 3

1.1.2. Language Categories 5

1.2. High Level Programming Languages 5

1.2.1. Syntax and Semantics of the Languages 6

1.3. The Structure of the Program 6

1.3.1. Lexical Structure 7

1.3.2. Syntactic Structure 8

1.3.3. Contextual Structure 9

1.4. Introduction of Translators 9

1.4.1. Assembler 10

1.4.2. Compilers 10

1.4.3. Interpreter 10

1.4.4. Loaders 11

1.4.5. Linkers 11

1.5. Need of Translators 11

1.5.1. Need of Compilers 11

1.5.2. Compiled Languages 12

1.5.3. Interpreted Languages 12

1.6. What is the Challenge in Compiler Design ? 13

1.7. Compiler Requirements 14

1.8. The Components of a Compiler 14

1.8.1. Modern Approach of Compiler Design 16

1.8.2. Lexical Analyzer (Scanner) 17

1.8.3. Syntax Analyzer (Parser) 18

1.8.4. Semantic Analyzer 19

1.8.5. Symbol Table Manager 20

1.8.6. Intermediate Code Generator 21

1.8.7. Code Optimization 21

1.8.8. Storage-Allocation 22

1.8.9. Code Generation 22

1.8.10. Peephole Optimization 23

1.8.11. Error Handler 23

1.9. The Number of Compiler Passes 24

1.9.1. Multi-pass Compilers 25

1.9.2. One-Pass Compilers 25

1.9.3. Advantages and Disadvantages for Both Single and Multipass Compilers 26

1.9.4. Difference Between Single Pass Compiler and Multi-Pass Compiler 26

1.10. Introduction to Bootstrapping (Compilers) 27

1.10.1. Notations 27

1.10.2. Bootstrapping (Compiling Compilers) 28
1.10.3. Porting 29
1.11. Lex 30
1.11.1. Lex File Format 30
1.11.2. Lex Predefined Variables 31
1.11.3. Lex Source Program 31
1.12. Introduction to Flex (Fast Lexical Analyzer Generator) 32
1.13. YAAC 34
1.13.1. YACC File Format 34
1.13.2. LEX and YAAC As Team 35
■ *Exercise* 36

2. INTRODUCTION TO FINITE AUTOMATA AND REGULAR EXPRESSION .... 37—84

2.1. Introduction to Automata 37
2.2. Alphabets 38
2.3. Strings 38
2.4. Languages 39
2.4.1. Concatenation of Languages 40
2.4.2. Languages in Set Forms 40
2.5. Deterministic Finite Automata 41
2.5.1. Processing of Strings by DFA 41
2.5.2. Simpler Notations for DFA's 42
2.5.3. The Language of a DFA 43
2.6. Nondeterminism 49
2.6.1. Why Nondeterminism 49
2.6.2. Extended Transition Function for NFA 53
2.7. The Language of NFA 54
2.8. Transformation of NFA to DFA 54
2.8.1. Procedure For Converting NFA to Equivalent DFA 55
2.9. NFA with ∈-transitions 58
2.10. Acceptance of a string by NFA with ∈-transitions 59
2.11. Conversion of NFA with ∈-transitions to NFA without ∈-transitions 59
2.12. Minimization of DFA's 62
2.12.1. Minimization Algorithm for DFA 63
2.13. Regular Expression 68
2.13.1. Definition of Regular Expression 68
2.13.2. Building Regular Expression 68
2.13.3. Inductive Step 69
2.14. Comparative Study of Regular Expression, Regular Sets and Finite Automata 74
2.15. Construction of FA for Regular Expression 74
2.16. Algebraic Laws for Regular Expressions 77
2.16.1. Associativity and Commutativity 78
2.16.2. Identities 78
2.16.3. Distributed Law 78
2.16.4. The Idempotent Law 79
2.16.5. Laws for Closure 79
2.16.6. Identities for Regular Expression 79
■ *Exercise* 83

3. LEXICAL ANALYSIS ..... 85—93

3.1. Introduction 85
3.2. Reasons for Decouple the Scanner from the Parser 86
3.3. The String Table 87
3.4. Strategies in Lexical Analyzer Construction 87
 3.4.1. Issues in Lexical Analyzer 88
 ■ *Exercise* 93

SECTION-B ..... 95—210

4. SYNTAX ANALYSIS ..... 97—134

4.1. Grammars 97
 4.1.1. Derivations and Their Variations 98
 4.1.2. Classic BNF Notation for Productions 99
4.2. Chomsky Hierarchy 99
 4.2.1. Unrestricted (Type-0) Grammars 101
 4.2.2. Context Sensitive Grammar 102
 4.2.3. Context-Free Grammars 102
 4.2.4. Regular Grammars 103
4.3. Left Most and Right Most Derivations 104
4.4. Parse Tree 104
 4.4.1. Definition 104
 4.4.2. The Yield of Parse Tree 104
4.5. Phrase Structure and Lexical Structure 115
4.6. Syntax Diagrams 116
4.7. Reduction of Context Free Grammar 117
 4.7.1. Eliminating Use Less Symbols 117
 4.7.2. Removal of Unit Production 119
 4.7.3. Removal of ∈-productions and Nullable Non-terminal 122
 4.7.4. Cycle-free Grammar 124
4.8. Ambiguity in Grammars and Languages 125
 4.8.1. Removing Ambiguity from Grammars 127
 4.8.2. Inherent Ambiguity 130
4.9. Syntax Analysis 131
4.10. Role of Parsers 132
 ■ *Exercise* 134

5. TOP-DOWN PARSING TECHNIQUES ..... 135—161

5.1. Top-Down Parsing 135
 5.1.1. Backtracking a Problem of Top-down Parsing 137
5.2. Issues of CFG for the Programming Languages 139
 5.2.1. Ambiguity 139
 5.2.2. Left Recursion and its Removal 141
 5.2.3. Left-Factoring 143
5.3. Recursive-Decent Parsing 145
5.4. Top-down Predictive Parsers 148
 5.4.1. First and Follow 149

5.4.2. Following Algorithm can be used to Compute First (X) for a Grammar Symbol X 150
5.4.3. Algorithm to Calculate the Follow (X) for a Non-terminal X 150
5.4.4. Constructing M-table/Predictive Parsing Table 151
5.5. LL (1) Grammars 153
5.5.1. Advantages and Disadvantages of LL(1) Parsing 154
■ *Exercise* 160

6. BOTTOM-UP PARSING TECHNIQUES .... 162—210

6.1. Introduction to Bottom-Up Parsing 162
6.1.1. Shift-Reduce Parsing 163
6.2. Handle of a Right Sentential Form 163
6.3. Shift Reduce Parsers 165
6.3.1. Shortcomings of Shift-reduce Parsing 167
6.3.2. Types of Shift-Reduce Parsers 168
6.4. Operator Precedence Parsers 168
6.4.1. Operator Precedence Relation 169
6.4.2. Developing Some Other Operator-Precedence Relation 171
6.4.3. Operator Precedence Parsing 173
6.4.4. Advantages and Disadvantages of Operator Precedence Parsing 177
6.5. LR Parsers 178
6.5.1. The LR Parsing Algorithm 178
6.5.2. LR Grammars 180
6.5.3. Augmented Grammar 181
6.5.4. Constructing the SLR Parsing Tables (or Tables for LR(0) Parsers) 181
6.5.5. Limitation of SLR Parser 195
6.5.6. Construction of LR(1) Parsing Table 197
6.5.7. LALR Parsing 204
6.5.8. Conflicts in Shift-Reduce Parsing 208
6.6. A Comparison of Predictive Parser with Shift-Reduce Parsers 208
■ *Exercise* 209

SECTION-C .... 211—257

7. SYNTAX-DIRECTED TRANSLATION .... 213—239

7.1. Introduction of Syntax-Directed Translation 213
7.2. Basic Definition of Attributes and Semantic Rules 215
7.3. Two-Pass Vs One Pass Compilation and Syntax Directed Translation 216
7.4. Syntax-Directed Definition 217
7.4.1. Synthesized Attributes 217
7.4.2. Inherited Attributes 218
7.4.3. Attribute Grammars 219
7.5. Syntax Tree 220
7.5.1. Structure of Syntax Trees 221
7.5.2. Construction of Syntax Tree 222
7.5.3. Annotated Parse Tree 223
7.5.4. Directed Acyclic Graph for Expression 224
7.5.5. A Syntax Directed Definition For Constructing Syntax Trees 224

7.6. S-Attributed Definitions 226
7.6.1. L-Attributed Definitions 226
7.6.2. Translation Schemes 227
7.7. Implementation Issues of Syntax-directed Translation 229
7.7.1. Top-down Translation 229
7.8. Bottom-Up Evaluation of S-Attributed Definitions 233
7.8.1. Triggering Semantic Actions in a LL(1) Parser 234
7.9. Implementation of Syntax-Directed Translation 235
7.9.1. Semantic Actions 236
■ *Exercise* 239

8. SYMBOL TABLE MANAGEMENT .... 240—250

8.1. Introduction to Symbol Table 240
8.2. Information Caching 241
8.3. What Should be Stored ? 242
8.4. Symbol Table Organization 244
8.5. Scoping 247
8.5.1. Scope-by-Number 248
8.5.2. Scope-by-Location 249
■ *Exercise* 250

9. ERROR DETECTION AND RECOVERY .... 251—257

9.1. Introduction to Error Handling 251
9.2. Categories and Goals of Error Handling 251
9.3. Error Recovery Strategies 252
9.4. Recovery from Lexical Phase Errors 252
9.5. Recovery from Syntax Error 253
9.6. Error Recovery in Predictive Parsing 254
9.6.1. Panic-Mode Error Recovery in LL(1) Parsing 254
9.7. Error Recovery in LR Parsing 255
9.8. Recovery from Semantic Errors 257

SECTION-D .... 259—322

10. INTERMEDIATE CODE GENERATION .... 261—287

10.1. Introduction to Intermediate Code Generation 261
10.2. Choosing an Intermediate Languages 262
10.2.1. Intermediate Languages: Design Issues 263
10.2.2. High-Level Intermediate Representations 263
10.3. Three-Address Code 264
10.3.1. Different Kinds of Three-Address Statement 264
10.3.2. Implementing Three-Address Instructions as Quadruples 266
10.3.3. Implementing Three-Address Instructions as Triples 267
10.4. P-Code 268
10.5. Byte Code 268
10.6. Intermediate Code for Control Flow 269
10.6.1. Syntax-Directed Translation for Intermediate Code Generation 269

10.6.2. Intermediate Code Generation for Boolean Expression ... 272
10.6.3. Code Generation for Conditions ... 273
10.6.4. Code Generation for Assignment ... 275
10.6.5. Back Patching ... 275
10.6.6. Code Generation for Conditions in Single Pass ... 277
10.6.7. Intermediate Code Generation for Case Statement ... 278
10.6.8. Code Generation for Function Calls ... 279
10.6.9. Accessing Array Elements ... 280
10.6.10. The Grammar for Array Reference ... 282
10.6.11. The Translation Scheme for Addressing Array Elements ... 282
■ *Exercise* ... 286

11. CODE OPTIMIZATION AND CODE GENERATION ... 288—322

11.1. Introduction to Code Optimization ... 288
11.2. Optimization Strategy ... 289
11.3. Loop Optimization ... 289
11.3.1. Analysis of Program for Loop Optimization ... 289
11.3.2. Loop Detection ... 293
11.3.3. Reducible Flow Graph ... 293
11.3.4. Pre-Headers ... 295
11.4. Global Data-Flow Analysis ... 295
11.5. Elimination of Induction variables ... 300
11.6. Elimination of Common Subexpression ... 300
11.6.1. Elimination of Local Common Subexpression ... 301
11.7. Loop Unrolling ... 303
11.8. Introduction to Code Generation ... 303
11.9. Machine Architecture Currently in Use ... 304
11.10. Major Issues of Code Generation ... 305
11.11. A Machine Model and Code Generation ... 307
11.12. A Simple Code Generator ... 308
11.13. Code Generation From DAG ... 310
11.14. Register Allocation ... 312
11.14.1. Basic Blocks ... 313
11.14.2. Spilling ... 313
11.14.3. Coalescing ... 314
11.14.4. Live Range Splitting ... 314
11.14.5. Interference Graph ... 314
11.14.6. Register Allocation Approaches ... 315
11.15. Algebraic Properties ... 319
11.16. Peephole Optimization ... 319
■ *Exercise* ... 320

APPENDIX : QUICK REVIEW OF COMPILER DESIGN ... 323—341

EXAMINATION PAPERS ... 342—343

SOLVED EXAMINATION PAPER ... 344—350

INDEX ... 351—354

SECTION-A

- Introduction to Compilers
- Introduction to Finite Automata and Regular Expression
- Lexical Analysis

SYLLABUS

Introduction to Compilers: Compilers and Translators, Need of Translators, Structure of Compiler: Its Different Phases, Compiler Construction Tools.

Lexical Analysis : Role of Lexical Analyzer, Design of Lexical Analyzer, Regular Expressions, Specification and Recognition of Tokens, Input Buffering, A Language Specifying Lexical Analyzer. Finite Automata, Conversion from Regular Expression to Finite Automata, and Vice-versa, Minimizing Number of States of DFA, Implementation of Lexical Analyzer.

1

Chapter

INTRODUCTION TO COMPILERS

INSIDE THIS CHAPTER

1.1. Introduction to Programming Languages **1.2.** High Level Programming Langauges **1.3.** The Structure of the Program **1.4.** Introduction of Translators **1.5.** Need of Translators **1.6.** What is the Challenge in Compiler Design ? **1.7.** Compiler Requirements **1.8.** The Components of a Compiler **1.9.** The Number of Compiler Passes **1.10.** Introduction to Bootstrapping (Compilers) **1.13.** Lex (Lexical Analyzer Generator) **1.14.** Introduction to Flex (Fast Lexical Analyzer Generator) **1.15.** YAAC.

1.1. INTRODUCTION TO PROGRAMMING LANGUAGES

We can easily analyse it, that every concept of designing of computer softwares is inspired by the life. The same thought is applied in the development of programming languages. **Like Natural Languages (Hindi, English, ...) Programming Languages are also notations, with which common man can communicate to the computers**.

Coded language used by programmers to write instructions that a computer can understand to do what the programmer (or the computer user) wants. The most basic (called low-level) computer language is the machine language that uses binary ('1' and '0') code which a computer can run (execute) very fast without using any translator or interpreter program, but is tedious and complex. The high-level languages (such as Basic, C, Java) are much simpler (more 'English-like') to use but need to use another program (a compiler or an interpreter) to convert the high-level code into the machine code, and are therefore slower. There are dozens of programming languages and new ones are being continuously developed. Also called computer language.

According to **Nico Habermann** *"An ideal language allows us to express what is useful for the programming task and at the same time makes it difficult to write what leads to incomprehensible or incorrect programs"*.

According to **Robert Harper** *"Good language makes it easier to establish, verify, and maintain the relationship between code and its properties"*.

1.1.1. The History of Computer Programming Languages

In 1945, John Von Neumann developed two important concepts that directly affected the path of computer programming languages. The first was known as "shared program technique". The technique states that the actual computer hardware should be simple and need not to be hand-wired for each program. Instead, complex instructions should be used to control the hardware, allowing it to be reprogrammed much faster.

The second concept was extremely important for development of programming languages. Von Neumann called it "Conditional Control Transfer". This idea gave rise to the notation of subroutines, or small block of code that could be jumped to in any order, instead of a single set of chronologically ordered steps for the computer to take.

In 1949, a few year, after Von Neumann's work, the language short code appeared. It was the first computer language for electronic devices and it required the programmer to change its statements in to 0's and 1's by hand. In 1957, the first of the major languages appeared in the form of FORTRAN. Its name stands for FORmula TRANslating system. The language was designed at IBM for scientific computing. Though FORTRAN was good at handling numbers, it was not so good at handling input and output, which mattered most to business computing. Business computing started to take off in 1959 and because of this COBOL was developed. COBOL statements also have a very english-like grammar, making it quite easy to learn. All of these features were designed to make it easier for the average business to learn and adopt it.

In 1958, John McCarthy of MIT created the LIST Processing (or LISP) language. It was designed for Artificial Intelligence (AI) research. Because it was designed for such a highly specialized field, its syntax has rarely seen before or since.

The Algol language was created by a committee for scientific use in 1958. It's major contribution is being the root of the tree that has led to such languages as Pascal, C, C++ and Java. It was also the first language with a formal grammar, known as Backus-Naur Form or BNF.

Pascal was begun in 1968 by Niklaus Wirth. Its development was mainly out of necessity for a good teaching tool. Wirth later created a successor to pascal, modula-2, but by the time it appeared, C was gaining popularity and users at a rapid pace. C was developed in 1972 by Dennis Ritchie while working at Bell Labs in New Jersy.

In the late 1970's and early 1980's, a new programming method was being developed. It was known as object oriented programming, or OOP. Bjarne developed extensions to C known as "C with classes." This set of extensions developed into the full-feature langauge C++, which was released in 1983.

In 1994, the Java project team changed their focus to the web. The next year, Netscape licensed Java for use in their internet browser, Navigator. In 1964, BASIC was developed by the John Kemeny and Thomas Kurtz. Microsoft has extended BASIC in its Visual Basic (VB) product. In 1987 Perl was developed by Larry Wall, it has very strong text matching functions which make it ideal for gateway designing for a web interface.

Programming language have been under development for years and will remain so for many years to come. They got their start with a list of steps to wire a computer to perform a task. These steps eventually found their way into software and began to acquire newer and better features. The first major languages were characterized by the simple fact that they were intended for one purpose only, while the languages of today are differentiated by the way they are programmed in, as they can be used almost for any purpose. And perhaps the language of tomorrow will be more natural with the invention of quantum and biological computers.

1.1.2. Language Categories

Programming languages are often categorized into four categories:

(*i*) Imperative languages
(*ii*) Functional languages
(*iii*) Logical languages or Declarative languages
(*iv*) Object oriented language.

The basic architecture of computers has a crucial effect on language design. Most of the popular languages of the past 35 years have been designed around the prevalent computer architecture, called the Von Neumann architecture, after one of its originators, John Von Neumann. These languages are called **imperative languages**. Programmers using imperative languages must deal with the management of variables and assignment of values to them. The results of this are increased efficiency of execution but laborious construction of programs.

The Functional Programming paradigm, which is based on mathematical functions, is the design basis for one of the most important nonimperative styles of languages. This style of programming is supported by functional or applicative, programming language. LISP is first functional programming language. The objective of the design of a functional programming language is to mimic mathematical functions to the greatest extent possible. This results in an approach to problem solving that is fundamentally different from methods used with imperative languages.

Programming that uses a form of symbolic logic as a programming language is often **called logic programming**, and **languages based on symbolic logic are called logic programming languages or declarative languages**. Prolog is example of logical language.

Procedure-oriented programming, which was the most popular software development paradigm in the 1970's, focuses on subprograms and sub-program libraries. Data are set to sub-programs for computations. Data-oriented programming focuses on abstract data types. **Languages that support data-oriented programming are often called object-oriented-based languages**. A language that is object oriented must provide support for three key languages features; abstract data types, inheritance and a particular kind of dynamic binding.

1.2. HIGH LEVEL PROGRAMMING LANGUAGES

Definition Advance computer programming language that is not limited by the type of computer or for one specific job and more easily understood.

In other words, a programming language designed to suit the requirement of the programmer and independent of the internal machine code of any particular computer, is called high level programming language. For example BASIC, FORTRAN, C++, Java and Visual Basic are high level programming languages.

In contrast, **low-level language**, such as assembly languages closely reflect the machine codes of specific computers and, therefore, described as **machine-oriented languages**. Unlike low-level languages, high-level languages are relatively easy to learn because the instructions bear a close resemblance to everyday language, and because the programmer does not require a detailed knowledge of the internal working of the computer. Each instruction in a high-level language is equivalent to several machine-code instructions. High-level programs are, therefore, more compact than equivalent low-level programs. However, each high-level instruction must be translated into machine code, by either a compiler or an interpreter, before it can be executed by a computer. **High-level languages are designed to be portable,** that is programs written in a high-level language can be run on any computer that has a compiler or Interpreter for those particular languages.

1.2.1. Syntax and Semantics of the Languages

It is very important for any programming language that it is easily understandable to the programmers and they are able to determine how expressions, statements and program module of the language are formed. Programmers must be able to write the software by the help of manual of programming language.

Definition

"So syntax of a programming language is the form of it's expressions, statements and program modules".

And the semantic of a programming language is the meaning given to the various syntactic structure.

For example in C we can take the decision by the following statement:

```
If (Conditional expression)
{
    ................
    Statements ;
    ...............
    ...............
}
```

The meaning of above syntactical structure is that if conditional expression have true value the statements of the blocks will be executed.

We have seen that, like any other natural language computer languages behave as communication media:

(*i*) human to machine

(*ii*) machine to machine, and

(*iii*) machine to human.

Unlike humans, machines are not fault tolerant, hence programming is often viewed as a "Complex task".

Before switching on the design of compiler, it is very much required to know about the structure of the programs.

1.3. THE STRUCTURE OF THE PROGRAM

We can analyse a computer program on four levels:

(*i*) Lexical structure level

(*ii*) Syntactic structure level
(*iii*) Contextual structure level
(*iv*) Semantic structure level.

1.3.1. Lexical Structure

The lexical level of any program is lowest level. At this level computer programs are viewed as simple sequence of lexical items called tokens. In fact at this level programs are considered as different groups of strings that make sense. Such a group is called a token. A token may be composed of a single character or a sequence of characters.

We can classify tokens as being either:

Identifier

Definition In a particular language names chosen to represent data items, functions and procedures, etc. are called identifiers.

Considerations for the identifier may differ from language to language, for example, some computer language may support case sensitivity and others may not. Number of characters in identifiers are also dependent on the design of the computer language.

Keywords

Definition Keywords are the Names chosen by the language designer to represent facts of particular language constructs which can not be used as identifiers (sometimes referred to as reserved words).

Let us see example of language C

```
int id ;
```

here id is identifier.

```
Struct id1 {
    int a;
    char b;
          }
```

here struct is keyword and id1 is identifier.

Operators

Definition In some languages, special "keywords" used to identify operations to be performed on operands, that is math operators.

For example in langauge C:

(*a*) Arithmathematical operators (+, –, ∗, /, %)
(*b*) Logical operators (>, = >, <, < =, = =, ! , ! =)

Separators

Definition These are punctuation marks used to group together sequences of tokens that have a "unit" meaning.

When outputting text it is often desirable to include punctuation, where these are also used (within the language) as separators, we must precede the punctuation character with what is called an **escape character** (usually a back slash '\').

Literals

Definition Some languages support literals which denote direct values.

For example:

(*i*) Numbers that is 1, –123, 3.14, 6.02.

(*ii*) Characters, that is 'a'

(*iii*) String, that is "Some text".

Comments

Definition We know that a good program is one that is understandable. We can increase understandability by including meaningful comments into our code.

Comments are omitted during processing. The start of a comment is typically indicated by a key word (that is comment) or a separator, and may also be ended by a separator.

1.3.2. Syntactic Structure

Definition The syntactic level describes the way that program statements are constructed from tokens. This is always very precisely defined in terms of a context-free grammars.

The best known examples are BNF (Backus Naur Form) or EBNF (Extended Backus Naur Form). Syntax may also be described using a syntax tree.

Let us see an EBNF example:

```
      <Sum>      ⇒  <operand>
                    <operator>
                    <operand>
   <operand>     ⇒  <number> / (<sum>)
   <operator>    ⇒  + / -
```

The syntax tree for above EBNF will be as follows:

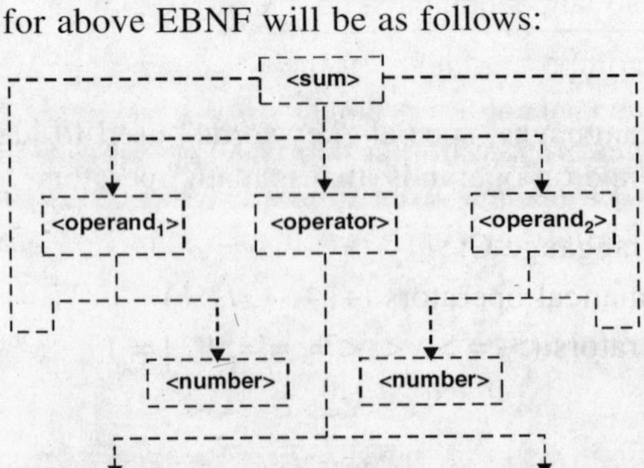

Fig. 1.1.

1.3.3. Contextual Structure

Definition The contextual level of analysis is concerned with the "Context" in which program statements occur. Program statements usually contain identifier whose value is dictated by earlier statements (especially in the case of the imperative or object oriented paradigms). Consequently the meaning of a statement is dependent on "what has gone before", that its context. Context also determines whether a statement is "legal" or not (context conditions – a data item must "exist before it can be used).

Let us see an example of language C.

```
int C = a + b ; }       not valid, since a and b are not defined
int  a;
int b ;                 This is valid.
int C = a + b ;
```

1.4. INTRODUCTION OF TRANSLATORS [MDU 2008, 2009, 2010]

Let us consider the translation from French to English. When translating from French to English, the translator must comprehend each French sentence and produce a corresponding sentence in English. Human translator receives a text that is written in the source language (here it is French) and he decomposes it in sentences and words using his knowledge of the language. This "knowledge" is stored in his "internal memory", that is his brain, and it consists of the vocabulary of the language (used to recognize the words) and the grammar of the language (used to analyze the correctness of the sentences). Once the translator recognizes all the words and understands the meaning of the sentence, he starts the process of translating.

TIPS

For any translator, the most difficult part comes from the fact that he or she can not translate word by word because the two language do not have the same sentence pattern's and in spite of this the meaning of the sentence in both languages must be the same.

A translator in computer science follows same sequence of operation that is performed by a human translator. Fortunately for the construction of translators, for the computer languages are far simpler than any spoken language, hence the task of writing a program that translates from source language to the target language is far more simple than translating from French to English. So finally we can define translators as follows:

Definition "Translators are the software programs, which can translate programs written into source language to target language".

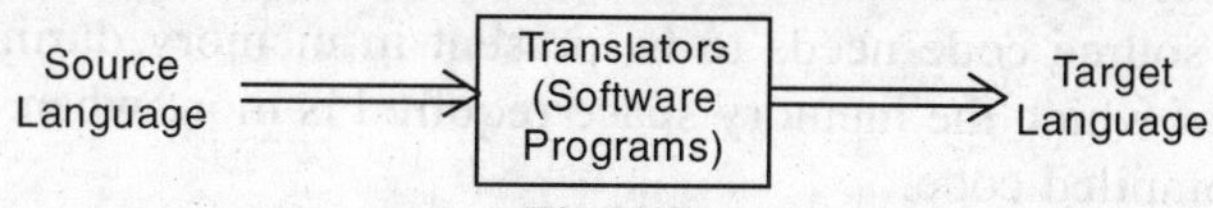

Fig. 1.2.

Let us discuss some translators, such as Assemblers, Interpreters and Compilers.

1.4.1. Assembler

Definition An Assembler is a translator for an assembly language of a computer.

An assembly language is the lowest level programming language for a computer. It is peculiar to a certain computer system and is hence machine-dependent.

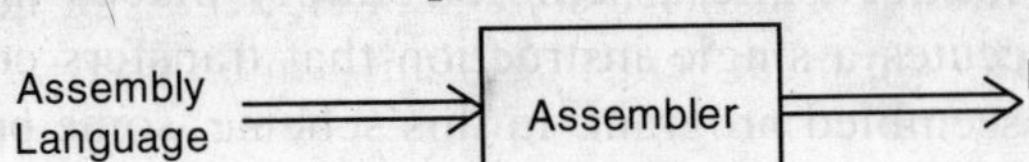

Fig. 1.3.

1.4.2. Compilers **[MDU 2009]**

Definition Compilers are system programs that translates an input program in a high level language into its machine language equivalent Checks for errors optimizes the code prepares the code for execution.

Compilers are the translators, which translate a program written in high-level langauge (like C++, FORTRAN, COBOL, PASCAL) into machine code for some computer architecture (such as on Intel pentium architecture). The generated code can be later executed many times against different data each time.

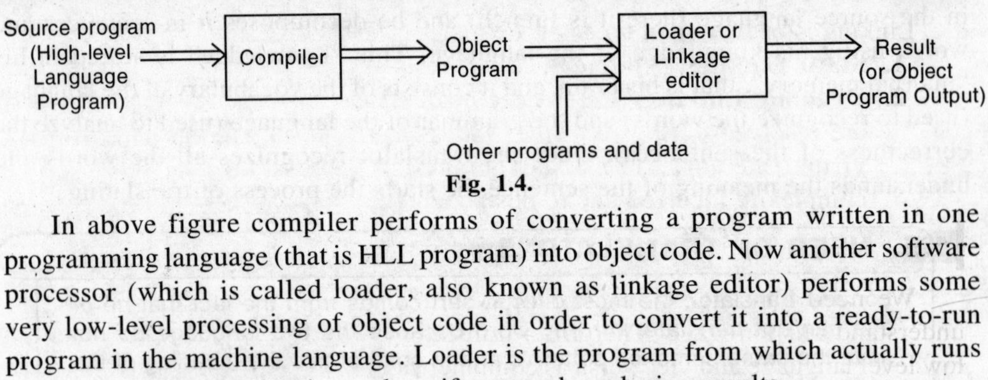

Fig. 1.4.

In above figure compiler performs of converting a program written in one programming language (that is HLL program) into object code. Now another software processor (which is called loader, also known as linkage editor) performs some very low-level processing of object code in order to convert it into a ready-to-run program in the machine language. Loader is the program from which actually runs on the computer, reading input data, if any, and producing results.

1.4.3. Interpreter

An interpreter reads an source program written a in high-level programming language as well as data for this program, and it runs the program against the data to produce some results.

(*i*) Is a systems program that looks at and executes programs on a line-by-line basis rather than producing object code.

(*ii*) Whenever the programs have to be executed repeatedly, the source code has to be interpreted every time. In contrast the compiled programs create an object code and this object code will be executed.

(*iii*) Entire source code needs to be present in memory during execution as a result of which the memory space required is more when compared to that of a compiled code.

(*iv*) Used in test environments as overhead of compilation is not there.

1.4.4. Loaders

Definition Loader is system programs. Loader loads the binary code in the memory ready for execution. Transfer the control to 1st instruction. Loaders are responsible for locating program in the main memory every time it is being executed.

There are various loading schemes:

1. Assemble-and-go loader: The assembler simply places the code into memory and the loader executes a single instruction that transfers control to the starting instruction of the assembled program. In this scheme, some portion of the memory is used by the assembler itself which would otherwise have been available for the object program.

2. Absolute Loader: Object code must be loaded into the absolute addresses in the memory to run. If there are multiple subroutines, then each absolute address has to be specified explicitly.

3. Relocating Loader: This loader modifies the actual instructions of the program during the process of loading a program so that the effect of the load address is taken into account.

1.4.5. Linkers

Definition Linking is the process of combining various pieces of code and data together to form a single executable that can be loaded in memory.

Linking is of two main types:

1. **Static Linking:** All references are resolved during loading at linkage time.
2. **Dynamic Linking:** References made to the code in the external module are resolved during run time. Takes advantage of the full capabilities of virtual memory. The disadvantage is the considerable overhead and complexity incurred due to postponement of actions till run time.

1.5. NEED OF TRANSLATORS

We need language translator in system software because computer can't understand our language. Therefore, translator converts high level language into low level language and vice versa as computer can only understand in binary format.

1.5.1. Need of Compilers

For performing different task on the computer, a compiler is communication medium that translate source language into target language. Compiler allows the user to perform customized task on machine.

Every Microprocessor has its own Machine Code. Our extremely simple microprocessor had its own machine code, Intel would have its own code and so would Motorola. For human being it is almost impossible to remember the machine code and to develop even a small application using machine codes. That's where the high level languages come in.

In C/C++, to add two numbers you would write the following code:

```
int i;
int j;
```

```
int k;
k = i + j;
```

It is far more viable to write applications in a high level language like C/C++ then to write the same application in machine language. But the problem is that the Computer would not understand anything but the machine language—what we need is some sort of a translator that will take the High level C/C++ code and translate it into machine code. Such a "translator" is called a Compiler.

The compiler is a program that takes in the C/C++ file(s) as an input and outputs an executable file, that can then be directly run on the host computer.

1.5.2. Compiled Languages

1. One of the biggest advantages of compiled languages is their execution speed. A program written in C/C++ runs 30-70% faster, then an equivalent program written in Java.
2. Compiled code also, takes less memory as compared to an interpreted program.
3. On the down side—a compiler is much more difficult to write than an interpreter.
4. A compiler does not provide much help in debugging a program.
5. The executable compiled code is much bigger in size than an equivalent interpreted code *e.g.*, a C/C++ .exe file is much bigger than an equivalent Java.class file.
6. Compiled programs are targeted towards a particular platform and hence are platform dependent.
7. Compiled programs do not allow security to be implemented within the code that is a compiled program can access any area of the memory, and can do whatever it wants with PC (most of the viruses are made in compiled languages).
8. Due to lose security and platform dependence—a compiled langauge is not particularly suited to be used to develop Internet or web-based applications.

1.5.3. Interpreted Languages

1. Interpreted language provides excellent debugging support. A Java programmer only spends a few minutes fixing a "Null pointer exception", because Java runtime not only specifies the nature of exception but also gives the exact line number and function call sequence where the exception occurred. This facility is something that a compiled language can never provide.
2. Another advantage is that Interpreters are much easier to build, then a compiler.
3. One of the biggest advantages of Interpreters is that they make platform-independence possible.
4. Interpreted language also, allows high degree of security - something badly needed for an Internet application.
5. An intermediate language code size is much smaller than a compiled executable code.
6. Platform independence and tight security are the two most important factors that make an interpreted language ideally suited for Internet and web-based applications.

7. Interpreted languages have some serious drawbacks. The interpreted applications take up more memory and CPU resources. This is because in order to run a program written in interpreted language; the corresponding interpreter must be run first. Interpreters are sophisticated, intelligent and resource hungry programs and they take up lot of CPU cycles and RAM.
8. Due to interpreted application's decode-fetch-execute cycle; they are much slower than compiled programs.
9. Interpreters also, do lot of code-optimization, security violation checking at run-time; these extra steps take up even more resources and further slow the application down.

1.6. WHAT IS THE CHALLENGE IN COMPILER DESIGN ?

Compiler writing is not an easy job. It is very challenging and you must have lot of knowledge about various fields of computer science. Let us discuss some of challenges:

(a) Many Variations

(*i*) Many programming languages (FORTRAN, C++, Java)

(*ii*) Many programming paradigms (that is, object-oriented, functional, logical)

(*iii*) Many computer architectures (that is MIPS, SPARC, Intel, Alpha)

(*iv*) Many operating systems (that is Linux, Solaris, Windows)

(b) Qualities of a Good Compiler (in Order of Importance)

(*i*) The compiler itself must be bug-free.

(*ii*) It must generate correct machine code.

(*iii*) The generated machine code must run fast.

(*iv*) The compiler itself must run fast (Compilation time must be proportional to program size).

(*v*) The compiler must be portable (that is modular, supporting, separate compilation).

(*vi*) It must print good diagnostics and error messages.

(*vii*) The generated code must work well with existing debuggers.

(*viii*) Must have consistent and predictable optimization.

(c) Building a Compiler Requires Knowledge of

(*i*) Programming languages (parameter passing, variable scoping, memory allocation etc.)

(*ii*) Theory of automata, context free languages etc.

(*iii*) Algorithms and data structure (hash tables, graph algorithms, dynamic programming etc.)

(*iv*) Computer Architecture (Assembly programming)

(*v*) Software Engineering.

(d) Instruction Parallelism

Handling of out of order execution and branch prediction is a challenge in compiler design.

(e) Parallel Algorithms

Incorporating the concept of grid computing and multi-core computers.

1.7. COMPILER REQUIREMENTS

As we will be implementing several compilers, it is important to understand which requirement compilers should satisfy. We discuss in each case to what extent it is relevant to this course.

(i) **Correctness:** Correctness is absolutely paramount. A buggy compiler is next to useless in practice. Since, we cannot formally prove the correctness of your compilers, we use extensive testing. This testing is end-to-end, verifying the correctness of the generated code on sample inputs.

(ii) **Efficiency:** In a production compiler, efficiency of the generated code and also efficiency of the compiler itself are important considerations.

The early emphasis on correctness has consequences for your approach to the design of the implementation. Modularity and simplicity of the code are important for two reasons: first, your code is much more likely to be correct, and, second, you will be able to respond to changes in the source language specification from lab-to-lab much more easily.

(iii) **Interoperability:** Programs do not run in isolation, but are linked with library code before they are executed, or will be called as a library from other code. This puts some additional requirements on the compiler, which must respect certain interface specifications. Your generated code will be required to execute correctly in the environment on the lab machines. This means that you will have to respect calling conventions early on (for example, properly save callee-save registers) and data layout conventions later, when your code will be calling library functions.

(iv) **Usability:** A compiler interacts with the programmer primarily when there are errors in the program. As such, it should give helpful error messages. Also, compilers may be instructed to generate debug information together with executable code in order help users debug runtime errors in their program.

(v) **Retargetability:** At the outset, we think of a compiler of going from one source language to one target language. In practice, compilers may be required to generate more than one target from a given source (for example, x86-64 and ARM code), sometimes at very different levels of abstraction (for example, x86-64 assembly or LLVM intermediate code).

1.8. THE COMPONENTS OF A COMPILER [MDU 2009]

A compiler is composed of several components also called phases, each performing one specific task. Each of these components performs at a slightly higher language level than the previous. The compiler must take an arbitrarily-formed body of text, and translate it into a binary stream that computer can understand.

The complete compilation procedure can be divided into six phases and these phases can be regrouped in two parts, as follows:

1. Analysis

In this part source program is broken into constituent pieces and creates an intermediate representation. Analysis can be done in three phases :

(*a*) Lexical Analysis

(*b*) Syntax Analysis

(*c*) Semantic Analysis.

2. *Synthesis*

Synthesis constructs the desired target program from intermediate representation. Synthesis can be done in following three phases:

(*a*) Intermediate code generation

(*b*) Code optimization

(*c*) Code generator.

TIPS

Every phase of compilation can intract with a special data structure called symbol table and with error handler.

Now let us draw the block diagram of compiler.

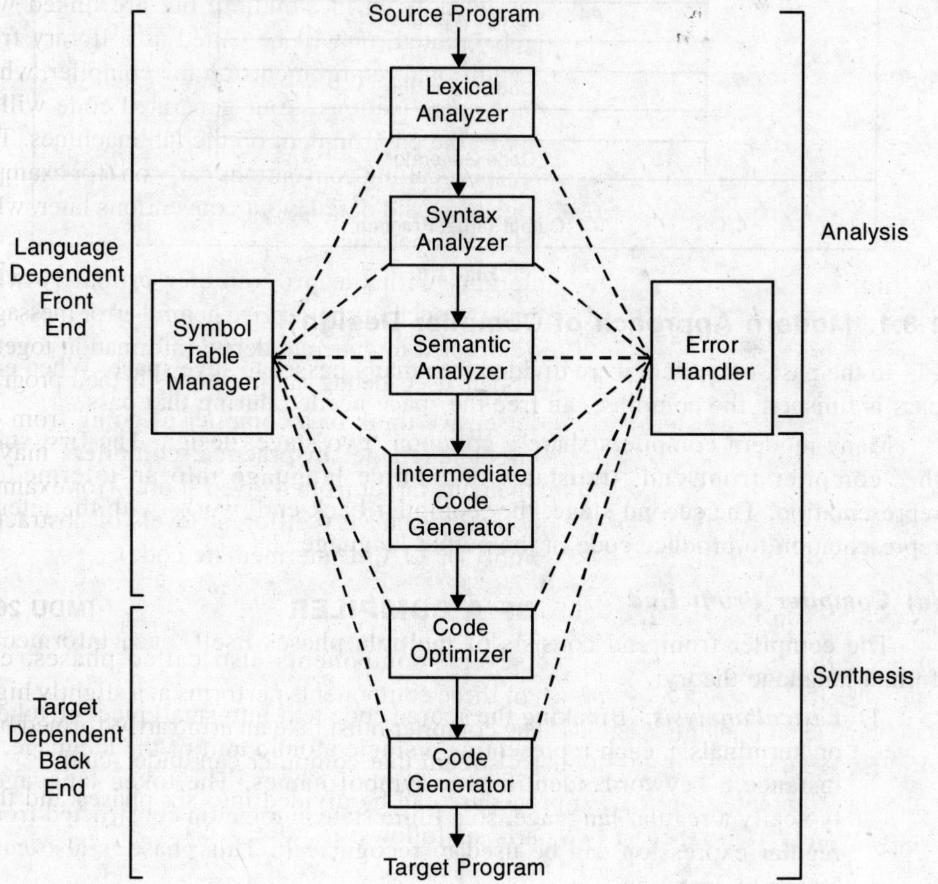

Fig. 1.5.

Sometimes we divide synthesis as follows:

(*a*) Intermediate code generation

(*b*) Code optimization
(*c*) Storage allocation
(*d*) Code generation.

Here we are looking storage allocation as a different phase, but in previous assumption it was the part of code generation.

Now block diagram can be drawn as follows:

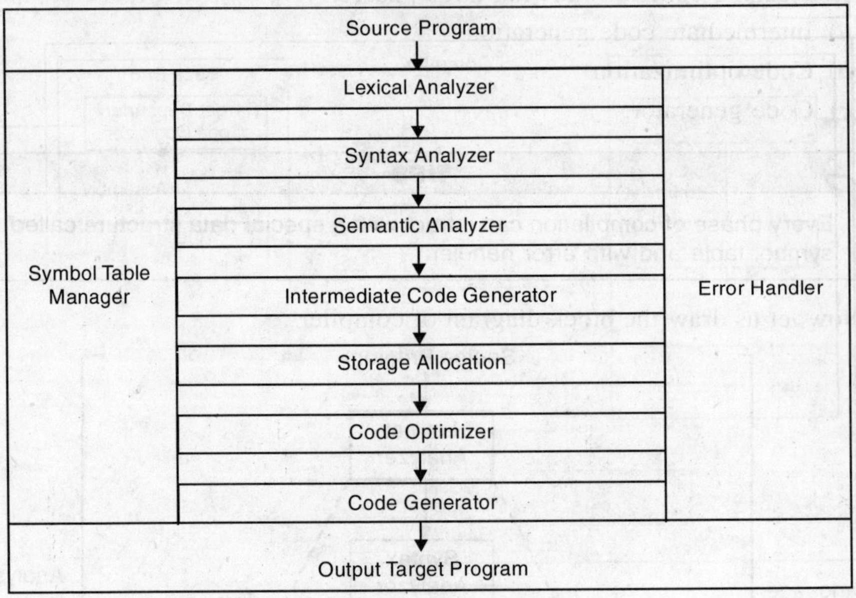

Fig. 1.6.

1.8.1. Modern Approach of Compiler Design

In the past, compilers were divided into many passes to save space. When each pass is finished, the compiler can free the space needed during that pass.

Many modern compilers share a common 'two stage' design. The first stage, the 'compiler front end' translates the source language into an intermediate representation. The second stage, the 'compiler back end' works with the internal representation to produce code in the output language.

(a) Compiler Front End

The compiler front end consists of multiple phases itself, each informed by formal language theory:

1. *Lexical analysis:* Breaking the source code text into small pieces ('tokens' or 'terminals'), each representing a single atomic unit of the language, for instance a keyword, identifier or symbol names. The token language is typically a regular language, so a finite state automaton constructed from a regular expression can be used to recognize it. This phase is also called lexing or scanning.
2. *Syntax analysis:* Identifying syntactic structures of source code. It only focuses on the structure. In other words, it identifies the order of tokens and understand hierarchical structures in code. This phase is also called parsing.

3. Semantic analysis is to recognize the *meaning* of program code and start to prepare for output. In that phrase, type checking is done and most of compiler errors show up.
4. Intermediate language generation—an equivalent to the original program is created in an intermediate language.

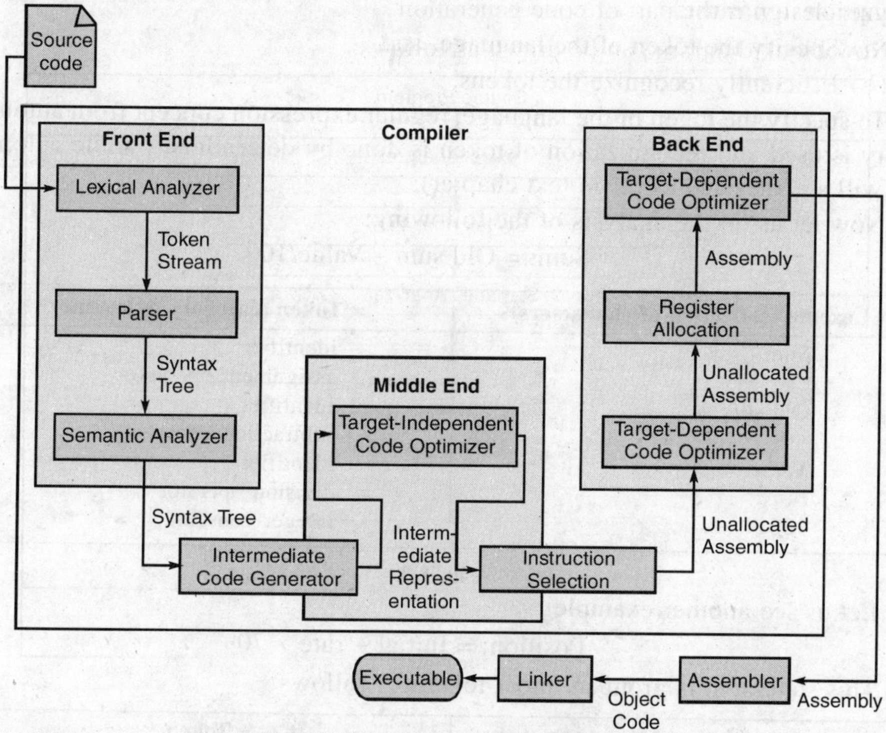

Fig. 1.7. Structure of a typical compiler.

(b) Compiler Back End

While there are applications where only the compiler front end is necessary, such as static language verification tools, a real compiler hands the intermediate representation generated by the front end to the back end, which produces a functional equivalent program in the output language. This is done in multiple steps:

1. *Optimization:* The intermediate language representation is transformed into functionally equivalent but faster (or smaller) forms.
2. *Code generation:* The transformed intermediate language is translated into the output language, usually the native machine language of the system. This involves resource and storage decisions, such as deciding which variables to fit into registers and memory and the selection and scheduling of appropriate machine instructions.

1.8.2. Lexical Analyzer (Scanner)

Definition

This module has the task of separating the continuous string of characters into distinctive groups that make sense. Such a group is called a token. A token may be composed of a single character or a sequence of characters (this sequence of characters is called lexeme).

In order to perform this task, the lexical analyzer must know the key words, identifiers, operators, delimiters and punctuation symbols of the language to be implemented. At the same time as forming characters into symbols the lexical analyser should deal with (multiple) spaces and remove comments and any other characters not relevant to the later stages of analysis. So now we can say that lexical analyzer design must:

(*a*) Specify the token of the language, and

(*b*) Efficiently recognize the tokens.

To specify the token of the language, regular expression concept from automata theory is used and recognization of token is done by deterministic finite automata (we will see these concepts in text chapter).

Now let us do the analysis of the following:

Sum: = Old sum – Value/100

Lexeme (collection of characters)	Token (category of lexeme)
Sum	identifier
: =	assignment operator
Old sum	Identifier
–	subtraction operator
Value	identifier
/	division operator
100	integer constant.

Fig. 1.8.

Let us see another example

Position: = Initial + rate * 70

This statement is grouped into 7 tokens as follows:

Lexeme	Token
position	identifier
: =	assignment operator
initial	identifier
+	addition operator
rate	identifier
*	multiplication operator
70	integer constant

Fig. 1.9.

After recognizing the token lexical analyser pass them to syntax analyser. Another way of looking at it is that the lexical analyser usually has no context to work with. As it is processing one symbol it has no knowledge of any of the symbols that preceded or will follow this symbol.

There are widely available tools that can be used for producing a lexical analyser for a langauge (we will see some of them).

1.8.3. Syntax Analyzer (Parser)

Definition Syntax analyser is the module in which the overall structure is identified, and involves an understanding of the order in which the symbols in a program may appear.

We know that it was the scanner's duty to recognize individual words or tokens of the language. The lexical analyser does not, however, recognize if these words have been used correctly. The main task of parser is to group the tokens into sentences, that is, to determine if the sequence of tokens that have been extracted by the syntax analyser are in the correct order or not. In other words, until the syntax analyser (parser) is reached the tokens have been collected with no regard to the whole context of the program as whole. The parser analyzes the context of each token and groups the token in declarations, statements and control statement.

TIPS

In the process of analyzing each sentence, the parser builds abstract tree structures.

A tree is a useful representation of a program segment because it facilitates transformations of the program that may lead to possible minimization of the machine instructions that are needed to carry out the required operations that is optimization. Let us see the procedure by an example:

```
Sum: = Old sum - Value/100
```

There are 7 tokens, sum (identifier) ': = ' (assignment operator), old sum (identifier), ' – ' (subtraction operator), value (identifier), / (division operator) and 100 (integer constant) as output of lexical analysis. Now these tokens work as input to the parser.

Parser will produce following syntax tree:

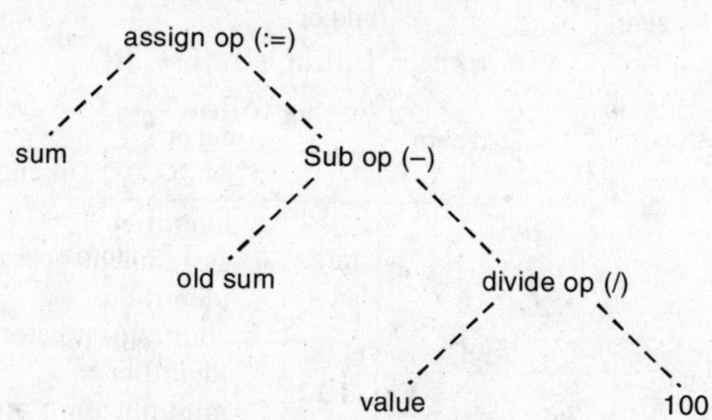

Fig. 1.10. Syntax tree as output of parser.

So we can say that syntax analysis is not only the key phase of the analysis stage of compilation but provides the framework for the compiler as a whole. It drives the lexical analysis phase and builds the structure upon which semantic analysis is performed. Syntax analysis also provides a framework for a whole range of source code analysis tools, including measuring tools of various sorts, cross-references, layout tools and so on.

1.8.4. Semantic Analyzer

Some features of programming cannot be checked in a single left to right scan of the source code without setting up arbitrary sized tables, to provide access to information that may be an arbitrary distance away, when it required. For example, information concerning the type and scope of variables falls into this category.

Definition The semantic analyzer gathers type information and checks the tree produced by the syntax analyzer for semantic errors.

Let us see the statement; and consider rate is real.

```
Sum: =Old sum + Rate & 60 ;
```

Here in this statement the semantic analyzer might add a type conversion node, say inttoreal, to the syntax tree to convert the integer to real quantity.

Output of syntax analysis:

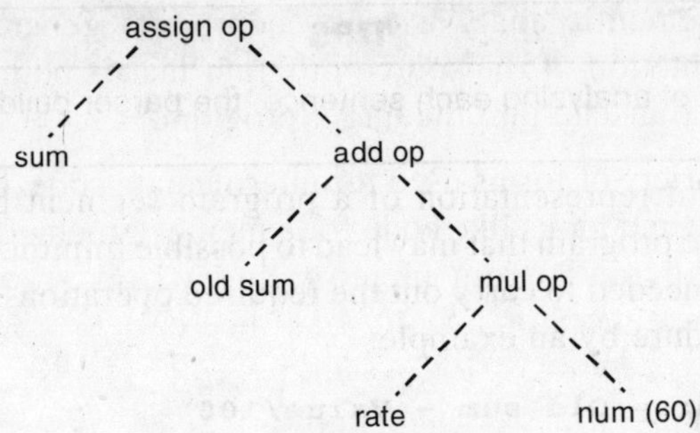

Fig. 1.11.

Now semantic analyzer comes in the picture and output will be as follows:

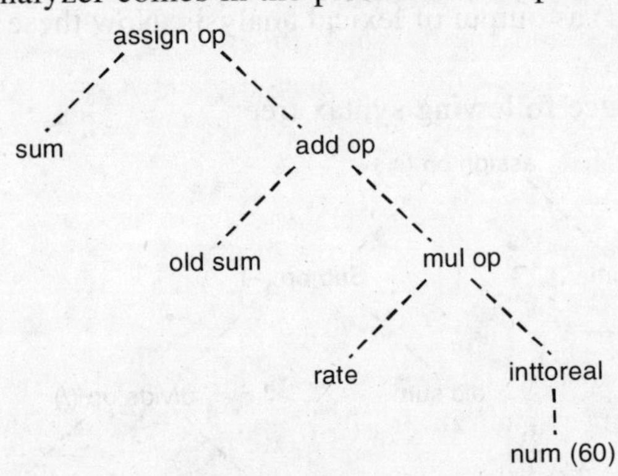

Fig. 1.12.

1.8.5. Symbol Table Manager

Though the parser determines the correct usage of tokens, and whether or not they appeared in the correct order, it still did not determine whether or not the program said anything that made sense. This is type of checking that occurs at the semantic level. In order to perform this task, the compiler makes use of a detailed system of lists, known as the symbol tables. The compiler needs a symbol table to record each identifier and collect information about it.

TIPS

For a variable the symbol table might record its type-expression (integer, float etc.), its scope (where the variable can be used), and its location in run-time storages, etc. For a procedure the symbol table might record the type-expression of its arguments and the type expression of its returned value.

The symbol table manager has a FIND function that returns a pointer to the descriptor for an identifier (descriptor is a record which contains all the information about identifier) when given its lexeme. Compiler phases use this pointer to read and/or modify information about the identifier. FIND returns a NULL pointer if there is no record for a given lexeme. The INSERT function in the symbol table manager inserts a new record into the symbol table when given its lexeme.

1.8.6. Intermediate Code Generator

Definition After semantic analysis many compilers generated an Intermediate representation of the source program that is both easy to produce and easy to translate into the target program.

There are a variety of forms used for intermediate code. One such form is called three-address code and looks like code for a memory-memory machine where every operation reads operands from memory and writes results into memory. The intermediate code generator usually has to create temporary location to hold intermediate results.

For example see following statement.

```
Sum: = Old sum + Rate & 60 ;
```

After the compiler broke this stream of text down into tokens, parsed the tokens and created the tree and finally performed the necessary type checking, it might create a sequence of three-address instructions. Three address instructions contain no more than, three operands (registers) per instruction, and in addition to an assignment, contain only one other operator. Three-address instructions also assume on unlimited supply of registers.

$$\begin{aligned}
&\texttt{temp}_1 \ := \ \texttt{inttoreal (60)}\\
&\texttt{temp}_2 \ := \ \texttt{rate \& temp}_1\\
&\texttt{temp}_3 \ := \ \texttt{old sum + temp}_2\\
&\texttt{Sum: = } \ \texttt{temp}_3
\end{aligned}$$

where $temp_1$, $temp_2$ and $temp_3$ are the names of three temporary locations created by the intermediate code generator.

1.8.7. Code Optimization

Definition The structure of the tree that is generated by the parser can be rearranged to suit the needs of the machine architecture or the tree can be restructured in order to produce an object code that run faster. What results is a simplified tree that may be used to generate the object code.

An example of such an optimization is the following FOR Loop:

```
for (int i = 0 ; i  < 2000 ; i + +)
   {
      x: = y + 3 ;
      q: = a * i ;
   }
```

In this loop there are instructions that are not related at all to the variables 'a' of the for statement, such as x: = y + 3;. In the above program this instruction is executed unnecessarily 2000 times. The object code can be generated such that the machine code for that line is moved outside the for's body, (either before or after the for command). The resulting object code is more efficient at run time because the loop has been significantly tightened, and the above mentioned instruction is executed only once, instead of 2000 times. **This is called loop optimization.**

In intermediate code generation phase, we have seen three-address instructions as follows:

```
temp₁  : =  inttoreal (60)
temp₂  : =  rate & temp₁
temp₃  : =  Old sum + temp₂
Sum: =  temp₃
```

We can optimize the code, by applying abstract algebraic rules to it. A subsequent optimization of the above code might be:

```
temp₁: = rate & inttoreal (60)
Sum: = Old sum + temp₁.
```

This is sometimes called local optimization. Once the intermediate code has been optimized, the compiler can generate the final code in its machine-readable form.

1.8.8. Storage-Allocation

Every constant and variable appearing in the program must have storage space allocated for its value during the storage allocation phase. This storage space may be one of the following types:

(*a*) **Static Storage:** If the life time of the variable is life time of the program and the space for its value once allocated cannot later be released. This kind of allocation is called static storage allocation.

(*b*) **Dynamic Storage:** We follow dynamic storage allocation if the life time is a particular block or function or procedure in which it is allocated so that it may be released when the block or function or procedure in which it is allocated is left.

(*c*) **Global Storage:** If its life time is unknown at compile time and it has to be allocated and deallocated at run time. The efficient control of such storage usually implies run-time overheads.

TIPS

After space is allocated by the storage allocation phase, an address, containing as much as is known about its location at compile time, is passed to the code generator for its use.

1.8.9. Code Generation

This is the part of the compiler where native machine code is actually generated. If the target machine has registers the target program. For example for statement

```
Sum: = Old sum + Rate & 60
```

Now we can understand that sum, Old sum and rate are identifier, let us say them id_1, id_2, and id_3.

that is $id_1 := id_2 + id_3 * 60$

```
MOVF   id3, R2
MULF   # 60.0, R2
MOVF   id2, R1
ADDF   R2, R1
MOVF   R1, id1
```

During this step, the compiler has to map address names from the three-address intermediate code onto the very finite amount of registers had by the machine. The final result is a piece of code that is (mostly) executable.

None of the remaining phases prior the actual execution of the program are a part of the compiler's responsibility. What remains to be done is to link the program and build a binary executable. In this phases, various libraries as well as the main executable portion of the program are combined to form a fully-fledged application. There are various types of linking, most of which are done prior to run time. The final phase that a program must pass through prior to execution is for it to be loaded in to memory. In this phase, the addresses in the binary code are translated from logical addresses, and any final run time binding are performed.

1.8.10. Peephole Optimization

The object code can be further optimized by a peephole optimizer. This process is present in a more sophisticated compiler. The purpose is to produce a more efficient object program that can be executed in a shorter time. Optimization is performed at a local level and it takes advantage of certain operator properties such as commutativity, associativity, and distributivity. To see a summary of the entire process of compilation from start to finish, refer to following figure.

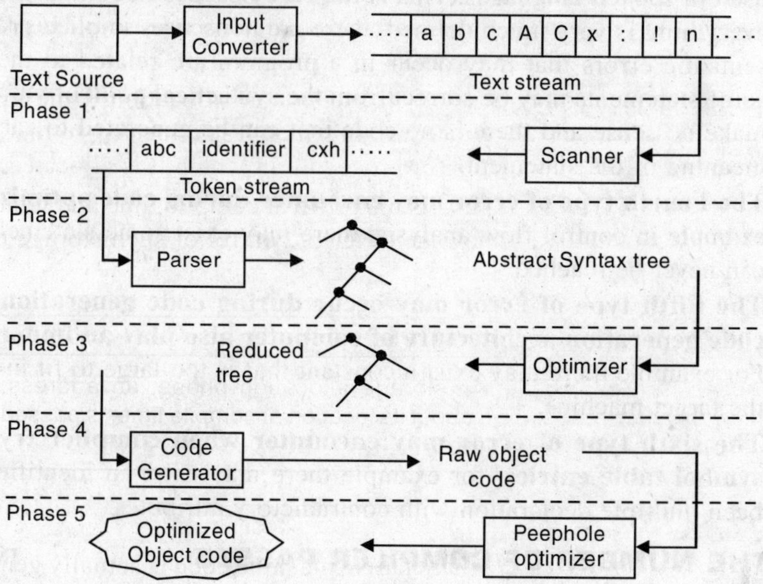

Fig. 1.13.

1.8.11. Error Handler

Detection and reporting of errors in source program is main function of compiler. Error may occur at any phase of compilation. A good compiler must determine the line number of program exactly, where the errors have occurred.

Let us see the various errors which may occur at different levels of compilation.

(*i*) **The first of these are lexical (scanner) errors.** Some of the most common types here consist of illegal or unrecognized characters, mainly, caused by typing errors. A common way for this to happen is for the programmer to type a character that is illegal in any instance in the language, and is never used. Finally the quite commonly, another type of error that the scanner may detect is an unterminated character or string constant. This happens whenever the programmer types something in quotes and forget the trailing quote. Again, these are mostly typing errors.

(*ii*) **The second class of error is syntactic in nature**, and is caught by the parser. These errors are among the most common. The really difficult part is to decide from where to continue the syntactical analysis after an error has been found. What happens if the parser is not carefully written, or if the error detection and recovery scheme is sloppy, the parser will hit one error and "mess up" after that, and cascade spurious error messages all throughout the rest of the program. In the case of an error, what one would like to see happening, is to have the compiler skip any improper tokens, and continue to detect errors without generating error messages that are not really an error but a consequence of the first error. This aspect is so important that some compilers are categorized based on how good their error detection system is.

(*iii*) **The third type of error is semantic in nature.** The semantics that are used in computer languages are by far simpler than the semantics that are used in spoken languages. This is the case because in computer languages everything is very much defined, there are nonsenses implied or used. The semantic errors that may occur in a program are related to the fact that some statements may be correct from the syntactical point of view, but they make no sense and there is no code that can be generated to carry out the meaning of the statement.

(*iv*) **The Fourth type of error may encounter during code optimization**, for example in control flow analysis, there may exist some statements, which can never be reached.

(*v*) **The Fifth type of error may occur during code generation, since in code generation architecture of computer also play an important role.** For example there may exist a constant that is too large to fit in a word of the target machine.

(*vi*) **The sixth type of error may encounter when compiler try to make symbol table entries**, for example there may exist an identifier that has been multiple declaration with contradictory attributes.

1.9. THE NUMBER OF COMPILER PASSES [MDU 2008]

We have seen the structure of compiler. The structure of the complier is divided into six phases. Each phase have it's own unique job to perform. The output of one phase is used as input to another phase. We have already seen that phases can be viewed as analysis phase and synthesis phase. On the basis of regrouping of phase compilers may multipass or one pass compilers.

(*i*) A pass is a complete traversal of the source program, or a complete traversal of some inter representation of the source program.
(*ii*) A pass can correspond to a "phase" but it does not have to!
(*iii*) Sometimes a single "pass" corresponds to several phases that are interleaved in time.
(*iv*) What and how many passes a compiler does over the source program is an important decision.

1.9.1. Multi-pass Compilers [MDU 2008]

We have already studied a compiler that scan the input source once, produces a first modified form, then scans the first-modified form and produce a second-modified from and so on, until the object form is produced. **Such a compiler is called a multipass compiler. In the case of the multi-pass compiler each function of the compiler can be performed by one pass of the compiler.** For instance the first pass can read the input source, scan and extract the tokens and store the result in an output file. The second pass can read the file which was produced in the first pass, do the syntactical analysis by building a syntactical tree and associate all the information relating to each node of the tree. The output of the second pass, then is a file containing the syntactical tree. The third pass, can read the output file produced by the second pass and perform the optimization by restructuring the tree structure.

Definition A multi-pass compiler makes several passes over the program. The output of a preceding phase is stored in a data structure and used by subsequent phases.

Dependency diagram of a typical Multi Pass Compiler:

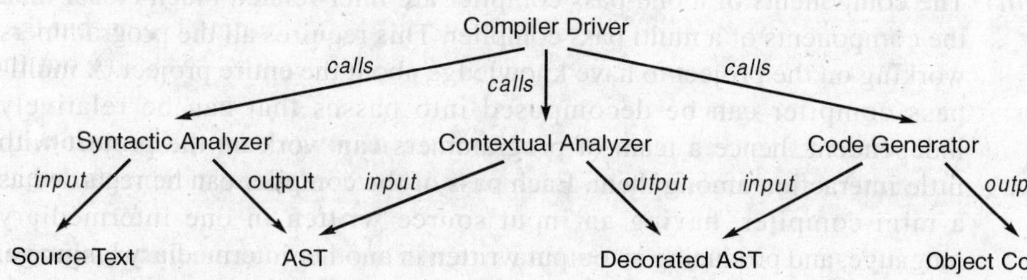

Fig. 1.14.

1.9.2. One-Pass Compilers

In a one-pass compiler, when a line source is processed it is scanned and the tokens are extracted. Then the syntax of the line is analyzed and the tree structure and some tables containing information about each token are built. Finally, after the semantical part is checked for correctness, the code is generated. The same process is repeated for each line of code until the entire program is compiled. Usually the entire compiler is built around the parser, which will call procedures that will perform different functions.

Definition A single pass compiler makes a single pass over the source text, parsing, analyzing and generating code all at once.

Dependency diagram of a typical single pass compiler:

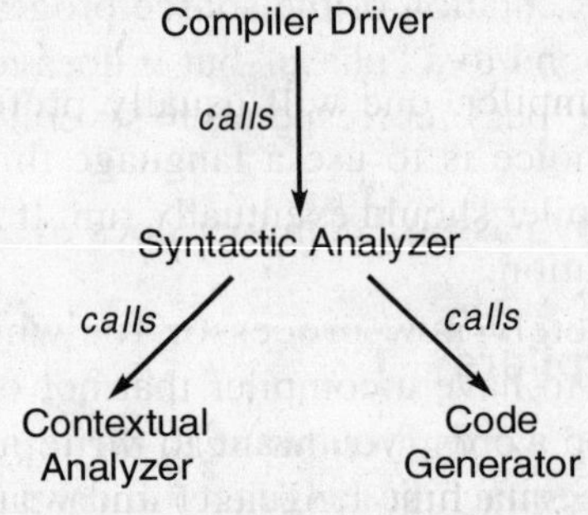

Fig. 1.15.

1.9.3. Advantages and Disadvantages for Both Single and Multi-pass Compilers [MDU 2008]

(*i*) A one pass compiler is fast, since all the compiler code is loaded in the memory at once. It can process the source text without the overhead of the operating system having to shut down one process and start another. Also, the output of each pass of the multi-pass compiler is stored on disk, and must be read in each time the next pass starts.

(*ii*) On the other hand, a one pass tends to impose some restrictions upon the program: constants, types, variables, and procedures must be defined before they are used. A multi-pass compiler does not impose this type of restrictions upon the user. Operation that can not be performed because of the lack of information can be deferred to the next pass of the compiler when the text has been traversed and the needed information made available.

(*iii*) The components of a one-pass compiler are inter-related much closer than the components of a multi pass compiler. This requires all the programmers working on the project to have knowledge about the entire project. A multi-pass compiler can be decomposed into passes that can be relatively independent, hence a team of programmers can work on the project with little interaction among them. Each pass of the compiler can be regarded as a mini-compiler, having an input source written in one intermediary langauge, and producing an output written in another intermediary language.

1.9.4. Difference Between Single Pass Compiler and Multi-Pass Compiler [MDU 2008]

(*i*) A **one-pass compiler** is a compiler that passes through the source code of each compilation unit only once. A **multi-pass compiler** is a type of compiler that processes the source code or abstract syntax tree of a program several times.

(*ii*) A one-pass compilers is faster than multi-pass compilers

(*iii*) A one-pass compiler has limited scope of passes but multi-pass compiler has wide scope of passes.

(*iv*) Multi-pass compilers are sometimes called **wide compilers** where as one-pass compiler are sometimes called **narrow compiler.**

(*v*) Many programming languages cannot be represented with single pass compilers, for example **Pascal** can be implemented with a single pass compiler where as languages like **Java** require a multi-pass compiler.

1.10. INTRODUCTION TO BOOTSTRAPPING (COMPILERS)

[MDU 2008, 2010]

When writing a compiler, one will usually prefer to write it in a high-level language. A possible choice is to use a language that is already available on the machine where the compiler should eventually run. It is, however, quite common to be in the following situation:

You have a completely new processor for which no compilers exist yet. Nevertheless, you want to have a compiler that not only target this processor, but also runs on it. In other words, you want to write a compiler for a language A, targeting langauge B (the machine language) and written language B.

The most obvious approach is to write the compiler in language B. But if B is machine language, it is a horrible job to write any non-trivial compiler in this language. Instead, it is customary to use a process called "bootstrapping", referring to the seemingly impossible task of pulling oneself up by the bootstraps.

The idea of bootstrapping is simple: You write your compiler in language A (but still let it target B) and then let it compile itself. The result is a compiler from A to B written in B.

It may sound a bit paradoxical to let the compiler compile itself. In order to use the compiler to compile a program, we must already have compiled it, and to do this we must use the compiler. In a way, it is a bit like the chicken-and-egg paradox. We shall shortly see how this apparent paradox is resolved, but first we will introduce some introduce some useful notation.

1.10.1. Notations

We will use a notation designed by H. Bratman. The notation is hence called "Bratman diagrams" or, because of their shape, "T-diagrams".

Let us see some notations as follows:

(*i*) A compiler written in language H that translates language S into language T.

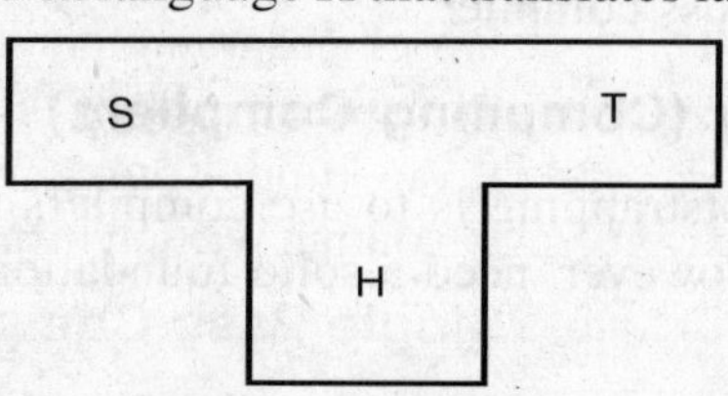

(*ii*) T-Diagram can be combined in two basic ways.

Now let us see the first T-diagram Combination as follows:

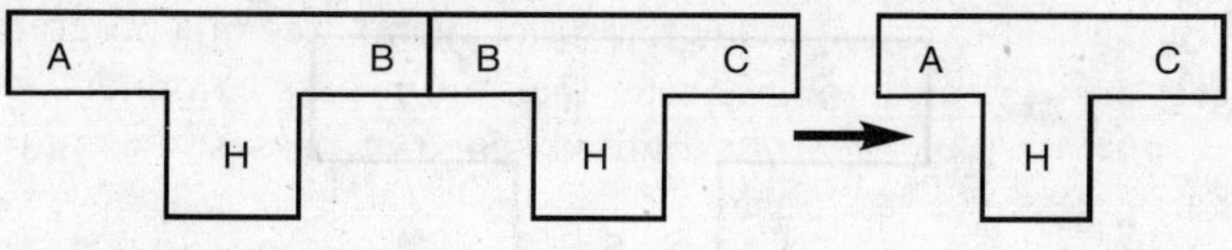

(*iii*) Two compilers run on the same machine H

- First from A to B
- Second from B to C
- Result from A to C on H

Now let us see the Second T-diagram Combination as follows:

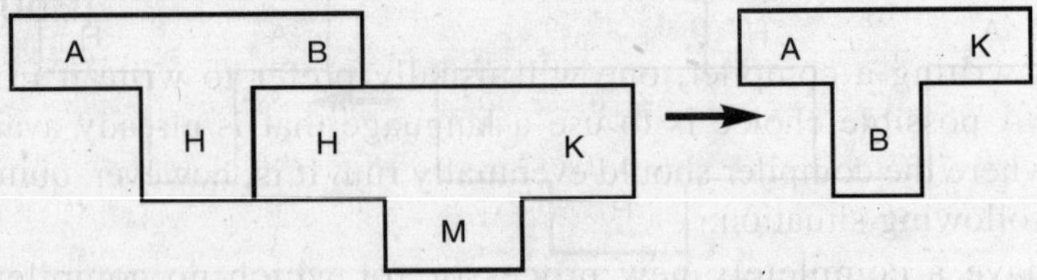

(*i*) Translate implementation language of a compiler from H to K

(*ii*) Use another compiler from H to K

Now let us study the following Scenario-1 as follows:

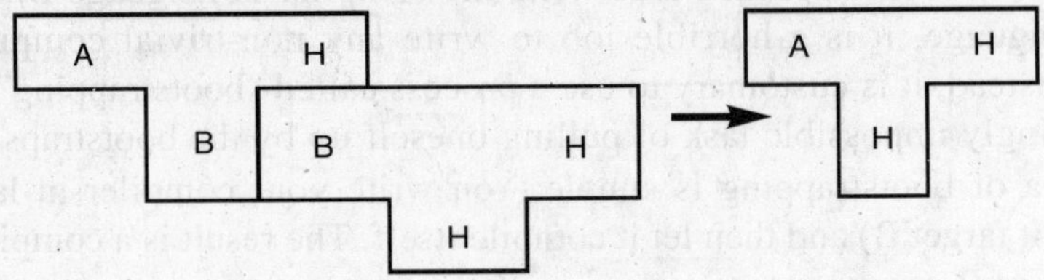

(*i*) Translate a compiler from A to H written in B

- Use an existing compiler for language B on machine H

Now let us study the following Scenario-2 as follows:

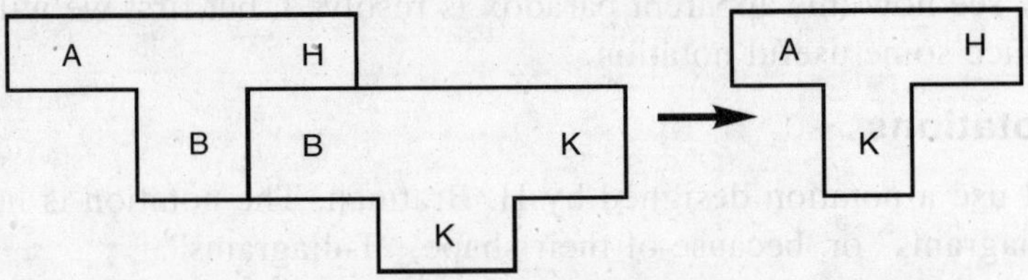

(*ii*) Use an existing compiler for language B on different machine K

- Result in a cross compiler

1.10.2. Bootstrapping (Compiling Compilers)

The basic idea in bootstrapping is to use compilers to compile themselves or other compilers. We do, however, need a solid foundation in form of a machine to run the compilers on.

Process of Bootstrapping

(*i*) Write a compiler in the same language.

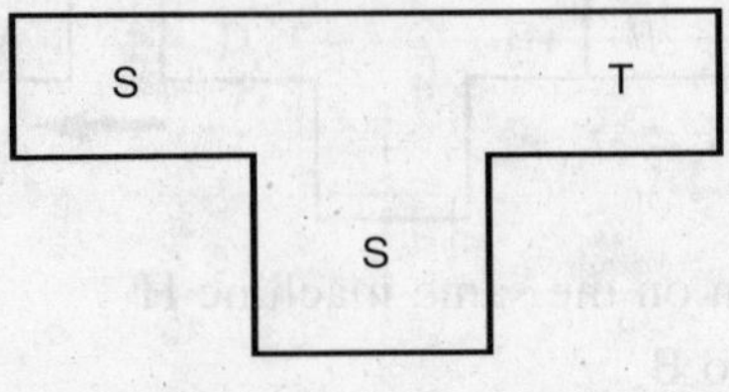

(*ii*) No compiler for source language yet.

(*iii*) Porting to a new host machine.

First step in bootstrap

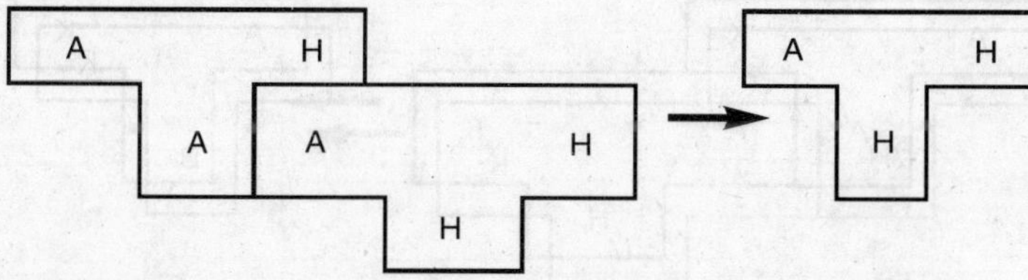

(*i*) "quick and dirty" compiler written in machine language H

(*ii*) Compiler written in its own language A

(*iii*) Result in running but inefficient compiler

The Second step in bootstrap

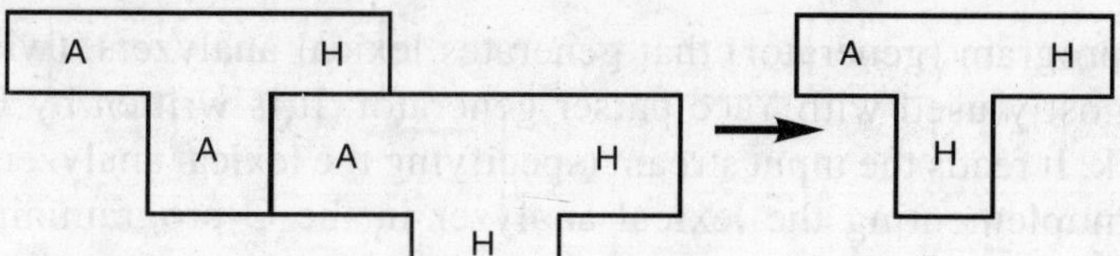

(*i*) Running but inefficient compiler

(*ii*) Compiler written in its own language A

(*iii*) Result in final version of the compiler

The above bootstrapping process relies on an existing compiler for the desired language, albeit running on a different machine. It is, hence, often called "half bootstrapping". When no existing compiler is available, *e.g.*, when a new langauge has been designed, we need to use a more complicated process called "full bootstrapping".

1.10.3. Porting

In computer science, **porting** is the process of adapting software so that an executable program can be created for a computing environment that is different from the one for which it was originally designed (*e.g.* different CPU, operating system, or third party library). The term's also used for when software/hardware is changed to make them usable in different environments.

Software is portable when the cost of porting it to a new platform is less than the cost of writing it from scratch. The lower the cost of porting software, relative to its implementation cost, the more portable it is said to be.

The step 1 in porting

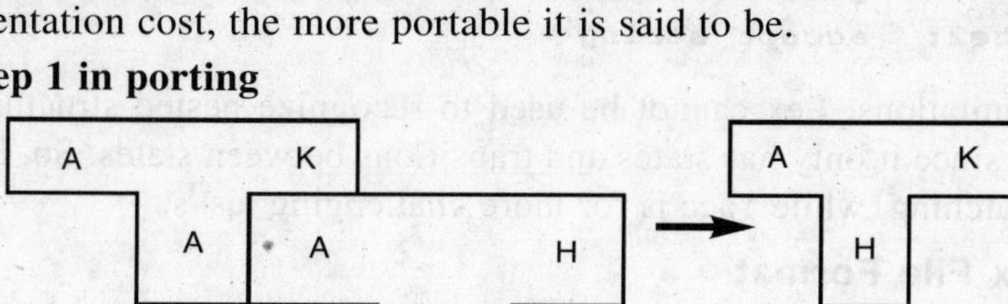

(*i*) Original compiler

(*ii*) Compiler source code retargeted to K

(*iii*) Result in Cross Compiler

The step 2 in porting

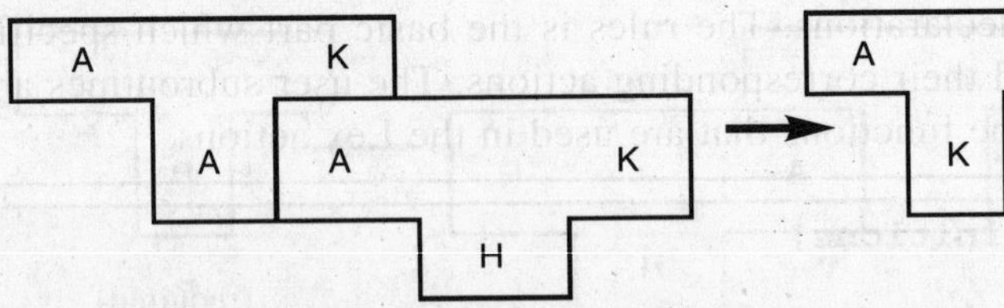

(*i*) Cross compiler
(*ii*) Compiler source code retargeted to K
(*iii*) Result in Retargeted Compiler

1.11. LEX

Lex is a program (generator) that generates lexical analyzers, (widely used on Unix). It is mostly used with Yacc parser generator. It is written by Eric Schmidt and Mike Lesk. It reads the input stream (specifying the lexical analyzer) and outputs source code implementing the lexical analyzer in the C programming language. Lex will read patterns (regular expressions); then produces C code for a lexical analyzer that scans for identifiers.

Regular expressions are translated by lex to a computer program that mimics an FSA. A simple pattern is **letter(letter | digit)*. This pattern matches a string of characters that begins with a single letter followed by zero or more letters or digits.**

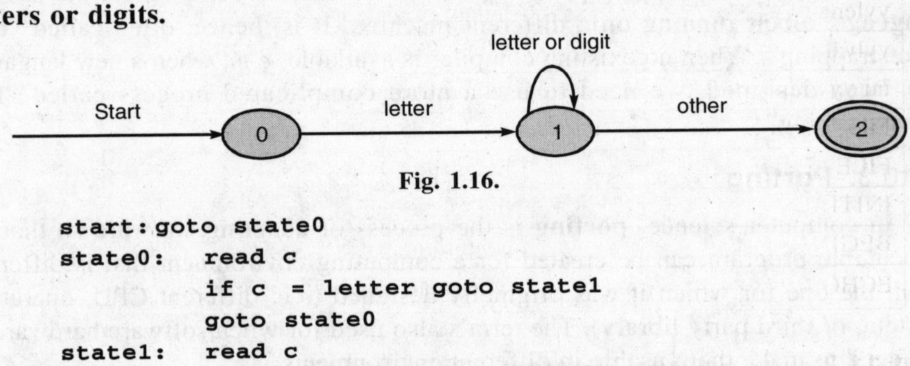

Fig. 1.16.

```
start: goto state0
state0:  read c
         if c  = letter goto state1
         goto state0
state1:  read c
         if c = letter goto state1
         if c = digit goto state1
         goto state 2
state2:  accept string
```

Some limitations, Lex cannot be used to recognize nested structures such as parentheses, since it only has states and transitions between states. So, Lex is good at pattern matching, while Yacc is for more challenging tasks.

1.11.1. Lex File Format

Lex source is separated into three sections by %% delimiters. The general format of Lex source is

Definitions consists of any external 'C' definitions used in the lex actions or subroutines. For example, all preprocessor directives like #include, #define macros etc. These are simply copied to the lex.yy.c file. The other type of definitions are

Lex definitions which are essentially the lex substitution strings, lex start states and lex table size declarations. The rules is the basic part which specifies the regular expressions and their corresponding actions. The user subroutines are the function definitions of the functions that are used in the Lex actions.

```
{definitions}
%%                          (required)
{transition rules}
%%                          (optional)
{user subroutines}
```

- The absolute minimum Lex program is thus

```
%%
```

Fig. 1.17.

1.11.2. Lex Predefined Variables

Name	Function
int yylex (void)	call to invoke lexer, returns token
char *yytext	pointer to matched string
yyleng	length of matched string
yylval	value associated with token
int yywrap (void)	wrapup, return 1 if done, 0 if not done
File *yyout	output file
FILE *yyin	input file
INITIAL	initial start condition
BEGIN	condition switch start condition
ECHO	write matched string

1.11.3. Lex Source Program

Lex source is a table of regular expressions and corresponding program fragments.

```
digit     [0-9]
letter    [a-zA-Z]
%%
{letter}({letter}|{digit})* printf("id: %s\n", yytext);
\n                          printf("new line\n");
%%
main () {
        yylex ();
}
```

Fig. 1.18.

The table is translated to a C program (lex.yy.c) which

- Reads an input stream
- Partitioning the input into strings which match the given expressions and
- Copying it to an output stream if necessary.

The functioning of Lex is as follows:

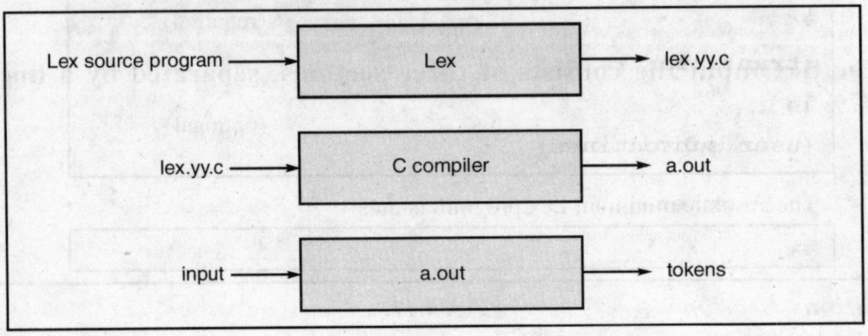

Fig. 1.19.

Let us see the Lex program for the regular expression is letter(letter|digit)*

```
digit      [0-9]
letter [A-Za-z]
%{
    int count;
%}
%%
    /* match identifier */
{letter} ({letter) | {digit}) *    count++;
%%
int main(void) {
    yylex();
    printf("number of identifiers = %d\n" count);
    return 0;
}
```

Whitespace must separate the defining term and the associated expression. Code in the definitions section is simply copied as-is to the top of the generated C file and must be bracketed with "%{" and "%}" markers. Substitutions in the rules section are surrounded by braces ({letter}) to distinguish them from literals.

1.12. INTRODUCTION TO FLEX (FAST LEXICAL ANALYZER GENERATOR)

Flex is a tool for generating scanners. A scanner, sometimes called a tokenizer, is a program which recognizes lexical patterns in text. The flex program reads user-specified input files, or its standard input if no file names are given, for a description of a scanner to generate. The description is in the form of pairs of regular expressions and C code, called rules. Flex generates a C source file named, "lex.yy.c", which defines the function yylex(). The file "lex.yy.c" can be compiled and linked to produce an executable. When the executable is run, it analyzes its input for occurrences of text matching the regular expressions for each rule. Whenever it finds a match, it executes the corresponding C code.

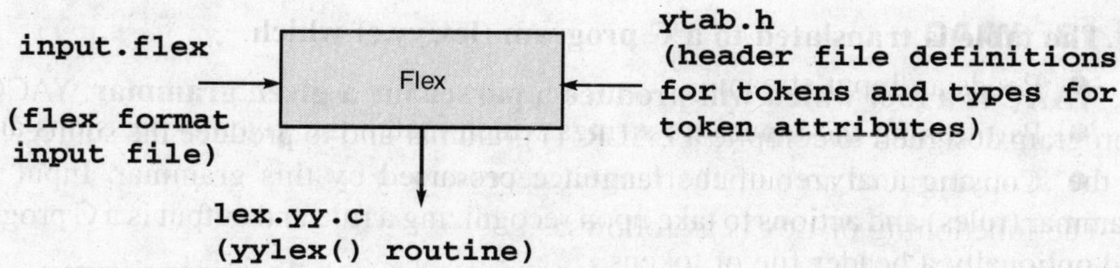

Fig. 1.20.

The flex input file consists of three sections, separated by a line with just %% in it:

```
definitions
%%
rules
%%
user code
```

Definitions Section

The definitions section contains declarations of simple name definitions to simplify the scanner specification. Name definitions have the form:

```
name definition
```

For example:

```
DIGIT   [0-9]
ID      [a-z][a-z0-9]*
```

Rules Section

The rules section of the flex input contains a series of rules of the form:

```
pattern action
```

For example:

```
{ID} printf( "An identifier: %s\n", yytext );
```

If action is empty, the matched token is discarded.

Action

The user code section is simply copied to *lex.yy.c* verbatim. The presence of this section is optional; if it is missing, the second %% in the input file may be skipped. In the definitions and rules sections, any indented text or text enclosed in %{ and %} is copied verbatim to the output (with the %{}'s removed).

Let us see a simple example as follows:

```
%{
    int num_lines = 0, num_chars = 0;
%}
%%
\n      ++num_lines; ++num_chars;
.       ++num_chars;
%%
main()  {
    yylex();
    printf( "# of lines = %d, # of chars = %d\n",
                num_lines, num_chars );
}
```

1.13. YAAC

YAAC is a tool which will produce a parser for a given grammar. YACC is a program designed to compile a LALR(1) grammar and to produce the source code of the syntactic analyzer of the language produced by this grammar. Input is a grammar (rules) and actions to take upon recognizing a rule and output is a C program and optionally a header file of tokens

Let us see some facts about Yaac as follows:

Input: A CFG and a translation scheme – file.y

Output: A parser file.tab.c (bison) or y.tab.c (yacc)

- An output file file.output containing the parsing tables (when invoked with –v option)
- A file file.tab.h containing declarations (if invoked with –d option)
- The parser called yyparse()
- Parser expects to use a function called yylex() to get tokens

The Basic Operational Sequence is as follows:

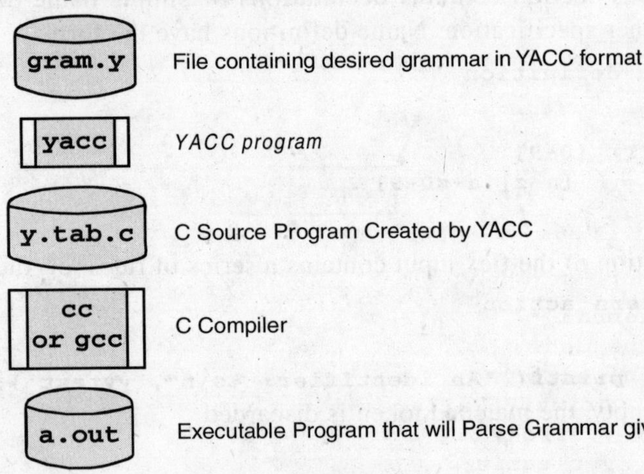

Fig. 1.21.

1.13.1. YACC File Format

The general format for the YACC file is very similar to that of the Lex file.

Definitions:

```
%%
```

Rules

```
%%
```

Supplementary Code: In Declarations apart from the legal 'C' declarations there are few Yacc specific declarations which begins with a %sign.

1. Definitions: It is Information about tokens:

(*a*) **Token names:**

(*i*) declared using '**%token**'

(*ii*) single-character tokens don't have to be declared

(*iii*) any name not declared as a token is assumed to be a nonterminal.

(*b*) start symbol of grammar, using '**%start**' [optional]

(*c*) operator info:

(*i*) precedence, associativity

(*d*) stuff to be copied verbatim into the output (e.g., declarations, **#includes**): enclosed in %{ … }%

Let us see the following example:

```
%{
#include <stdio.h>
#include <stdlib.h>
%}
%start  line
%token  CHAR, COMMA, FLOAT, ID, INT, SEMI
%%
```

Rules: The rules are written in the form of grammar as follows:

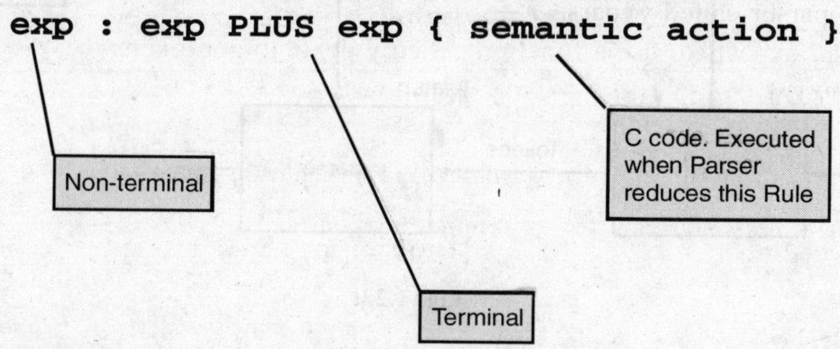

Fig. 1.22.

Consider the example as follows:

```
line: /* lambda */
      | line decl
      | line error {
              printf("Failure:-(\n");
              yyerrok;
              yyclearin;
              }
      ;
decl:  type ID list { printf("Success!\n"); } ;
list:  COMMA ID list
      | SEMI
      ;
type:  INT | CHAR | FLOAT
      ;
%%
```

1.13.2. LEX and YAAC As Team

Parser (Yaac) assumes the existence of a function 'int yylex()' that implements the scanner (lex). Lex return value indicates the type of token found and other values communicated to the parser using yytext, yylval (see man pages).

Yacc determines integer representations for tokens:

(*i*) Communicated to scanner in file y.tab.h

- use "yacc-d" to produce y.tab.h

(*ii*) Token encodings:

- "end of file" represented by '0';
- a character literal: its ASCII value;
- other tokens: assigned numbers ≥ 257.

The process of Yaac and Lex team work is as follows:

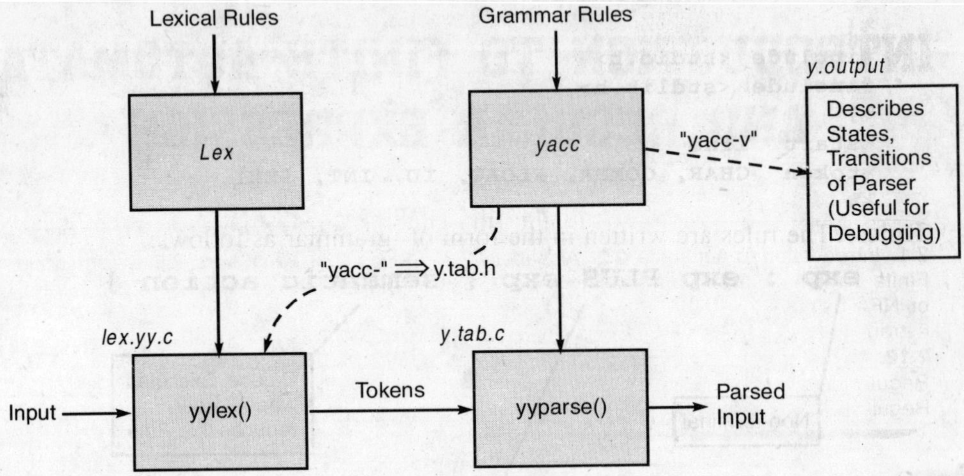

Fig. 1.23.

EXERCISE

1. What are phases of a compiler ? Explain the function of each phase in brief.
2. Symbol table is necessary for compiler construction, justify your statement with example.
3. What is a cross compiler ? How is boot-strapping of a compiler done to a second machine ?
4. What is a translator ? Discuss the role of various phases of the compiler in the translation of source program to object code.
5. Suppose you have a working *C* compiler on machine *A*. Discuss the steps you would take to create a working compiler for another language *C* on machine *B*.
6. Discuss the action taken by every phase of the compiler on the following string :

$$A = B * C + D/E$$

7. Write the difference between syntax and semantics of the languages, also give the example in C language.
8. Analyse the programs of any language on the basis of their different programming structure.
9. Discuss the advantages and disadvantages for single and multipass compilers.
10. Discuss the role of machine architecture in compiler design, also give the pro and cons of virtual machine and real machine architecture.

□□□

2

Chapter

INTRODUCTION TO FINITE AUTOMATA AND REGULAR EXPRESSION

INSIDE THIS CHAPTER

2.1. Introduction to Automata **2.2.** Alphabets **2.3.** Strings **2.4.** Languages **2.5.** Deterministic Finite Automata **2.6.** Nondeterminism **2.7.** The Language of NFA **2.8.** Transformation of NFA to DFA **2.9.** NFA with ∈-transitions **2.10.** Acceptance of a string by NFA with ∈-transitions **2.11.** Conversion of NFA with ∈-transitions to NFA without ∈-transitions **2.12.** Minimization of DFA's **2.13.** Regular Expression **2.14.** Comparative Study of Regular Expression, Regular Sets and Finite Automata **2.15.** Construction of FA for Regular Expression **2.16.** Algebraic Laws for Regular Expressions

2.1. INTRODUCTION TO AUTOMATA

These are machines (automata) that mimic a computer by providing a medium, which executes a finite number of instructions sequentially according to an algorithm, accepting valid input and producing an output if the input is accepted. All these machines have an input tape, a tape reading head, a register recording the state of the machine which is one of the finite states in which it can exist, and a set of instructions that govern the transition from one state to another.

$P_0, P_1, P_2, P_3, P_4, P_5$ are states in finite control system, x and y are input symbols. At regular interval the automation

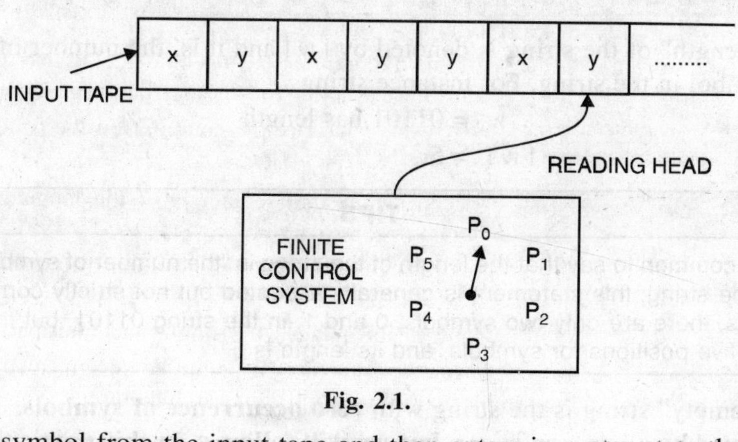

Fig. 2.1.

reads one symbol from the input tape and then enters in a new state that depends only on the current state and the symbol just read. After reading an input symbol, reading head moves one square to the right on the input tape, so that on the next move, it will read the symbol in next tape square. This process is repeated again and again. The automation then indicates its an approval or disapproval of what it has read by the state it is at the end:

TIPS

If it winds up *in one of a set of final states* the input strings is considered to be accepted. The language accepted by the machine is the set of strings, it accepts.

2.2. ALPHABETS

Everything in mathematics is based on symbols. This is also true for automata theory of computation. The symbols are generally letters and digits.

Definition "Alphabets are defined as a finite set of symbols"

An example of an alphabet is a set of decimal numbers.

$$\Sigma = \{0, 1, 2, 3, 4, 5, 6, 7, 8, 9\}$$

Here Σ is representation for the alphabet.

Another example is a binary alphabet that is:

$$\Sigma = \{0, 1\}$$

Similarly $\Sigma = \{A, B, ... Z\}$ is a alphabet.

$\Sigma = \{a, b, ... z\}$ is also a alphabet}.

2.3. STRINGS

Definition "A string (or sometimes word) is a finite sequence of symbols selected from some alphabet". For example, if Σ $\{a, b\}$ is an alphabet then *abab* is string over alphabet Σ. A string is generally denoted by w.

The "length" of the string is denoted by $| w |$ and it is, the number of positions for the symbol in the string. For instance string

$$w = 01101 \text{ has length}$$

$$| w | = 5.$$

TIPS

It is common to say that the length of the string is "the number of symbols" in the string; this statement is generally accepted but not strictly correct. Thus, there are only two symbols, 0 and 1, in the string 01101, but there are five positions for symbols, and its length is 5.

The "empty" string is the string with **zero occurrence of symbols.** This string is represented by $\in$, or e or λ, or γ **but we will follow $\in$ in this particular book.**

The set of strings, including the empty, over an alphabet Σ is denoted by Σ^*, that is suppose $\Sigma = \{0, 1\}$ then Σ^* is a set containing empty and all combinations of 0's and 1's. Thus

$$\Sigma^* = \{\in, 0, 1, 01, 10, 000, 010, 0000, 1111000 ...\}$$

Similarly $\{0\}^* = \{\in, 0, 00, 000 ...\}$,

$\{1\}^* = \{\in, 1, 11, 111 ...\}$

Note that $\in$ is in Σ^*, regardless of what alphabet Σ is.

That is $\in$ is only string whose length is 0.

if $\Sigma = \{0, 1\}$, then $\Sigma^1 = \{0, 1\}$, $\Sigma^2 = \{00, 01, 10, 11\}$,

$\Sigma^3 = \{000, 001, 010, 011, 100, 101, 110, 111\}$

and so on. Note that there is slight confusion between Σ and Σ^1. The formal is an alphabet; its members 0 and 1 are symbols. The latter is a set of strings; its members are the strings 0 and 1, each of which is of length 1. We shall not try to use separate notations for the two sets, relying on context to make it clear whether {0, 1} or similar sets are alphabets or set of strings.

Sometimes, we wish to exclude the empty string from the set of strings. The set of non-empty strings from alphabet Σ is denoted by Σ^+. Thus, two appropriate equivalence are:

$$\Sigma^+ = \Sigma^1 \cup \Sigma^2 \cup \Sigma^3 \cup ...$$

$$\Sigma^* = \Sigma^+ \cup \{\in\}.$$

➤ **Concatenation of strings.** Let w_1 and w_2 be two strings. Then $w_1 w_2$ denotes the concatenation of w_1 and w_2, that is, the string formed by making a copy of w_1 and followed it by a copy of w_2.

if $w_1 = abc$

$w_2 = xyz$

then $w_1 w_2 = abc\ xyz$

➤ In the set form string w_2 is defined as:

$$w_2 = \begin{cases} \in & \text{for } i = 0 \\ w^{i-1}\ w & \text{for } i \geq 1 \end{cases}$$

that is if $w = aba$

then $w^3 = (aba)^3$

$w^3 = aba\ aba\ aba$

$= (aba)^2\ aba.$

➤ **Reverse of the string.** Reverse of the string can be achieved by flipping over last symbols:

if $w = abc$

then $w^R = cba$, where w^R is reverse of w, $|w| = |w^R|$

2.4. LANGUAGES

Definition "A set of strings all of which are chosen from some Σ^*, where Σ is a particular alphabet, is called a language". If Σ is an alphabet, and $L \subseteq \Sigma^*$, then L is said to be language over alphabet Σ.

Notice that a language over Σ need not include strings with all the symbols of Σ, so once we have established that L is a language over Σ, we also know it is a language over any alphabet which is a superset of Σ.

An example is english, where the collection of legal english words is a set of strings over the alphabet that consists of all the letters. Another example is C or any other programming language, where the legal programs are a subset of the possible strings that can be formed from the alphabet of the language. This alphabet is a subset of the ASCII characters. The exact alphabet may differ slightly among different programming languages, but generally includes the upper-and lower-case letters; the digits, punctuation, and mathematical symbols.

However, there may be some other languages that appear when we study automata. Some examples are as follows:

(*a*) The language of all strings consisting of n 0's followed by n 1's, for some $n \geq 0$: {∈, 01, 0011, 000111, ...}

(*b*) The set of strings over 0's and 1's with equal number of each: {∈, 01, 10, 0011, 0101, 1001 ...}.

(*c*) The set of binary numbers whose value is a prime: {10, 11, 101, 111, 1011, ...}

(*d*) Σ^* is a language for any alphabet Σ.

(*e*) ϕ, the empty language, is a language over any alphabet.

(*f*) {∈}, the language consisting of only the empty string, is also a language over any alphabet. Notice that $\phi \neq \{\in\}$; the former has no strings and latter has one string.

The only important constraint on what can be a language is that all alphabet are finite. Thus languages, although they can have an infinite number of strings, are restricted to consist of strings drawn from one fixed, finite alphabet.

2.4.1. Concatenation of Languages

Concatenation of languages can be defined as:

if L_1 and L_2 are two languages, their concatenation

$$L = L_1 . L_2 \text{ where } L = \{w/w = xy, x \in L_1, y \in L_2\}.$$

That is, the string in $L_1 L_2$ are formed by choosing a string in L_1 and followed by a string in L_2, in all possible combinations.

2.4.2. Languages in Set Forms

It is very common to describe a language using a "set-form":

{*w*/some logical view about *w*}.

This expression is read "the set of words w such that (whatever is said about w to the right of the vertical bar)." For example:

(*a*) {w/w consists an equal number of 0's and 1's}

(*b*) L = {w/w is a binary integer that is prime}

It is common to replace w by some expression with parameters and describe the strings in the language by stating conditions on the parameters. Here are some examples :

$\{a^n b^n / n \geq 1\}$. Read "the set of a to the n, b to the n such that n is greater than or equal to 1," this language consists of the strings {*ab*, *aabb*, *aaabbb*, ...}. Notice that as with alphabet, we can raise a single symbol to a power n in order to represent n copies of that symbol.

2.5. DETERMINISTIC FINITE AUTOMATA

We know that computer is deterministic, by which we mean that, on reading one particular input instruction, the machine converts itself from the state it was into some particular other state, where the resultant state is completely determined by the prior state and the input instruction. Some sequence of instruction may lead to success and some may not. Success is entirely.

Finite automata is called "finite" because number of possible states and number of letter in the alphabet are both finite, and "automation" because the change of the state is totally governed by the input. It is deterministic, since, what is next is automatic not will-full, just as the motion of the hands of the clock is automatic, while the motion of hands of a human is presumably the result of desire and thought.

When we analyse above discussion mathematically the following formal version is obtained.

Definition A deterministic finite automata is a quintuple

$$M = (Q, \Sigma, \delta, q_0, F)$$

where

- Q : Is a non-empty finite set of states present in the finite control. $(q_0, q_1, q_2, ...)$
- Σ : Is a non-empty finite set of input symbols which can be passed to the finite state machine. $(a, b, c, d, e, ...)$
- q_0 : Is a starting state, one of the state in Q.
- F : Is a non-empty set of final states or accepting states, set of final states belongs to Q.
- δ : Is a function called transition function that takes two arguments—a state and a input symbol, it returns a *single* state.

Let 'q' is the state and 'a' be input symbol passed to the transition function as :

$$\delta(q, a) = q'$$

q' is the output of the function *i.e.*, a single state only, q' may be q.

Some people prefer to call this just a finite acceptor because its sole job is to accept certain input strings and rejects others. It does not do anything like printing machine. Some time it is also referred as language recognizer because FA merely recognizes whether the input string is in the language as we can also recognize that some body is speaking *Russian* or *English* without necessarily understanding what it means.

2.5.1. Processing of Strings by DFA

Suppose $a_1, a_2, a_3, ..., a_n$ is a sequence of input symbols. We start out with deterministic finite automata having $q_0, q_1, q_2, q_3, ..., q_n$ states where q_0 is the initial state and q_n is the final state and transition function and processed as

$$\begin{aligned}
\delta(q_0, a_1) &= q_1 \\
\delta(q_1, a_2) &= q_2 \\
\delta(q_2, a_3) &= q_3 \\
\vdots \quad & \quad \vdots \\
\delta(q_{n-1}, q_n) &= q_n
\end{aligned}$$

Input $a_1, a_2, ..., a_n$ is said to be **"accepted"** since q_n is the member of the final states, and if not then it is **"rejected"**. By adopting this approach we can define the language of a particular DFA, and can also check that particular string is accepted or rejected by given FA.

TIPS

"So a string is said to be accepted by the given FA only when it is processed by transition function d in such a manner that it ends at the final state".

2.5.2. Simpler Notations for DFA's

There are two preferred notations for describing automata:

1. Transition Diagram,
2. Transition Table.

1. Transition Diagram Notations. A transition diagram for DFA, $M = (Q, \Sigma, \delta, q_0, F)$ is a graph defined as follows:

(*a*) For each state in Q there is a node represented by the circle.

(*b*) For each state q in Q and each input symbol a in Σ, let $\delta(q, a) = P$. Then transition diagram has an arc from node q to node P, labelled a. If there are several input symbols that cause transition from q to P, then the transition diagram can have one arc, labelled by the list of these symbols as:

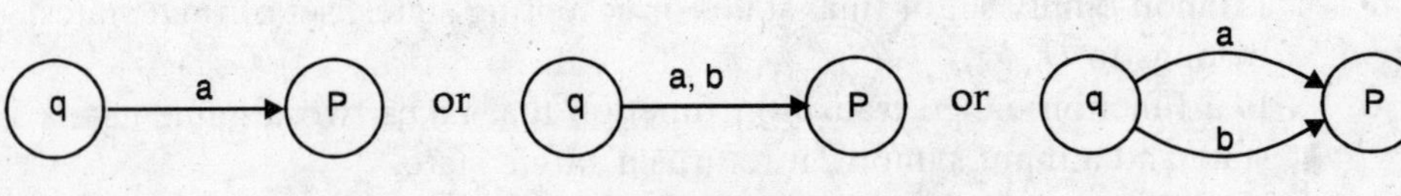

Fig. 2.2.

(*c*) If any state q in Q is the starting state then it represented by the circle with arrow as

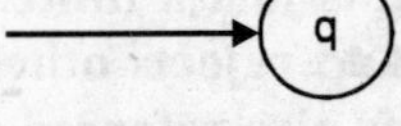

Fig. 2.3.

(*d*) Nodes corresponding to accepting states (those are in F) are marked by a double circle. States not in F have a single circle.

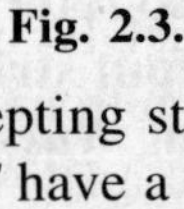

Fig. 2.4.

2. Transition Tables. A transition table is a Conventional, tabular representation of a function like δ that takes two arguments and returns a state. The rows of the table correspond to the states, and the columns correspond to the input. The entry for one row corresponding to state q and the column corresponding to input a is the state $\delta(q, a)$. For example:

δ/Σ	a	b
$\rightarrow q_0$	q_1	q_2
q_1	q_2	q_0
$*q_2$	q_2	q_2

Fig. 2.5.

q_0 is the starting state and q_2 is the final state,

$$Q = \{q_0, q_1, q_2\}$$
$$F = \{q_2\}$$
$$\delta(q_0, a) = q_1, \delta(q_0, b) = q_2$$
$$\delta(q_1, a) = q_2, \delta(q_1, b) = q_0$$
$$\delta(q_2, a) = q_2, \delta(q_2, b) = q_2$$

Let us draw the transition diagram for the transition table shown in Fig. 2.6.

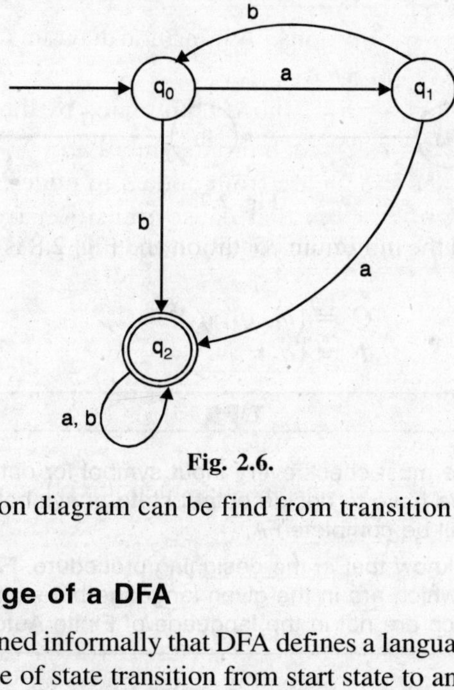

Fig. 2.6.

So clearly transition diagram can be find from transition table and vice-versa is also true.

2.5.3. The Language of a DFA

It have been explained informally that DFA defines a language. The set of all strings that result in a sequence of state transition from start state to an accepting state.

Now we can define the language of a DFA $M = (Q, \Sigma, \delta, q_0, F)$. This language is denoted by $L(M)$, and defined by

$$L(M) = \left\{w / \hat{\delta}(q_0, w) \text{ is in } F\right\}$$

* $\hat{\delta}$ (extended transition function) will be explained afterwords.

That is the language of M is the set of strings (w) that take the start q_0 to one of the accepting states.

TIPS

If L is $L(M)$ for some deterministic finite automata, then we say L is a regular language.

Now for practical understanding of FA, let us discuss the some examples.

Example 2.1. Design a FA that accepts set of strings such that every string ends in 00, over alphabet

$$\{0, 1\} \text{ i.e., } \Sigma = \{0, 1\}$$

Solution. Let $M = (Q, \Sigma, \delta, q_0, F)$ be the DFA

q_0 : initial state.

Σ : {0, 1} is given.

So according to the problem we have to design a FA which accepts strings $w = w'\,00$.

Where w' can be any combination of 0's and 1's for example 011100, 111100, 00, 100, 01010100 etc.

Now first we will decide the approach to design FA. It is not an easy task to think FA as a whole. So first we have to fulfill minimum condition *i.e.,* every string end in 00.

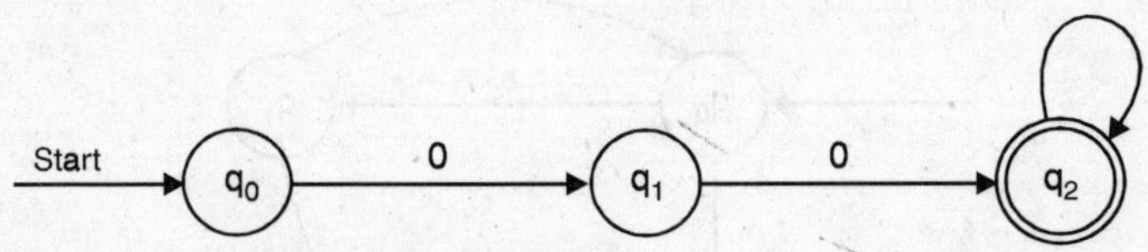

Fig. 2.7.

So Fig. 2.7 satisfied the minimum condition and Fig. 2.8 is the transition diagram of required FA.

$$Q = \{q_0, q_1, q_2\}$$
$$F = \{q_2\}$$

TIPS

Being the DFA we must check every input symbol for output state from every state. So we have to decide output state at symbol 1 from q_0, q_1 and q_2. Then it will be complete FA.

It is important to know that in the designing procedure, FA must accept all those strings which are in the given language but should not accept those strings which are not in the language of Finite Automata.

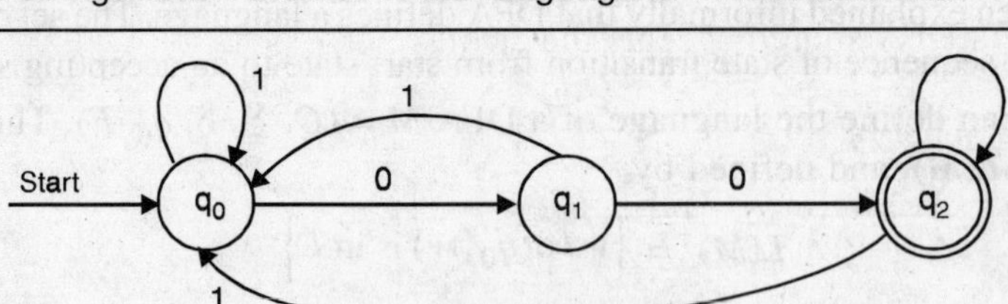

Fig. 2.8.

Facts in Designing Procedure of FA

On the basis of example 2.1, we can decide some *facts or observation* for FA.

(*a*) In the designing of the FA, first of all we have to analyse the set of strings *i.e.,* language which is accepted by the FA.

(*b*) Make sure that every state is check for the output state for every input symbol of Σ.

(*c*) Make sure that no state must have two different output state for a single input symbol.

(*d*) Make sure, that there is one initial state and at least one final state in transition diagram of FA.

Consideration of above discussed points in the designing, will minimize the error.

Example 2.2. Construct a FA that accepts set of strings where the number of 0's in every string is multiple of three over alphabet $\Sigma = \{0, 1\}$.

Solution. Multiple of three means numbers of 0's in the strings may be 0, 3, 6, 9, 12, ... in the number at any position.

Let FA be $M = (Q, \Sigma, \delta, q_0, F)$

Here it is necessary to consider acceptance of Null ($\in$) string since it contains number of 0's zero times which is divisible by three so it must be very clear that here initial state *i.e.*, q_0 is also final state. Here positions of 0's and 1's are irrelevant but number of 0's should be divisible by 3.

$$Q = \{q_0, q_1, q_2\}$$
$$F = \{q_0\}$$

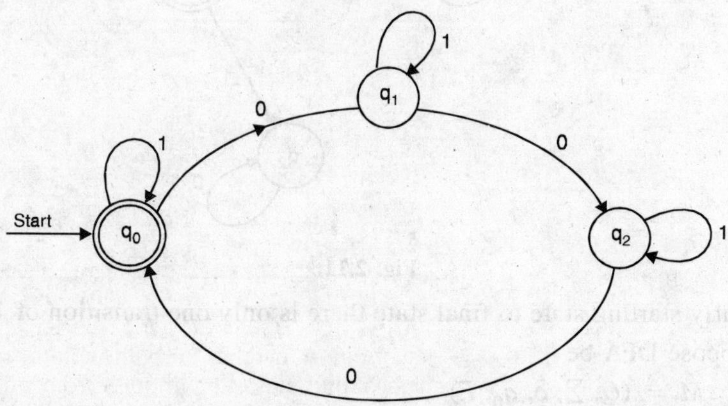

Fig. 2.9.

Example 2.3. Design a FA which accepts set of strings containing exactly four 1's in every string over alphabet $\Sigma = \{0, 1\}$.

Solution. DFA should accepts the strings 1111, 0101011, 0110110000 ..... .

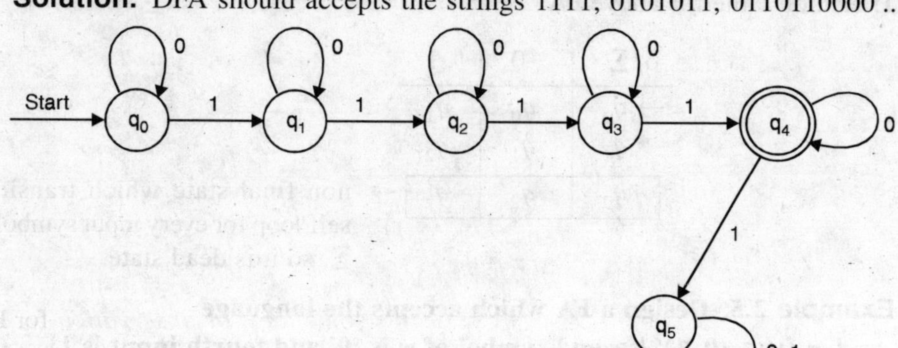

Fig. 2.10.

Let DFA be $M = (Q, \Sigma, \delta, q_0, F)$

$Q = \{q_0, q_1, q_2, q_3, q_4, q_5\}$

$F = \{q_4\}$

From q_0 to q_4 transitions four 1's are already there and q_4 is final state so 1 from q_4 should go on a trap path, that is why q_5 is there so q_5 is called trap or dead state.

"Dead states are those non final states which transits in itself for all the input symbol. So q_5 is the example of dead state."

Example 2.4. Design a FA that accepts strings containing exactly 1 over alphabet {0, 1}

Solution. String set according to the problem, which is accepted by the FA is like

1, 0001, 01000, 0001000, ...

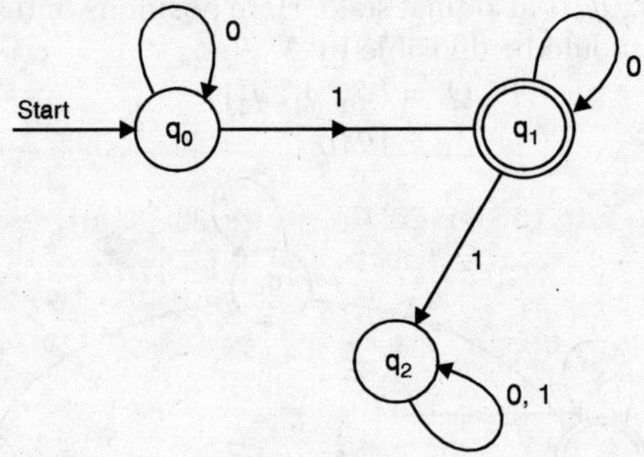

Fig. 2.11.

So clearly starting state to final state there is only one transition of 1.

Let suppose DFA be

$M = (Q, \Sigma, \delta, q_0, F)$

$Q = \{q_0, q_1, q_2\}$

$F = \{q_1\}$

q_2 is the dead state.

Transition table for above FA

δ/Σ	0	1
$\rightarrow q_0$	q_0	q_1
$* q_1$	q_1	q_2
q_2	q_2	q_2

$\rightarrow$ non final state which transit in self loop for every input symbol in Σ, so it is dead state.

Example 2.5. Design a FA which accepts the language

$L = \{w \in (0, 1)^*$/second symbol of w is '0' and fourth input is 1}.

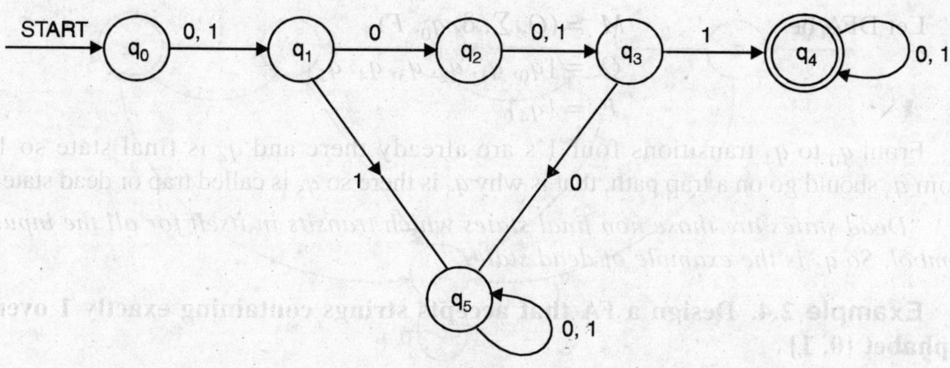

Fig. 2.12.

Solution. So after analysing the problem we find that every string of the language is like

0001, 1011, 10110000, ...

Let FA is $M = (Q, \Sigma, \delta, q_0, F)$

$Q = \{q_0, q_1, q_2, q_3, q_4, q_5\}$

$F = \{q_4\}$

q_5 is dead state.

Example 2.6. For the given FA write the language and also give the transition table

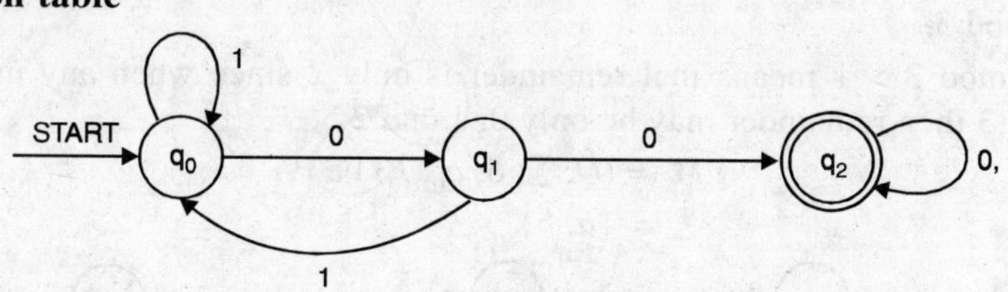

Fig. 2.13.

Solution. Let the language be L for FA shown in Fig. 2.13.

$L = \{w \in (0, 1)^*$/every string w of the language containing 00 as substring$\}$.

Transition table:

δ/Σ	0	1
$\rightarrow q_0$	q_1	q_0
q_1	q_2	q_0
$* \ q_2$	q_2	q_2

Example 2.7. Design FA for the language

$$L = \{(01)^i \ 1^{2j} \mid i \geq 1, J \geq 1\}.$$

Solution. By analysing language L, it is clear that FA will accepts strings start with any number of 01 (not empty) and end with even number of 1's.

Let FA be $M = (Q, \Sigma, \delta, q_0, F)$

q : initial state.

δ : is transition function.

$\Sigma = \{0, 1\}$ given.

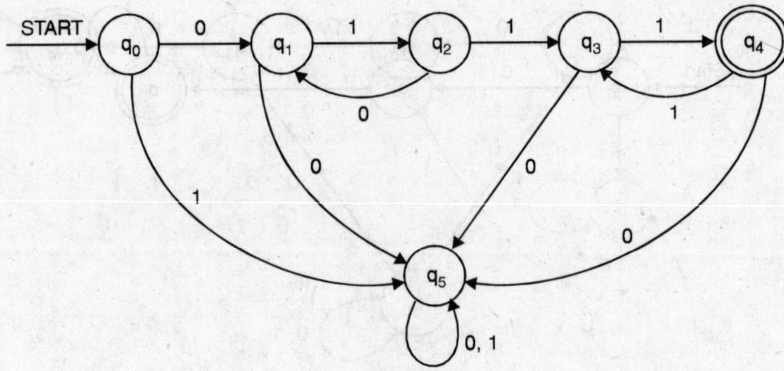

Fig. 2.14.

$$Q = \{q_0, q_1, q_2, q_3, q_4, q_5\}$$
$$F = \{q_4\}$$

q_5 is dead state.

Example 2.8. Design DFA for the language

$$L = \{w \in (a, b)^* / n_b(w) \bmod 3 > 1\}.$$

Solution. Let us first analyse the language L.

$n_b(w)$: number of b's in every string of language.

$L_b(w)$ mod 3 : number of b's in string is divided by three and remainder is result of $n_b(w)$ mod 3.

$n_b(w)$ mod 3 > 1 means that remainder is only 2 since when any number is divided by 3 then remainder may be only 0, 1 and 2.

Let $M = (Q, \Sigma, \delta, q_0, F)$ be FA

$$\Sigma = \{a, b\}$$

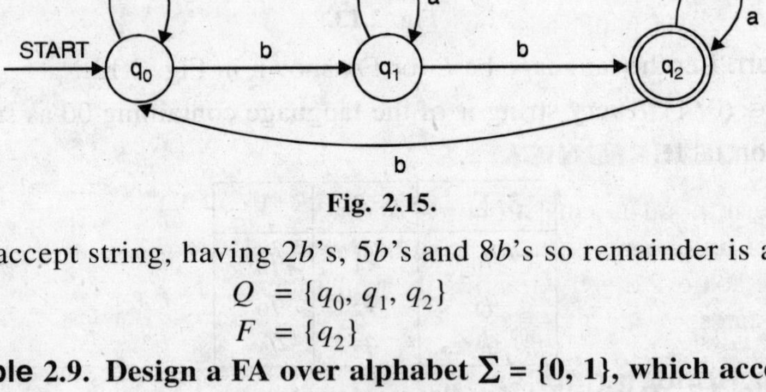

Fig. 2.15.

It will accept string, having $2b$'s, $5b$'s and $8b$'s so remainder is always two.

$$Q = \{q_0, q_1, q_2\}$$
$$F = \{q_2\}$$

Example 2.9. Design a FA over alphabet $\Sigma = \{0, 1\}$, which accepts the set of strings either start with 01 or end with 01.

Solution. By the analysis of problem, it is clear that FA will accepts the strings such as 01, 011111, 01000, 000101, 00011101, 111101..... .

Let FA $M = \{Q, \Sigma, \delta, q_0, F)$

$$\Sigma = \{0, 1\} \text{ given}$$

q_0 is initial state.

$$Q = \{q_0, q_1, q_2, q_3, q_4, q_5\}$$
$$F = \{q_2, q_5\}$$

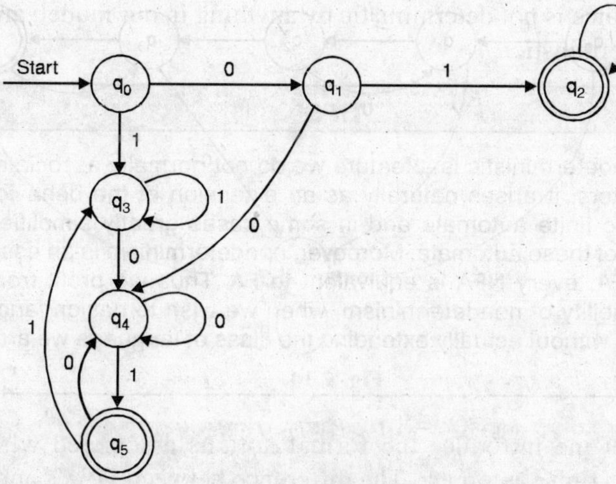

Fig. 2.16.

Example 2.10. Design a DFA for the language

$$L = \{w: n_a = 1, w \in (a, b)^*\}$$

Solution. Let DFA be $M = (Q, \Sigma, \delta, q_0, F)$

When we analyse the language, then it contains a set of string, every string contains exactly one '*a*' only, that is *a*, *bbb* ... *a*, *bbb* ... *abbb* ..., and *abbb*... . Then transition diagram will be as follows:

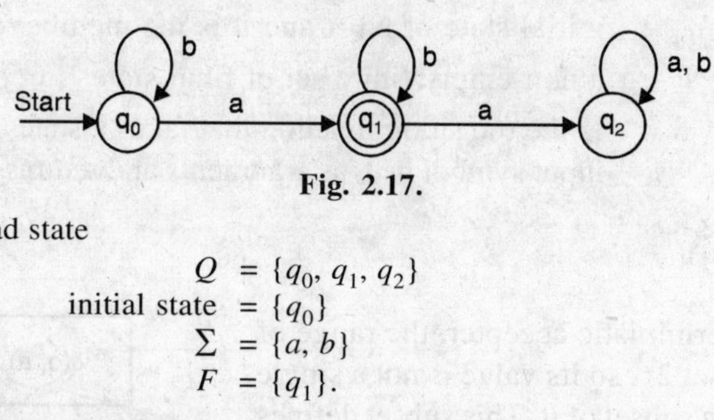

Fig. 2.17.

Here q_2 is dead state

$$Q = \{q_0, q_1, q_2\}$$
$$\text{initial state} = \{q_0\}$$
$$\Sigma = \{a, b\}$$
$$F = \{q_1\}.$$

2.6. NONDETERMINISM

Nondeterminism means a choice of moves for an automaton. Rather than prescribing a unique move in each situation, we allow a set of possible moves. Formally, we achieve this by defining the transition function so that its range is a set of possible states.

2.6.1. Why Nondeterminism **[MDU 2009]**

A "nondeterministic" finite automation (NFA) has the power to be in several states at once. This ability is often expressed as an ability to "guess", something about its output. Each NFA accepts a language that is also accepted by some DFA. That is the NFA's accept exactly the regular language, just as DFA's do. We can always convert a NFA to DFA. It is clear now, we can permit several possible "next states" for a given combination of current state and input symbol. The automation, as it reads the input string, may choose at each step to go into any one of these legal next states, the choice is not deterministic by anything in our model, and is therefore said to be nondeterminstic.

TIPS

Though nondeterministic is a feature we do not normally associate with real computers, it arises naturally as an extension of the behaviour of deterministic finite automata and in some cases greatly simplifies the description of these automata. Moreover, nondeterminism is an essential feature of FA, every NFA is equivalent to FA. Thus we profit from the greater flexibility of nondeterminism, when we wish to design language recognizers without actually extending the class of language we are able to describe.

Definition Now let me introduce the formal notions associated with nondeterministic finite automata. The difference between DFA's and NFA's will be pointed out as we do.

Let M be NFA then M is defined as

$$M = (Q, \Sigma, \delta, q_0, F)$$

where
- Q : is a non empty finite set of states in NFA.
- Σ : is a non-empty finite set of input symbols.
- q_0 : initial state of NFA and it is the member of Q.
- F : a non-empty finite set of final states in Q.
- δ : is the transition function that takes a states and Q and an input symbol in Σ as arguments and returns a subset of Q.

that is $Q \times (\Sigma \cup \{\in\}) \rightarrow 2^Q$

So in nondeterministic accepter, the range of δ is in the power set 2^Q, so its value is not a single element of Q, but a subset of it. This subset defines the set of possible states that can be reached by transition.

Let q be an state of Q and $a \in \Sigma$ then

We can transit from a state q on same input a to different states $q_0, q_1, ..., q_n$ in Q. This not possible in the case of FA.

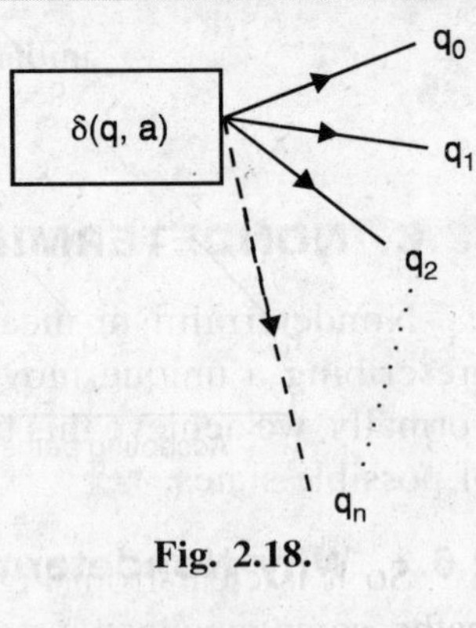

Fig. 2.18.

Example 2.11. Design a NFA for the language L = all strings over {0, 1} that have at least two consecutive 0's or 1's.

Solution. By the analysis it is clear that NFA will accepts the strings of the pattern <u>00</u>, <u>11</u>, 1010<u>00</u>, 10<u>11</u>00, ...

Let NFA be $M = (Q, \Sigma, \delta, q_0, F)$

$$\Sigma = \{0, 1\}$$

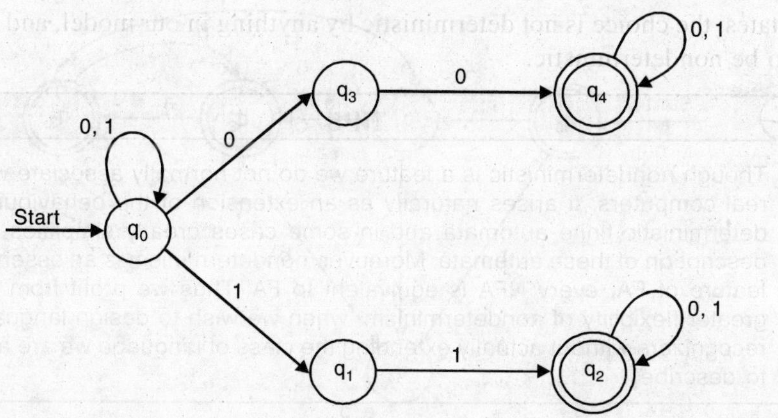

Fig. 2.19.

Note that state q_3 does not go anywhere for an input of 1. **We use the terminology that the path "dies" if in q_3 getting an input 1.**

The tree of states this NFA is in for the input 0100011.

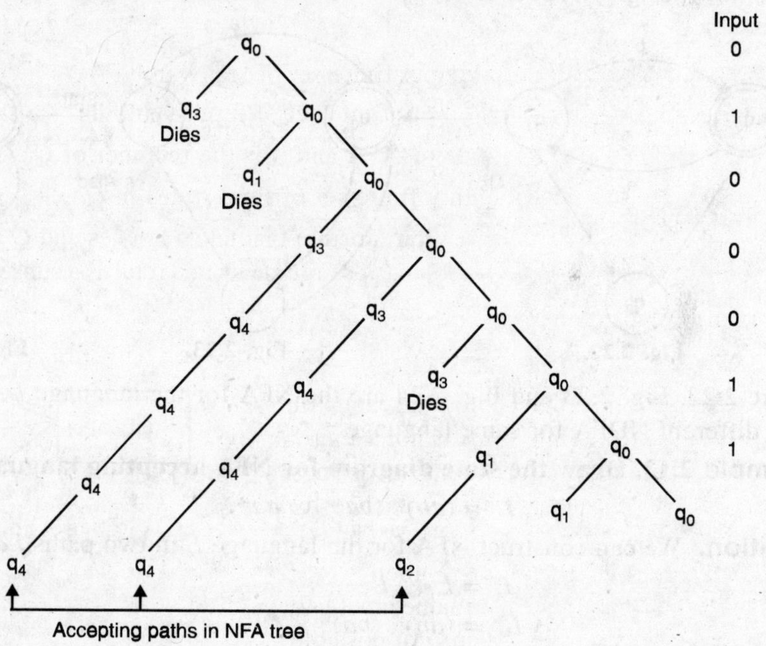

Fig. 2.20.

So it is clear from Fig. 2.20 that for a particular string there may be various paths, some may lead it to accepted states, and some may lead it to non accepted states and some may lead it to the blocked paths as $q_0 \to q_3$ as q_3 dies.

Example 2.12. Design a NFA for the language $L = (ab \cup aba)^*$.

Solution. Let us first analyse the given language *i.e.,*

$L = (ab \cup aba)^*$ means either *ab*, *aba* or any combination of *ab* and *aba* should be accepted by the NFA. null string ($\in$) is also accepted.

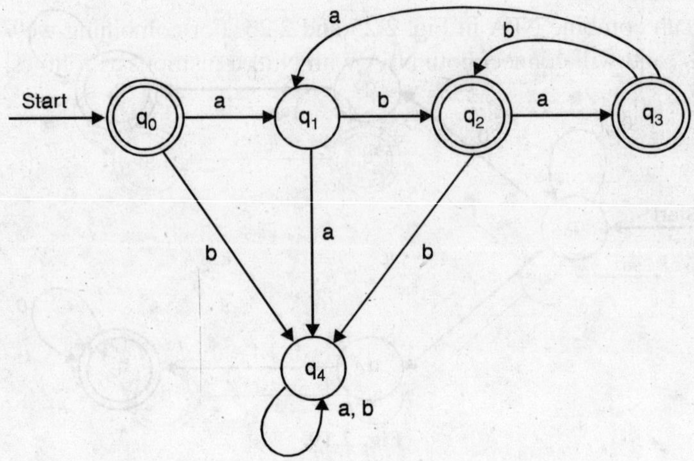

Fig. 2.21. DFA for the *L*.

If DFA for language *L*.

Let us now design NFA for *L*.

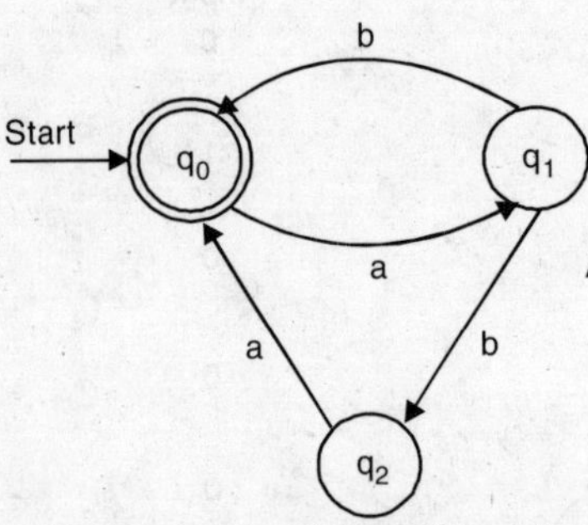

Fig. 2.22.

Also
⇒

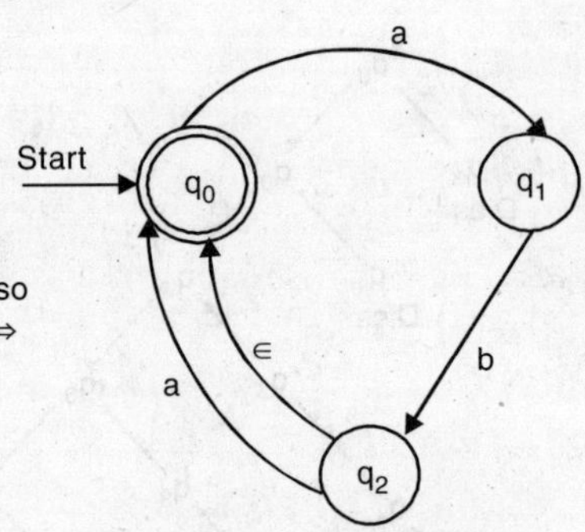

Fig. 2.23.

Also
⇒

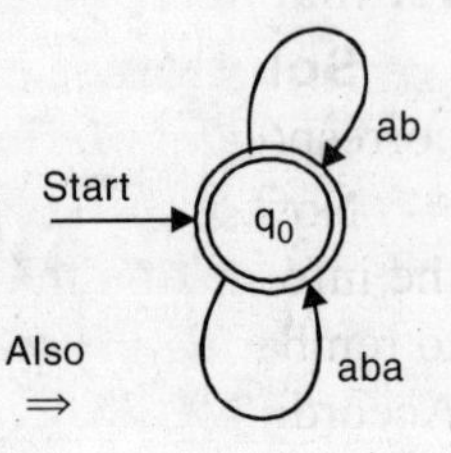

Fig. 2.24.

Figure 2.22, Fig. 2.23 and Fig. 2.24 are the NFA for the language *L*. Thus we can have different NDFA for same language.

Example 2.13. Draw the state diagram for NFA accepting language

$$L = (ab)^* (ba)^* \cup aa^*.$$

Solution. We can construct NFA for the language *L* in two paths *i.e.*,

$$L = L_1 \cup L_2$$

$$L_1 = (ab)^* (ba)^*$$

$$L_2 = aa^*$$

First we can construct NFA for L_1

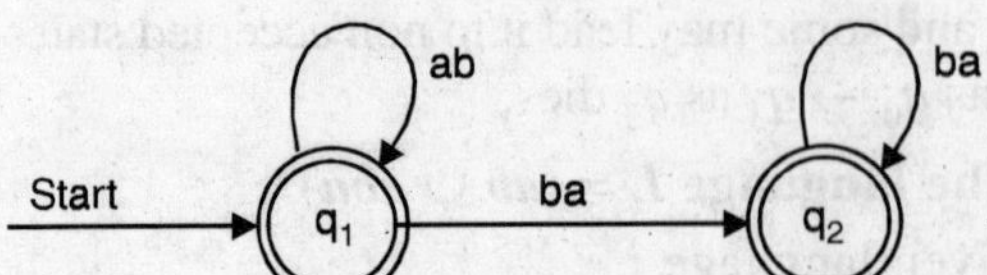

Fig. 2.25. NFA for L_1.

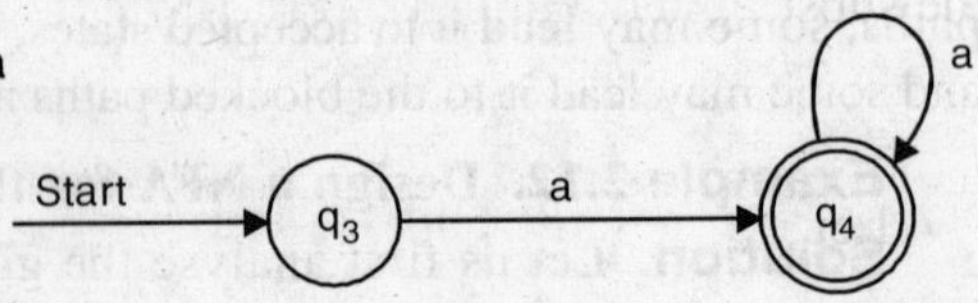

Fig. 2.26.

Secondly we construct NFA for L_2. (Fig. 2.26)

Now we can combine NFA in Fig. 2.25 and 2.26. For combining we will introduce another state q_0 and will connect both NFA with Null transitions as follows:

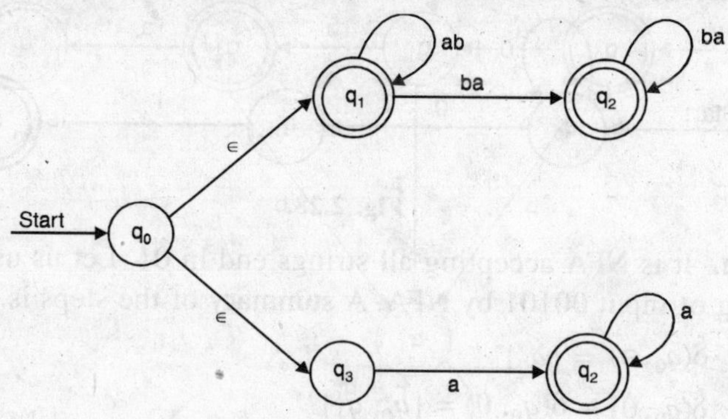

Fig. 2.27.

NFA in Fig. 2.27 is NFA for the language *L*.

Example 2.14. If L_1 and L_2 are languages over alphabet Σ, then design an FA that accept $L_1 - L_2$.

Solution. Let FA corresponding L_1 is $M_1 = (Q_1, \Sigma, \delta_1, q_1, F_1)$ and FA corresponding L_2 is $M_2 = (Q_2, \Sigma, \delta_2, q_2, F_2)$.

Now let us assume FA for $L_1 - L_2$ is *M*. *M* should be rember at each step both the information that 'M_1' rembers and information 'M_2' rembers. That is, it amount to rembering the ordered pair (p, q), where *P* is the current of M_1 and *q* is not in F_2. Accordingly, we construct:

$$M = (Q, \Sigma, \delta, q_0, F)$$

where
$$Q = Q_1 \times Q_2$$
$$q_0 = (q_1, q_2)$$

and the transition function is defined as follows. $\delta((p, q), a) = (\delta_1 (p, a), \delta(q, a)$ for any $p \to Q_1$, $q \to Q_2$, $a \in \Sigma$. If

$$F = \{(p, q): p \in F_1 \text{ and } q \notin F_2\},$$

then *M* accepts the language $L_1 + L_2$.

2.6.2. Extended Transition Function for NFA

As for DFA's, we need to extend the transition function δ of an NFA to a function $\hat{\delta}$ that takes a state and string and returns the set of states.

Basis. $\hat{\delta}(q,\in) = \{q\}$. That is without reading any input symbols, we are only in the same state.

Induction. Suppose *w* is of the form $w = xa$, where *a* is the last symbol and *x* is the substring containing rest of *w*. Let us suppose that

$$\hat{\delta}(q, x) = \{p_1, p_2, \dots p_k\},$$

Let
$$\bigcup_{i=1}^{k} \delta(p_i, a) = \{r_1, r_2, r_3, \dots, r_m\}$$

Then $\hat{\delta}(q, w) = \{r_1, r_2, \dots r_m\}$. Less formally, we compute $\hat{\delta}(q, w)$ by first computing $\hat{\delta}(q, x)$, and by then following any transition from any of these states that is labeled *a*. Let us see an example for better under standing of the concept.

Example 2.15. Consider NFA in Fig. 2.28, and describe the processing of string 00101 by it.

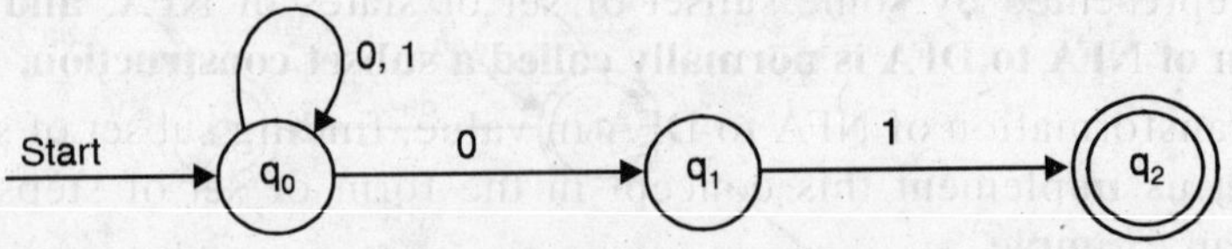

Fig. 2.28.

Solution. It is NFA accepting all strings end in 01. Let us use $\hat{\delta}$ to describe the processing of input 00101 by NFA. A summary of the steps is.

Step 1. $\hat{\delta}(q_0, \in) = \{q_0\}$

Step 2. $\hat{\delta}(q_0, 0) = \delta(q_0, 0) = \{q_0, q_1\}$

Step 3. $\hat{\delta}(q_0, 00) = \delta(q_0, 0) \cup \delta(q_1, 0) = \{q_0, q_1\} \cup \phi = \{q_0, q_1\}$

Step 4. $\hat{\delta}(q_0, 001) = \delta(q_0, 1) \cup \delta(q_1, 1) = \{q_0\} \cup \{q_2\} = \{q_0, q_2\}$

Step 5. $\hat{\delta}(q_0, 0010) = \delta(q_0, 0) \cup \delta(q_2, 0) = \{q_0, q_1\} \cup \phi = \{q_0, q_1\}$

Step 6. $\hat{\delta}(q_0, 00101) = \delta(q_0, 1) \cup \delta(q_1, 1) = \{q_0\} \cup \{q_2\} = \{q_0, q_2\}$.

Step 1 is the basis rule. We obtain step 2 by applying δ to the q_2, that is in the previous set, and get $\{q_0, q_1\}$ as a result. Step 3 is obtained by taking the union over two states in previous set of what we get when we apply δ to the input 0. That is, $\delta(q_0, 0) = \{q_0, q_1\}$, while $\delta\{q_1, 0\} = \phi$.

For step 4, we take union of $\delta(q_0, 1) = q_0$ and $\delta(q_1, 1) = \{q_2\}$ and step 5 and step 6 are obtained in similar way.

2.7. THE LANGUAGE OF NFA

Definition As we discussed, an NFA accepts a string *w* if it is possible to make any sequence of choice of next state while reading the characters of *w*, and go from the start state to any accepting state.

If *M* is a NFA

$$M = (Q, \Sigma, \delta, q_0, F) \text{ then}$$

$$L(M) = \left\{w / \hat{\delta}(q_0, w) \cap F \neq \phi\right\} \text{ is the Language of } M \text{ i.e.,}$$

NFA. *L*(*M*) is the set of strings *w* in Σ* such that $\hat{\delta}(q_0, w)$ contains at least one accepting state.

2.8. TRANSFORMATION OF NFA TO DFA

For every non deterministic finite automata there exist an equivalent deterministic finite automata. The equivalence is defined in terms of language acceptance. Since a NFA is a nothing but a finite automata in which zero, one or more transition on an input symbol is permitted, we can always construct a finite automata which will simulate all the moves of NFA on a particular input symbol in parallel, then get a finite automata in which there will be exactly one transition on every input symbol, hence it will be DFA equivalent to NFA.

Since the DFA equivalent of NFA is to simulate the moves in parallel state of a DFA will be combination of one or more states of NFA, hence every state of a DFA will be represented by some subset of set of states of NFA, **and therefore the transition of NFA to DFA is normally called a subset construction.**

So transformation of NFA to DFA in value, finding subset of set of states of NFA. Let us implement this concept in the form of set of steps as algorithm through an example.

2.8.1. Procedure For Converting NFA to Equivalent DFA

Let $M_2 = \{Q_2, \Sigma, q'_0, \delta_2, F_2\}$ be NFA that recoginzes the languages L. Then the DFA $M = \{Q, \Sigma, q_0, \delta, F\}$ that satisfies the following conditions recoginzes L:

$Q = 2^{Q_2}$, that is the set of all subsets of Q_2.

$q_0 = \{q'_0\}$

$\delta(q, a) = \bigcup_{p \in q} \delta(p, a)$ for each state q in Q and each symbol a in Σ and

$F = \{q \in Q \mid q \cap F_2 \neq \phi\}$.

To obtain a DFA $M = \{Q, \Sigma, \delta, q_0, F\}$ which accepts the same language as given NFA $M_2 = \{Q_2, \Sigma, q'_0, \delta_2, F_2\}$ does, we may proceed as follows :

Step 1. Initially $Q = \phi$.

Step 2. Put $\{q'_0\}$ into Q. $\{q'_0\}$ is the initial state of the DFA M.

Step 3. Then for each state q in Q do the following: add this new state, add $\delta(q, a) = \bigcup_{p \in q} \delta(p, a)$ to δ, where the δ on the right hand side is that of NFA M_2.

Step 4. Repeat step-3 till new states are there to add in Q, if there is new state found to add in Q, the process terminates. All the states of Q that contain accepting states of M_2 are accepting states of M.

Note: The states that are not reached from the initial state are not included in Q obtained by this procedure. Thus the set of states Q thus obtained is not necessarily equal to 2^{Q_2}.

Example 2.16. Convert the following NFA in to DFA.

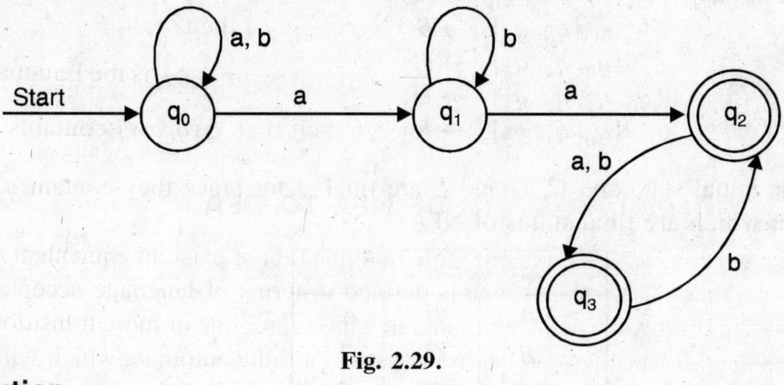

Fig. 2.29.

Solution.

Step 1. Seek all the transition from starting state q_0 for every symbol in Σ *i.e.*, (a, b). If you get a set of states for same input then consider that set as new single state as:

$$\delta(q_0, a) = \{q_0, q_1\}$$

in Fig. 2.29 q_0 transit to both q_0 and q_1 for input a.

$$\delta(q_0, b) = \{q_0\}$$

Step 2. In step 1 we are getting a new state $\{q_0, q_1\}$. Now repeat step 1 for this new state only *i.e.,* check all transitions of a and b (that is Σ) from $\{q_0, q_1\}$ as :

$$\delta(\{q_0, q_1\}, a) = \delta(q_0, a) \cup \delta(q_1, a) = \{q_0, q_1\} \cup \{q_2\}$$
$$= \{q_0, q_1, q_2\}$$
$$\delta(\{q_0, q_1\}, b) = \delta(q_0, b) \cup \delta(q_1, b) = \{q_0, q_1\}$$

Step 3. Repeat step 2 till you are getting any new state. All those states which consists at least one accepting state of given NFA as member state will be considered as final states.

$[q_0, q_1, q_2]$ is new state.

$$\delta(\{q_0, q_1, q_2\}, a) = \delta(q_0, a) \cup \delta(q_1, a) \cup \delta(q_2, a)$$
$$= \{q_0, q_1\} \cup \{q_2\} \cup \{q_3\}$$
$$= \{q_0, q_1, q_2, q_3\} \quad \text{(New state)}$$
$$\delta(\{q_0, q_1, q_2\}, b) = \delta(q_0, b) \cup (q_1, b)\ \delta(q_2, b)$$
$$= q_0 \cup q_1 \cup q_3 = \{q_0, q_1, q_3\} \quad \text{(New state)}$$
$$\delta(\{q_0, q_1, q_2, q_3\}, a) = \{q_0, q_1, q_2, q_3\} \quad \text{(Old state)}$$
$$\delta(\{q_0, q_1, q_2, q_3\}, b) = \{q_0, q_1, q_3, q_2\} \quad \text{(Old state)}$$
$$\delta(\{q_0, q_1, q_3\}, a) = \{q_0, q_1, q_2, q_3\} \quad \text{(Old state)}$$
$$\delta(\{q_0, q_1, q_3\}, b) = \{q_0, q_1, q_3, q_2\} \quad \text{(Old state)}$$

Let draw transition table

δ/Σ	a	b
$\to q_0$	$\{q_0, q_1\}$	$\{q_0\}$
$\{q_0, q_1\}$	$\{q_0, q_1, q_2\}$	$\{q_0, q_1\}$
$*\{q_0, q_1, q_2\}$	$\{q_0, q_1, q_2, q_3\}$	$\{q_0, q_1, q_3\}$
$*\{q_0, q_1, q_2, q_3\}$	$\{q_0, q_1, q_2, q_3\}$	$\{q_0, q_1, q_2, q_3\}$
$*\{q_0, q_1, q_3\}$	$\{q_0, q_1, q_2, q_3\}$	$\{q_0, q_1, q_2, q_3\}$

Let us say

$$q_0 \to A$$
$$\{q_0, q_1\} \to B$$
$$\{q_0, q_1, q_2\} \to C$$
$$\{q_0, q_1, q_2, q_3\} \to D$$
$$\{q_0, q_1, q_3\} \to E$$

A is initial state and C, D and E are final states since they contain q_2 and q_3 as member which are final states of NFA.

So

δ/Σ	a	b
$\to A$	B	A
B	C	B
$*C$	D	E
$*D$	D	D
$*E$	D	D

Let us draw transition diagram

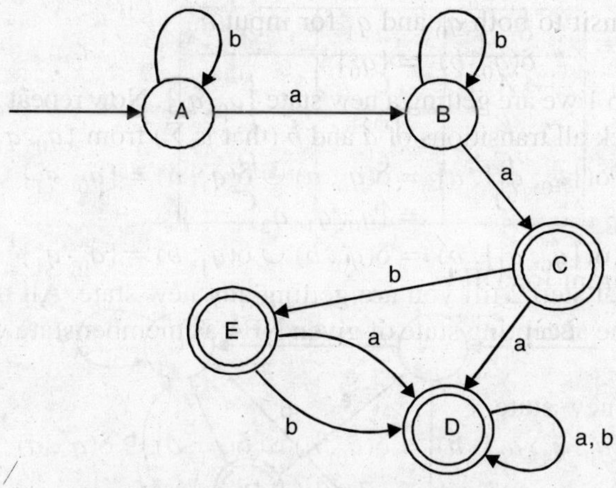

Fig. 2.30. DFA equivalent of given NFA.

It is a DFA.

Example 2.17. Given NFA is

δ/Σ	a	b
$\rightarrow q_0$	$\{q_0, q_1\}$	$\{q_2\}$
q_1	$\{q_0\}$	$\{q_1\}$
$*q_2$	ϕ	$\{q_0, q_1\}$

Convert it in to DFA.

Solution. Do the same process as in example 2.16.

$$\delta(q_0, a) = \{q_0, q_1\} \quad \text{(New)}$$

$$\delta(q_0, b) = \{q_2\} \quad \text{(New)}$$

$$\delta(\{q_0, q_1\}, a) = \delta(q_0, a) \cup \delta(q_1, a) = \{q_0, q_1\} \cup \{q_0\}$$

$$= \{q_0, q_1\} \quad \text{(Old)}$$

$$\delta(\{q_0, q_1\}, b) = \delta(q_0, b) \cup \delta(q_1, b) = \{q_1\} \cup \{q_1\}$$

$$= \{q_1, q_2\} \quad \text{(New)}$$

$$\delta\{q_2, a) = \{\phi\} \quad \text{(Empty set)}$$

$$\delta(q_2, b) = \{q_0, q_1\} \quad \text{(Old)}$$

$$\delta(\{q_1, q_2\}, a) = \delta(q_1, a) \cup \delta(q_2, a) = \{q_0\} \cup \{\phi\}$$

$$= \{q_0\} \quad \text{(Old)}$$

$$\delta(\{q_1, q_2\}, b) = \delta(q_1, b) \cup \delta(q_2, b) = \{q_1\} \cup \{q_0, q_1\}$$

$$= \{q_0, q_1\} \quad \text{(Old)}$$

Let say

$$q_0 \rightarrow A \quad \text{(Initial state)}$$

$$q_2 \rightarrow B \quad \text{(Final state)}$$

$$\{q_0, q_1\} \rightarrow C$$

$$\{q_1, q_2\} \rightarrow D \quad \text{(Final state)}$$

Transform table for DFA

δ/Σ	a	b
$\rightarrow A$	C	B
$*B$	$\{\phi\}$	C
C	C	D
$*D$	A	C

Transition diagram for DFA

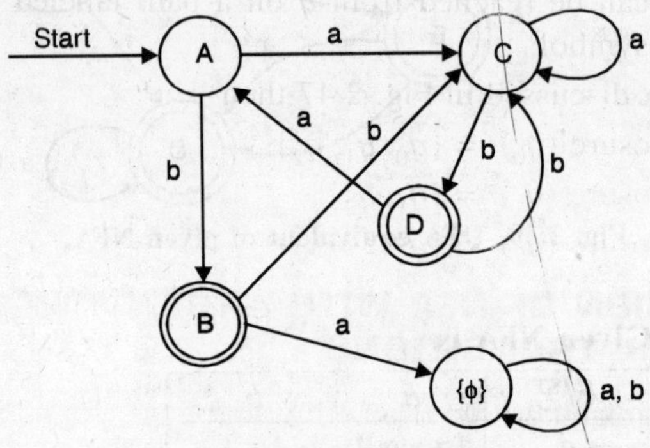

Fig. 2.31.

Here $\{\phi\}$ state is nothing but interpreted as if there is a input 'a' on state B then there is no path for it and machine will be hang.

2.9. NFA WITH ∈–TRANSITIONS

TIPS

If a FA is modified to permit transition without input symbols, along with zero, one or more transition on input symbols, then we get a NFA with ∈-transitions, because the transition made without symbols are called as ∈-transitions.

Consider the NFA given by following diagram.

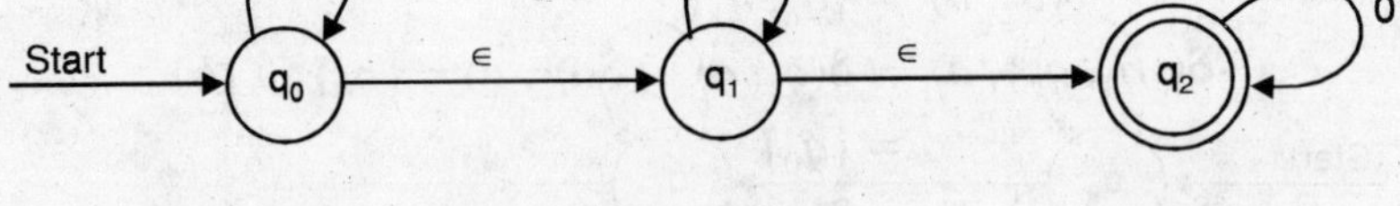

Fig. 2.32.

This is a NFA with ∈-transitions because it is possible to make transition from state q_0 to q_1 without consuming any of the input symbols. Being a finite Automata, NFA with ∈-transitions will also be denoted as five tuple.

$$M = (Q, \Sigma, \delta, q_0, F)$$

where Q, Σ, q_0 and F are having usual meaning, and transition function δ defines a mapping from $Q \times (\Sigma \cup \in)$ to 2^Q.

2.10. ACCEPTANCE OF A STRING BY NFA WITH ∈-TRANSITIONS

A string w in Σ^* will be accepted by NFA with ∈-transition, if there exist at least one path corresponding w, which starts in an initial state, and ends in one of the final states. Since this path may formed by ∈-transition as well non ∈-transitions to find out whether w is accepted or not by the NFA with ∈ moves, we require to defined a function ∈-closure (q), where q is a state of the automata.

The function ∈-closure (q) is defined as follows:

∈-closure (q) = set of all those states of the automata (NFA with ∈-transitions) which can be reached from q on a path labeled by ∈ *i.e.,* without consuming any input symbol.

Consider the NFA discussed in Fig. 2.47 then

$$\in\text{-closure } (q_0) = \{q_0, q_1, q_2\}$$
$$\in\text{-closure } (q_1) = \{q_1, q_2\}$$
$$\in\text{-closure } (q_2) = \{q_2\}$$

2.11. CONVERSION OF NFA WITH ∈-TRANSITIONS TO NFA WITHOUT ∈-TRANSITIONS

There is a conversion algorithm from NFA with ∈-transition *s* to *a* NFA without ∈-transitions !

We will learn this algorithm by solving an example.

Example 2.18. Consider the NFA with ∈-transition $M = (Q, \Sigma, \delta, q_0, F)$

$$Q = \{q_0, q_1, q_2\}$$
$$\Sigma = \{a, b, c\} \text{ and } \in \text{ moves.}$$
$$\text{Initial state} = \{q_0\}$$
$$F = \{q_2\}$$

Transition Table

δ/Σ	a	b	c	∈
$\rightarrow q_0$	$\{q_0\}$	$\{\phi\}$	$\{\phi\}$	$\{q_1\}$
q_1	$\{\phi\}$	$\{q_2\}$	$\{\phi\}$	$\{q_2\}$
$* q_2$	$\{\phi\}$	$\{\phi\}$	$\{q_2\}$	$\{\phi\}$

and transition diagram is following:

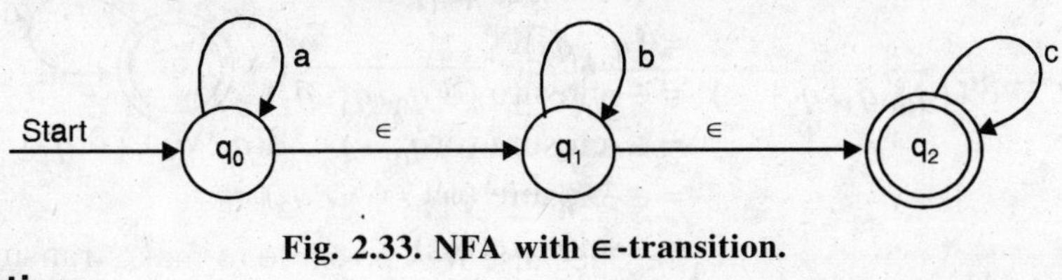

Fig. 2.33. NFA with ∈-transition.

Solution.

Step 1. Find the states of NFA without ∈-transition including initial state and final states as follows:

(*i*) The language accepted by the above NFA with ∈-transform is the set of strings over $\{a, b, c\}$ including the null string and all strings with any number of a's followed by any number of b's and follows by any number of c's.

(*ii*) There will be the same number of states but the names can be constructed by writing the state name as the set of states in the ∈-closure.

(*iii*) Initial state will be ∈-closure of initial state of NFA with ∈-transition in Fig. 2.33 as

$$\in\text{-closure }(q_0) = \{q_0, q_1, q_2\} \quad \text{(new initial state for NFA, without } \in \text{ transition)}$$

(*iv*) Rest of the states are

$$\in\text{-closure }(q_1) = \{q_1, q_2\} \quad \text{(New state)}$$
$$\in\text{-closure }(q_2) = \{q_2\} \quad \text{(New state)}$$

(*v*) The final states of NFA without ∈-transition are all those new states which contains final state of NFA with ∈-transition as member.

So $\{q_0, q_1, q_2\}$, $\{q_1, q_2\}$ and $\{q_2\}$ all are final states.

So if NFA without ∈-transition is

$$M' = (Q', \Sigma, \delta', q_0', F')$$
$$Q' = (\{q_0, q_1, q_2\}, \{q_1, q_2\}, \{q_2\})$$
$$q_0' = \{q_0, q_1, q_2\}$$
$$F' = (\{q_0, q_1, q_2\}, \{q_1, q_2\}, \{q_2\})$$

Step 2. Now we have to decide δ′ to find out the transitions as follows:

Very carefully consider each old machine transition in each row, you can ignore ϕ entries and ∈-transitions column. In the old machine $\delta(q_0, a) = q_0$ thus in new machine *i.e.*, in M', $\delta'(\{q_0, q_1, q_2\}, a) = \{q_0, q_1, q_2\}$ this is just because the new machine accepts the same language as the old machine and must at least have the same transitions for the new state names. We can allow interpret it as follows:

$$\begin{aligned}
\delta'(\{q_0, q_1, q_2\}, a) &= \in\text{-closure }(\delta(q_0, q_1, q_2), a)\\
&= \in\text{-closure }(\delta(q_0, a) \cup \delta(q_1, a) \cup \delta(q_1, a))\\
&= \in\text{-closure }(q_0 \cup \phi \cup \phi)\\
&= \in\text{-closure }(q_0)\\
&= \{q_0, q_1, q_2\}
\end{aligned}$$

$$\begin{aligned}
\text{Similarly } \delta'(\{q_0, q_1, q_2\}, b) &= \in\text{-closure }(\delta(q_0, q_1, q_2), b)\\
&= \in\text{-closure }(\delta(q_0, b) \cup \delta(q_1, b) \cup \delta(q_2, b))\\
&= \in\text{-closure }(\phi \cup q_1 \cup \phi)\\
&= \in\text{-closure }(q_1)\\
&= \{q_1, q_2\}.
\end{aligned}$$

$$\begin{aligned}
\delta'(\{q_0, q_1, q_2\}, c) &= \in\text{-closure }(\delta(q_0, q_1, q_2), c)\\
&= \in\text{-closure }(\delta(q_0, c) \cup \delta(q_1, c) \cup \delta(q_1, c)\\
&= \in\text{-closure }(\phi \cup \phi \cup q_2)\\
&= \in\text{-closure }(q_2)\\
&= \{q_2\}.
\end{aligned}$$

In the same fashion we can write

$$\delta'(\{q_1, q_2\}, a) = \phi$$
$$\delta'(\{q_1, q_2\}, b) = \{q_1, q_2\}$$
$$\delta'(\{q_1, q_2\}, c) = \{q_2\}$$

$$\delta'(q_2, a) = \phi$$
$$\delta'(q_2, b) = \phi$$
$$\delta'(q_2, c) = \{q_2\}$$

So transition table for NFA without $\in$-transition will be as follows:

δ'/Σ	a	b	c
$\rightarrow * \{q_0, q_1, q_2\}$	$\{q_0, q_1, q_2\}$	$\{q_1, q_2\}$	$\{q_2\}$
$* \{q_1, q_2\}$	ϕ	$\{q_1, q_2\}$	$\{q_2\}$
$* \{q_2\}$	ϕ	ϕ	$\{q_2\}$

Let us say $\{q_0, q_1, q_2\}$ as q_x

$\{q_1, q_2\}$ as q_y

and $\{q_2\}$ as q_z

So transition diagram is as follows:

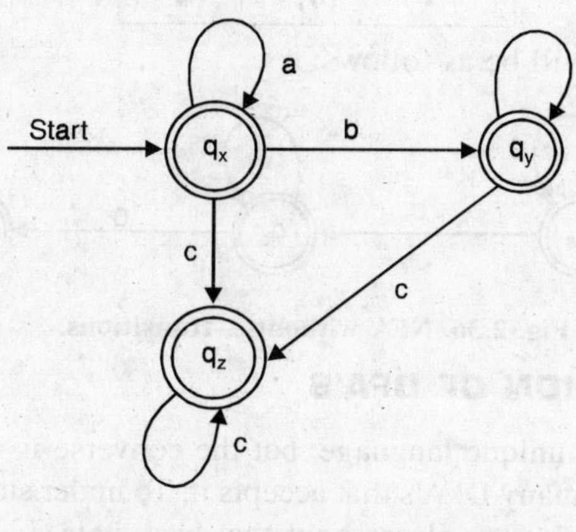

Fig. 2.34. NFA without $\in$-transitions.

Let us see one more example.

Example 2.19. Consider the NFA given by following diagram

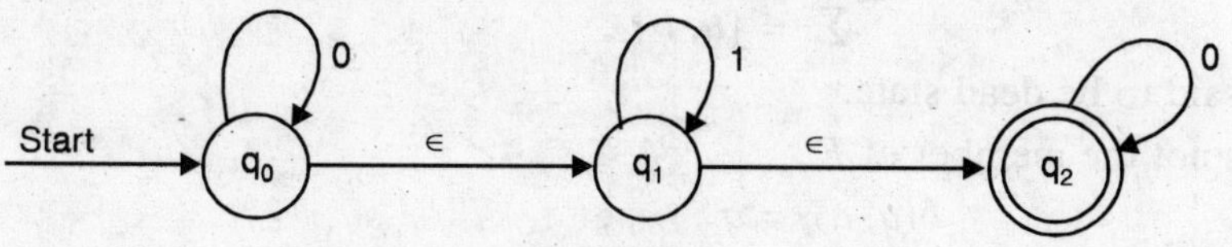

Fig. 2.35.

Find the equivalent NFA without $\in$-transitions.

Solution. Let NFA without $\in$-transition be

$$M' = \{Q', \Sigma, q_0', \delta', F')$$

So as we see in Example 2.18.

$$Q' = (\in\text{-closure } (q_0), \in\text{-closure } (q_1), \in\text{-closure } (q_2))$$
$$= (\{q_0, q_1, q_2\}, \{q_1, q_2\}, \{q_2\}) \cong (q_a, q_b, q_c)$$
$$\Sigma = \{0,1\}$$

$$F' = (q_a, q_b, q_c) \text{ as } q_a \rightarrow \{q_0, q_1, q_2\}$$
$$q_b \rightarrow \{q_1, q_2\}$$
$$q_c \rightarrow \{q_2\}$$

and
$$\delta'(q_a, 0) = q_a$$
$$\delta'(q_a, 1) = q_b$$
$$\delta'(q_b, 0) = q_c$$
$$\delta'(q_b, 1) = q_b$$
$$\delta'(q_c, 0) = q_c$$
$$\delta'(q_c, 1) = \phi$$

So transition table is as follows:

δ'/Σ	*0*	*1*
$\rightarrow * q_a$	q_a	q_b
$* q_b$	q_c	q_b
$* q_c$	q_c	ϕ

and transition diagram will be as follows:

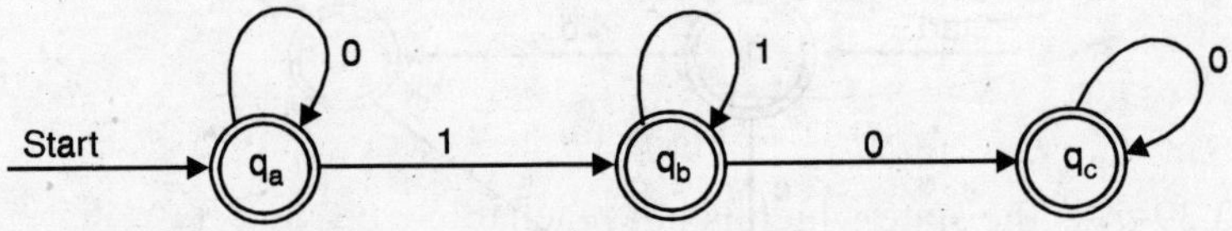

Fig. 2.36. NFA without $\in$-transitions.

2.12. MINIMIZATION OF DFA'S

Any DFA defines a unique language, but the converse is not true. For a given language, there may be many DFA's that accepts it. To under stand the minimization first of all one should be very clear about the dead state, inaccessible states and indistinguishable states, so let us discuss them first.

Dead state. All those non final states which transit to itself for all input symbols in Σ, are called Dead state.

Let $\Sigma = \{a, b\}$

then q is said to be dead state.

If q is not the member of F

and $\delta(q, a) = q$

i.e., $\delta(q, b) = q$

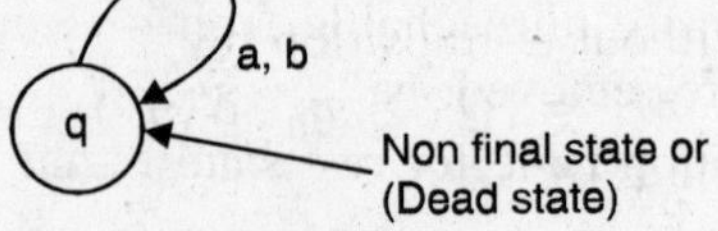

Fig. 2.37.

in the following Fig. 2.38 q_4 is dead state.

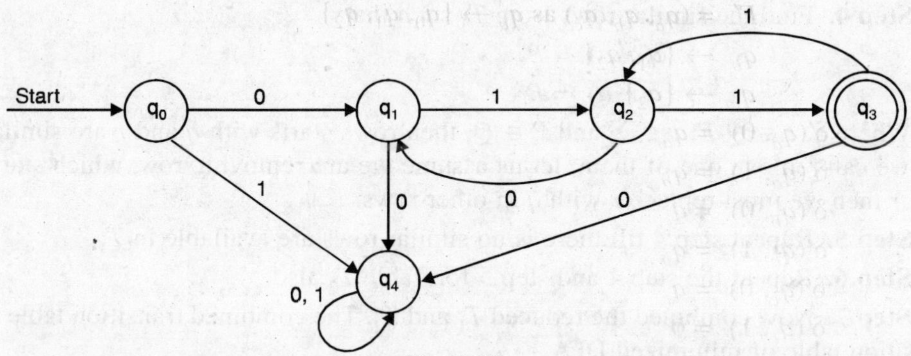

Fig. 2.38.

Inaccessible states. All those states which can never be reached from initial state are called inaccessible states. For example:

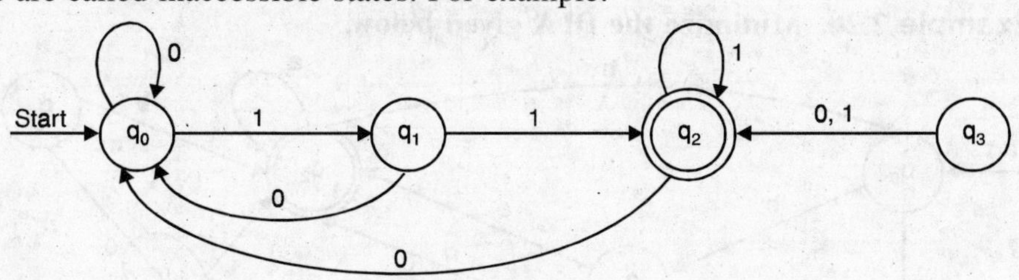

Fig. 2.39.

In Fig. 2.39 q_3 is the state which is not reachable from q_0 (initial state). So q_3 is the unreachable state.

Indistinguishable states. Two states q and r of DFA are called indistinguishable if

$$\hat{\delta}(q, w) \in F \ \Rightarrow \hat{\delta}(r, w) \in F$$

and $\quad \hat{\delta}(q, w) \notin F \ \Rightarrow \hat{\delta}(r, w) \notin F$

for all $\quad w \in \Sigma^*$.

Distinguishable states. Two states q_1 and q_2 of DFA are called distinguishable if $\hat{\delta}(q_1, w) \in F$ and $\hat{\delta}(q_2, w) \notin F$ or vice versa; for all $w \in \Sigma^*$.

2.12.1. Minimization Algorithm for DFA

One important result of finite automata, both theoretically and practically, is that for any regular language there is a unique DFA having the smallest number of states that accepts it. Let $M = (Q, \Sigma, \delta, q_0, F)$ be a DFA that accepts a language L. Then the following algorithm produces the DFA, denote it by M_1, that has the smallest number of states among the DFAs that accept L.

Minimization Algorithm for DFA:

Step 1. All states that are unreachable from the initial state via any set of transitions of the DFA M are removed.

Step 2. Draw the transition table for rest states, after removing the unreachable states.

Step 3. Split the transition table in two tables T_1 and T_2, where

T_1 = {contains all rows which starts with states from Q-F}

T_2 = {contains all rows which starts with states from F}

Step 4. Find the similar rows from T_1 such that:

$$\delta(q, a) = P$$
$$\delta(r, a) = P$$

Where $q, r \in Q\text{-}F$, $a \in \Sigma$ and $P \in Q$, then rows starts with q and r are similar and we can remove one of them, let us assume we are removing row, which starts with r then we must replace r with q in other rows.

Step 5. Repeat step 4 till there is no similar rows are available in T_1.

Step 6. Repeat the step 4 and step 5 for Table T_2 also.

Step 7. Now combined the reduced T_1 and T_2. The combined transition table is transition table of minimized DFA.

Note: Basically by applying above algorithm, we are removing non-reachable states, dead states and indistinguishable states.

Example 2.20. Minimize the DFA given below.

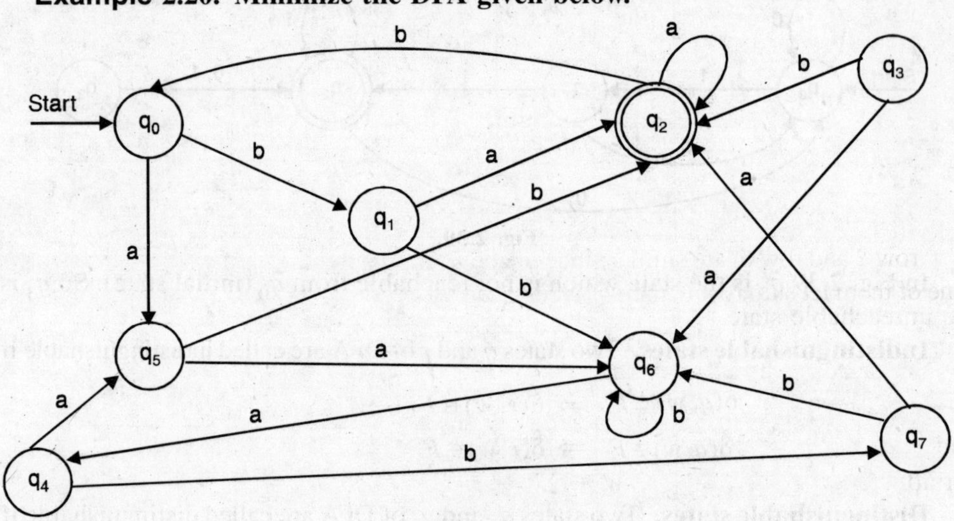

Fig. 2.40.

Solution. Let us discuss the solution step by step.

Step 1. Remove the unreachable states from the given DFA. The DFA given in Fig. 2.40, q_3 is unreachable state so remove it.

Step 2. Draw the transition table for rest states, after removing q_3. (unreachable states).

δ/Σ	a	b
$\rightarrow q_0$	q_5	q_1
q_1	q_2	q_6
$* q_2$	q_2	q_0
q_4	q_5	q_7
q_5	q_6	q_2
q_6	q_4	q_6
q_7	q_2	q_6

Step 3. Now divide rows of transition table in two sets as:

(*i*) One set contains only those rows, which starts from non-final states *i.e.*,

q_0	q_5	q_1
q_1	q_2	q_6
q_4	q_5	q_7
q_5	q_6	q_2
q_6	q_4	q_6
q_7	q_2	q_6

Set-1 (say)

(*ii*) Other set contains only those rows, which starts from final states.

q_2 q_2 q_0 **Set-2 (say)**

Step 4. Apply step 4 on both the sets individually. Now first consider set-1

q_0	q_5	q_1	$\rightarrow$	row-1
q_1	q_2	q_6	$\rightarrow$	row-2
q_4	q_5	q_7	$\rightarrow$	row-3
q_5	q_6	q_2	$\rightarrow$	row-4
q_6	q_4	q_6	$\rightarrow$	row-5
q_7	q_2	q_6	$\rightarrow$	row-6

row 2 and row 6, are similar since q_1 and q_7 transit to same states on a and b. So skip one of them let skip row 6, then q_7 is replaced by q_1 in the rest. Write the rest as follows :

q_0	q_5	q_1	$\rightarrow$	row-1
q_1	q_2	q_6	$\rightarrow$	row-2
q_4	q_5	q_1	$\rightarrow$	row-3
q_5	q_6	q_2	$\rightarrow$	row-4
q_6	q_4	q_6	$\rightarrow$	row-5

Step 5. Repeat step 4 on the rest of set 1 till you are finding similar rows and write rest, now this set is minimised.

q_0	q_5	q_1
q_1	q_2	q_6
q_4	q_5	q_1
q_5	q_6	q_2
q_6	q_4	q_6

as row-1 and row-3 are similar

Now minimised set is

q_0	q_5	q_1
q_1	q_2	q_6
q_5	q_6	q_2
q_6	q_0	q_6

See, now there are no two rows are similar.

Step 6. Do same process for set 2 as:

$$q_2 \quad q_2 \quad q_0$$

So only single row, already minimised.

Step 7. Now combine set 1 and set 2 as:

δ/Σ	*a*	*b*
$\rightarrow q_0$	q_5	q_1
q_1	q_2	q_6
q_5	q_6	q_2
q_6	q_0	q_6
$*q_2$	q_2	q_6

Now it is the transition table of minimised DFA.

Transition diagram of minimised DFA:

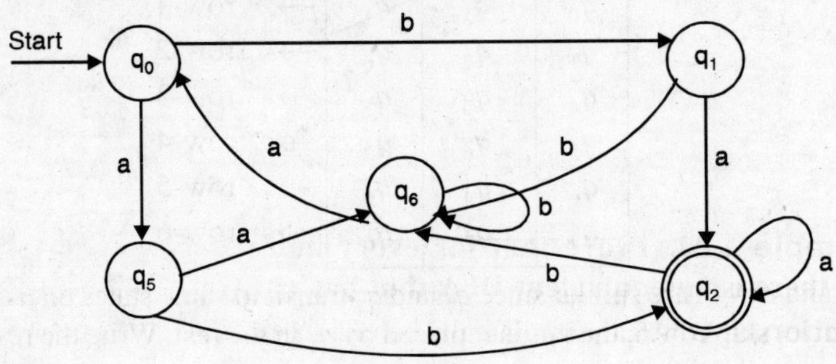

Fig. 2.41.

Example 2.21. Minimise the given DFA.

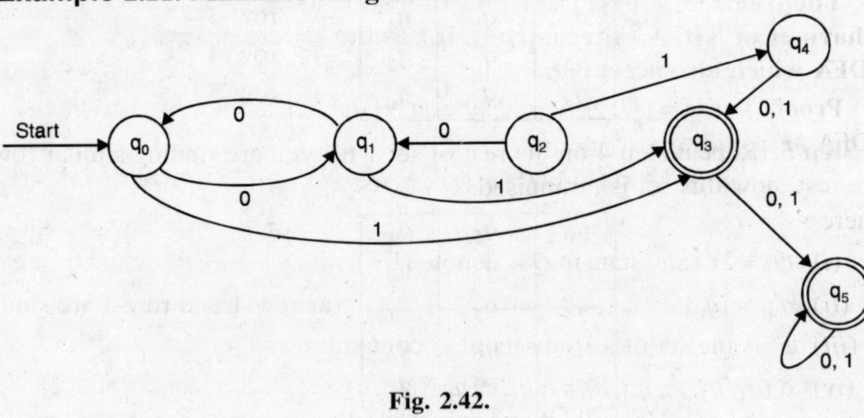

Fig. 2.42.

Solution. DFA given in Fig. 2.42 is having q_2 and q_4 as unreachable states, so remove them and draw the transition table for rest of the states.

	δ/Σ	*0*	*1*
set 1	$\rightarrow q_0$	q_1	q_3
	q_1	q_0	q_3
set 2	$*q_3$	~~q_5~~ q_3	~~q_5~~ q_3
	~~$*q_5$~~	~~q_5~~	~~q_5~~

Set 1 has no, two similar rows so we have to minimise only set 2. Now transition table for minimised DFA is as follows:

δ/Σ	*0*	*1*
$\rightarrow q_0$	q_1	q_3
q_1	q_0	q_3
$* \; q_3$	q_3	q_3

Transition diagram is as follows:

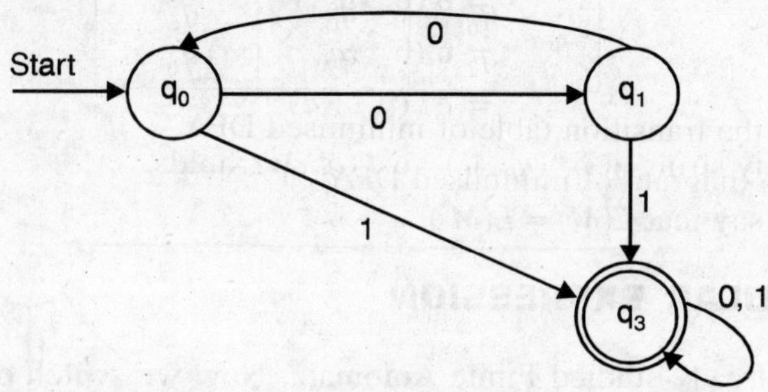

Fig. 2.43.

Example 2.22. Prove that for every non-deterministic finite automata (NDFA) there is an equivalent DFA that has only one final state.

Solution. A DFA can simulate the behaviour of NDFA by increasing the number of states. Any NDFA is more general machine without being more powerful. We now give a theorem on equivalence of DFA and NDFA.

Theorem: For every NDFA, there exists a DFA which simulates the behaviour of NDFA. Alternatively, if *L* is the set accepted by NDFA then exists a DFA which also accepts *L*.

Proof. Let $M = \{Q, \Sigma, \delta, q_0, F\}$ be an NDFA accepting language L. We construct a DFA M' as follows:

$$M' = \{Q', \Sigma, \delta', q_0', F'\}$$

where

(*i*) $Q_1 = 2^Q$ (any state in Q is denoted by $\{q_1, q_2, ..., q_i\}$ where $q_1, q_2, ..., q_j \in Q$);

(*ii*) $q'_0 = [q_0]$

(*iii*) F' is the set of all subsets of Q containing an element of F.

(*iv*) $\delta'(q_1, q_2, ..., q_i], a) = \delta(q_1, a) \cup \delta(q_2, a) \cup ... \cup \delta(q_i, a)$ equivalently $\delta'([q_1, q_2, ..., q_i]\ a) = [P_1 ... P_j]$ if and only if $\delta[\{q_1, q_2, ..., q_i\}, a) = \{P_1, P_2, ..., P_j\}$

if and only if $\delta(q_0, x) = \{q_1 ... q_i\}$ for all x in Σ

First we are going to prove by induction that $\delta^*(q_0, w) = \delta'^*(q_0', w)$ for any string w, when it is proven, it obviously implies that NDFA M and DFA M' accept the same strings.

Now we have a task to prove $\delta^*(q_0, w) = \delta'^*(q_0', w)$. This is going to be proven by induction on w.

Note: δ^* represents the transition function for processing strings.

Basic step: For $w = t$

$$\delta'^*(q'_0, t) = q'_0 \text{ by the definition of } \delta'^*$$
$$= q_0 \text{ by construction of DFA } M'$$
$$= \delta^*(q_0, t) \text{ by the definition of } \delta^*.$$

Inductive Step: Assume that $\delta^*(q_0, w) = \delta'^*(q_0', w)$ for an arbitrary string w. ... Induction hypothesis.

For the string w and an arbitrary symbol a in Σ,

$$\delta^*(q_0, wa) = \mathrm{U}_{P \in \delta^*(q_0, w)} \delta(P, a)$$
$$= \delta'(\delta^*(q_0, w), a)$$
$$= \delta'(\delta'^*(q_0', w), a)$$
$$= \delta^*(q_0', wa)$$

Thus for any string w $\delta^*(q_0, w) = \delta'^*(q_0', w)$ holds.

So we can say that $L(M) = L(M')$.

2.13. REGULAR EXPRESSION [MDU 2009]

We have already studied Finite Automata. Now we switch our attention from machine like description of language to an algebraic description: "the regular expression." We shall find that, regular expressions can define exactly the same languages that the various forms of automata describe: the regular languages.

However, regular expression offer something that automata do not: a declarative way to express the strings, we want to accept. Thus regular expression as the input language for many system. For example, lexical-analyser generators, such as Lex or Flex.

2.13.1. Definition of Regular Expression

Definition The set of regular expression is defined by the following rules:

1. Every letter of Σ can be made into a regular expression, null string, $\in$ itself is a regular expression.
2. If r_1 and r_2 are regular expression, then
   (*i*) (r_1) (*ii*) $r_1 r_2$
   (*iii*) $r_1 + r_2$ (*iv*) $r_1{}^*$
   (*v*) r_1^+ are also regular expression.
3. Nothing else is regular expression.

2.13.2. Building Regular Expression

The algebra of regular expressions follows this path using constants and variables that denote languages, and operators for three operations of union, dot and star. We can describe the regular expression recursively, as follows:

In this definition, we not only describe what are the legal regular expression, but for each regular expression r, we describe the language, it represents, which we denote as $L(r)$. We can consider some facts:

1. The constants $\in$ (null string) and ϕ (empty set) are regular expression, denoting the languages $\{\in\}$ and ϕ, respectively.

That is, $L(\in) = \{\in\}$, and $L(\phi) = \phi$.

2. If a is any symbol, then a is regular expression. This expression denotes the language $\{a\}$. That is $L(a) = \{a\}$.

3. A variable, usually capitalized and italic such as L is a variable, representing any language.

2.13.3. Inductive Step

There are four parts to the inductive step, one for each of three operators and one for the introduction of parantheses.

1. If r_1 and r_2 are regular expressions, then $r_1 + r_2$ is a regular expression denoting the union of $L(r_1)$ and $L(r_2)$.

That is, $L(r_1 + r_2) = L(r_1) \cup L(r_2)$

2. If r_1 and r_2 are regular expression, then $r_1 r_2$ is a regular expression denoting the concatenating of $L(r_1)$ and $L(r_2)$

That is, $L(r_1 r_2) = L(r_1) . L(r_2)$

3. If r is a regular expression then r^* is a regular expression denoting the closure of $L(r)$. That is,

$$L(r^*) = (L(r))^*$$

4. If r is a regular expression, then (r), a parenthesized r, is also a regular expression, denoting the same language as r.

Formally, $L((r)) = L(r)$.

Now let us discuss some examples of building regular expression for better understanding.

Example 2.23. Write the regular expression over alphabet (*a*, *b*, *c*) containing at least one *a* and at least one *b*.

Solution. Let us first analyse the language of regular expression.

Problem is very simple and the language will have set of strings, like *ab*, *abbbb*, *abab*, *aaaab*, *baaaa* .... So strings are 1̲ *a* 2̲ *b* 3̲, any combination of *a*'s, *b*'s and *c*'s is possible at the places 1, 2, and 3 any combination of *a*, *b*, *c* means $(a + b + c)^*$.

So regular expression is

$$r = (a+b+c)^* a(a+b+c)^* b(a+b+c)^* + (a+b+c)^* b(a+b+c)^* a(a+b+c)^*$$

If $r = r_1 + r_2$ (say)

then minimum string of r_1 is *ab* and minimum string of r_2 is *ba*.

Example 2.24. Write the regular expression for the set of strings of 0's and 1's whose tenth symbol from the right end is 1.

Solution. $\Sigma = \{0, 1\}$ given the set of strings, according to the required regular expression are 111̲000000000, 10011̲010101010 .... . So clearly we concern about only tenth position and it should be 1, rest 9 places from the right are either 0 or 1. Similarly positions 11th onward from right end may be either 0 or 1.

So regular expression is

$$r = (0+1)^* 1(0+1)(0+1)(0+1)(0+1)(0+1)(0+1)(0+1)(0+1)(0+1)$$

$$r = (0+1)^* 1(0+1)^9.$$

Example 2.25. Write the regular expression for the set of strings of an equal number of 0's and 1's such that in every prefix, the number of 0's differs from the number of 1's by at most 1.

Solution. Clearly the set of strings is having strings like 1010101010, 10101010, 0110, 1001 ...

So regular expression will be

$$r = (01 + 10)^*.$$

Example 2.26. Write a regular expression for the set of strings of 0's and 1's not containing 101 as a substring.

Solution. When we analyse the set of strings then it becomes clear that string may start with '0' and 0's may be repeated, wherever one is encountered then no single 0 must follow it so strings are like 0000010, 00000111111, 1001001 ....

Regular expression is $r = (0^*1^*00)^*\ 0^*\ 1^*$.

Example 2.27. Write the regular expression over alphabet $\{a, b\}$ for the set of strings with even number of a's followed by odd number of b's that is for the language

$$L = \{a^{2n} b^{2m+1} : n \geq 0, m \geq 0\}.$$

Solution. Clearly every string of language L will start with even number of a's. String should end with odd number of b's.

Regular expression is $r = (aa)^*\ (bb)^*\ b$.

Example 2.28. Write the regular expression for the language

$L = \{w \in \{0, 1\}^*: w$ has no pair of consecutive zeros$\}$.

Solution. Clearly whenever a 0 occurs, it must be followed by a 1. Such a substring may be proceeded and followed by any number of 1's, that is $(1^*011^*)^*$. But strings ending in 0 or consisting of all 1's are unaccounted here, that is $1^*(0 + \in)$.

So regular expression is $r_1 = (1^*011^*)^*\ (0 + \in) + 1^*(0 + \in)$

Another possible solution is $r_2 = (1 + 01)^*\ (0 + \in)$

"Although the two expression look different, both answers are correct, as they denote the same language. Generally, there are an unlimited number of regular expressions for any given language."

Example 2.29. Write the regular expression for the language

$$L = \{a^n b^m / n \geq 4, m \leq 3\}$$

Solution. Language contain the set of strings such that every string start with atleast 4 a's and end at the most 3 b's.

So regular expression will be $r = a^4\ a^*\ (\in + b + b^2 + b^3)$.

Example 2.30. Write the regular expression for the language

$$L = \{a^n\ b^m : (n + m) \text{ is even}\}$$

Solution. We know $(n + m)$ can be even in two cases

Case 1. n and m both are even.

Case 2. n and m both are odd.

Let regular expression for case 1 is r_1 then

$$r_1 = (aa)^* (bb)^*$$

Let regular expression for case 2 is r_2 then

$$r_2 = (aa)^* a (bb)^* b$$

Let regular expression for language L is r

then $r = r_1 + r_2$

$$r = (aa)^* (bb)^* + (aa)^* a (bb)^* b.$$

Example 2.31. Write the regular expression for the language

$$L = \{ab^n w: n \geq 3, w \in (a, b)^+\}$$

Solution. Clearly every string of language L starts with a followed by atleast three b's, followed by atleast one a or one b, that is strings are like ab^3a, *abbbbbba*, *abbbbbbbb*, *abbbb....a*

So regular expression is: $r = ab^3b^* (a + b)^+$

Here + is a positive closure *i.e.*, $(a + b)^+ = (a + b)^* - \in$.

Example 2.32. Write the regular expression for the language

$$L = \{w: | w | \bmod 3 = 0\}, w \in (a, b)^*\}$$

Solution. Let us first understand the meaning of $| w | \bmod 3 = 0$, when length of string belongs to the language L, is divided by 3 then reminder should be 0, that is length of w must be 0, 3, 6, 9 .... Clearly multiple of three.

Regular expression is: $r = ((a + b)^3)^*$

Example 2.33. Write down regular expression corresponding to each of the following languages over {0, 1}. The langauge of all strings in which every 0 is followed immediately by 11.

Solution. Let us analyse strings, as follows:

1, 1, 111, 011, 011011, 0111, 1011 etc.

So regular expression will be

$$r = (1 + 011)^*$$

Example 2.34. Find a regular expression corresponding to each of the following subset [0, 1]:

(i) The language of all strings containing at least two 0's.

(ii) The language of all strings containing at most two 0's.

(iii) The language of all strings ending with 1 and don't contain 00.

(iv) The language of all strings in which both the number of 0's and number of 1's are odd.

Solution. (*i*) $r = (0 + 1)^* 0 (0 + 1)^* 0 (0 + 1)^*$

(*ii*) $r = 1^* + 1^*01^* + 1^*01^*01^*$

It means string may contain all 1's, at most one 1 or atmost two 0's.

(*iii*) $r = (1 + 01)^* 1$

(*iv*) $r = (1(11)^* + 0(00)^*)$

Example 2.35. Find the regular expression representing the following sets:

(*i*) **The set of all strings over [0, 1] having at most one pair of 0's or at most one pair of 1's.**

(*ii*) **The set of all strings over [*a*, *b*] in which there are at least two occurrence of *b* between any two occurrences of *a*.**

Solution.

(*i*) Given t = [0, 1] at most means upto one pair of 0's and 1's.

$$0^* = \{\in, 0, 00, 000, 0000, \ldots\} \qquad \ldots(i)$$

$$1^* = \{\in, 1, 11, 111, 1111, \ldots.\} \qquad \ldots(ii)$$

$$01^* = \{0, 01, 011, 0111, 01111, \ldots\} \qquad \ldots(iii)$$

$$10^* = \{1, 10, 100, 1000, 10000, \ldots\} \qquad \ldots(iv)$$

Now we can find that which regular expression does not have a pair of 0's *i.e.*, 00.

Hence the set 1* and 01* (*i.e.*, from 2 and 3) not containing pair of 0's.

Now we can write as:

$$1^*+01^*$$

$$(1 + 01)^* \qquad \ldots(v)$$

Thus (1 + 01)* is regular expression that contain no pair of 0's. To find the string that contain exactly one pair of zero's, we simply write the string which have not a pair of 0's on both side of the pair of 0's *i.e.*, 00. Thus, the string which have exactly one pair 0's can be shown by

$$(1 + 01)^*\ 00\ (1 + 01)^* \qquad \ldots(vi)$$

Similarly, regular expression does not have pair of 1's *i.e.*, 0* and 10* (*i.e.*, from (*i*) and (*iv*)) we can write these two as

$$0^* + 10^*$$

$$(0 + 10)^* \qquad \ldots(vii)$$

Similarly, as above, the string that not have pair of 1's on the both side of the pair of 1's *i.e.*, 11. Thus this string which have exactly one pair of 1's:

$$(0 + 10)^*\ 11\ (0 + 10)^* \qquad \ldots(viii)$$

From (*v*), (*vi*), (*vii*) and (*viii*), we get a set of strings over {0, 1} having at most one pair of 0's or 1's *i.e.*,

$$r = (1 + 01)^* + (1 + 01)^*\ 00\ (1 + 01)^* + (0 + 10)^* + (0 + 10)^*\ 11\ (0 + 10)^*$$

(*ii*) When we say (*bb*)* *a* (*bb*)* it means that one *a* between even *b*. Then if we do (*a*(*bb*)* *a* *b*(*bb*)* *a*)* then obviously there will be at least two *b* between two '*a*' hence corresponding regular expression (RE) will be:

$$r = (a(bb)^*\ a(bb)^*a)^*$$

Example 2.36. Let Σ = {*a*, *b*} be an alphabet. Show that

(*i*) *a* * (*a* ∪ *b*)*

(ii) $aa^* (a \cup b)^* a$

(iii) $(a \cup b)^* (a \cup \phi)$

are regular expression over Σ.

Solution.

(*i*) $\Sigma = \{a, b\}$ so a^* and a are regular expression over Σ. $a \cup b$ is also regular expression by the definition of regular expression, therefore $(a \cup b)^*$ is regular expression over Σ. Hence by concatenation of two regular expressions $a^* (a \cup b)^*$ is a regular expression.

(*ii*) a, b and a^* are regular expression and hence aa^* is r.e. (regular expression) by concatenation $aa^* (a \cup b)$ is regular expression. This regular expression concatenates with the regular expression a produces regular expression $aa^* (a \cup b)^* a$ on Σ.

(*iii*) Since $a, b \in \Sigma$, ϕ, a, b are regular expressions. Hence $a \cup \phi$ and therefore $(a \cup \phi)^*$ is a regular expression. Again $(a \cup b)$ and $(a \cup b)^*$ is a regular expression. By concatenation $(a \cup b)^* (a \cup \phi)$ is a regular expression over Σ. Note that ϕ represents the empty set, that is the set with no strings.

Example 2.37. Find regular expression '*r*' over $\Sigma = \{a, b\}$ such that $L(r) = \{a^m b^m : m > 0\}$.

Solution. Given language is $L = \{a^m b\, a^n : m > 0 ; n > 0\}$

Regular expression, $r = aa^*\ baa^*$

Example 2.38. Exhibit the language *L* (a^*, $(a + b)$ in set notation.

Solution.

$$L(a^* . (a + b)) = L(a^*) . L(a + b)$$
$$= (L(a))^* . (L(a) + L(b))$$
$$= \{\in, a, aa, aaa, ...\} . \{a + b\}$$
$$= \{a, aa, aaa, ..., b, ab, aab, ...\}$$

Example 2.39. Find a regular expression for the language $L = \{w \in (0, 1)^* : w$ has no pair of consecutive zero's$\}$.

Solution. If language is set of strings which does not have any consecutive zero's as substring we can analyse that *L* as the repetition of string 1 and 01 only. So as per this logic regular expression is:

$$r = (1 + 01)^* \underline{(0 + \in)}$$

String may end in 0 or 1.

Example 2.40. Let $L = \{uvu : u, v \in (a, b)^*, | w | = 2\}$, find regular expression for the same.

Solution. We can have following case for $| w | = 2$.

(*i*) $u = aa$ (*ii*) $u = ab$

(*iii*) $u = ba$ (*iv*) $u = bb$

So regular expression is

$$r = aa(a + b)^* aa + ab(a + b)^* ab + ba(a + b)^* ba + bb(a + b)^* bb$$

2.14. COMPARATIVE STUDY OF REGULAR EXPRESSION, REGULAR SETS AND FINITE AUTOMATA [MDU 2009]

Regular expression	Regular set	Finite automata
ϕ	{ }	 **Fig 2.44.** q_f **is the final state.**
$\in$	{$\in$}	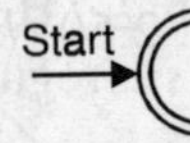 **Fig. 2.45.**
Every a in Σ is a regular expression.	{a}	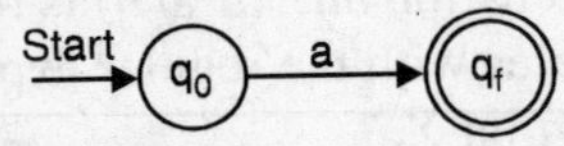 **Fig. 2.46.**
If r_1 and r_2 are regular expression then $(r_1 + r_2)$ or r_1/r_2 is regular expression.	$R_1 \cup R_2$ (where R_1 and R_2 are regular sets corresponding to r_1 and r_2 respectively).	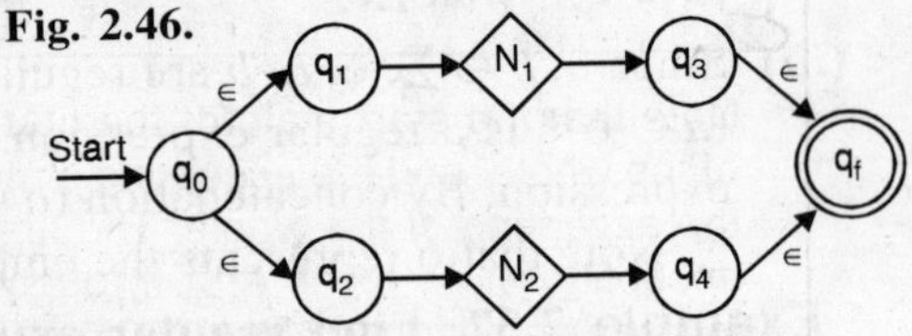 Where N_1 is FA accepting R_1 and N_2 is a finite automata accepting R_2. **Fig. 2.47.**
r_1r_2 is a regular expression.	R_1R_2 (where R_1 and R_2 are regular sets corresponding to r_1 and r_2 respectively).	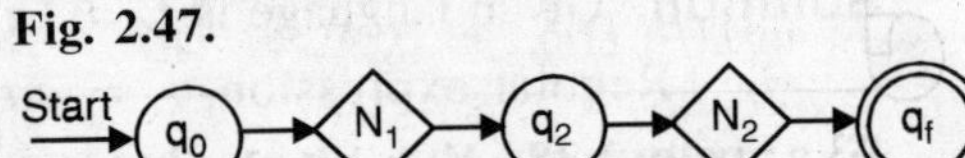 Where N_1 is finite automata for R_1 and N_2 is FA for R_2. **Fig. 2.48.**
r^* is a regular expression if r is regular expression.	R^* (where R is a regular set corresponding to r).	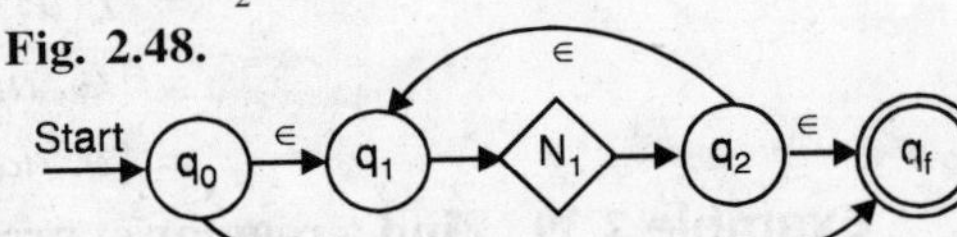 where N_1 is finite automata accepting R. **Fig. 2.49.**

2.15. CONSTRUCTION OF FA FOR REGULAR EXPRESSION [MDU 2009, 2010]

We can construct NFA with $\in$ transmission by considering following four facts.

1. The expression is $r + s$ for some smaller expression r and s. The following automation serves, where R is automation for r and S is automation for s. That is, starting at the new start state, we can go to the start state of either the automation for r or the automation for S. We can reach the accepting state of one of these automata, following a path labelled by some string in $L(r)$ and $L(s)$, respectively. Once we reach the accepting state of the automation for r or s, we can follow one of the $\in$-arcs to the accepting state of the new automation. The language for the automation in Fig. 2.50 (a) is $L(r) \cup L(s)$.

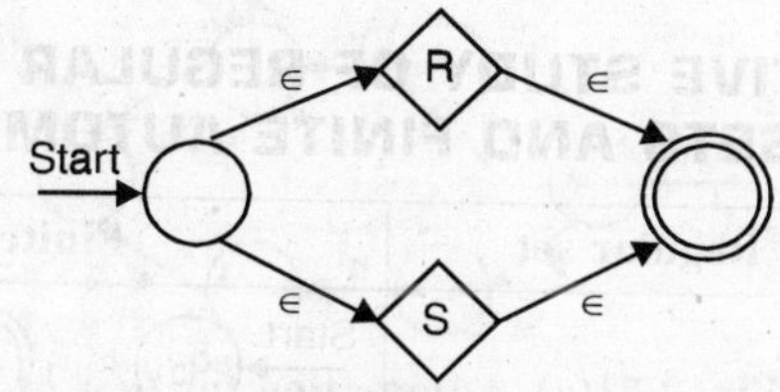

Fig. 2.50 (*a*).

2. The expression *rs* for some smaller expression *r* and *s*. The automation for the concatenation is show in Fig. 2.50 (*b*).

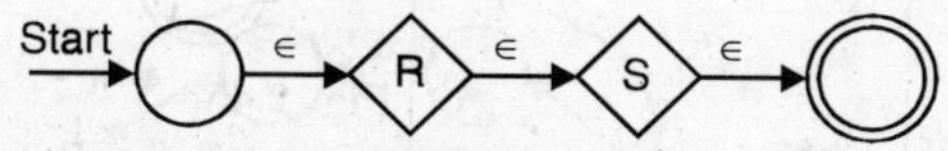

Fig. 2.50 (*b*).

TIPS

Note that the start state of the first automation becomes the start state of the whole, and the accepting state of the second automation becomes the accepting of the whole. The idea is that the only paths from start to accepting state to go first through the automation of *r*, where it must follow a path labelled by a string in *L*(*r*), and then through the automation for *s*, where it follows a path labelled by a string in *L*(*s*). Thus, the path in the automation of Fig. 2.50 (*b*) are all and only those labelled by strings in *L*(*R*) *L*(*s*).

3. The expression is r^* for some smaller expression *r*. Then we use automation of Fig. 2.50 (*c*). That automation allow us to go either:

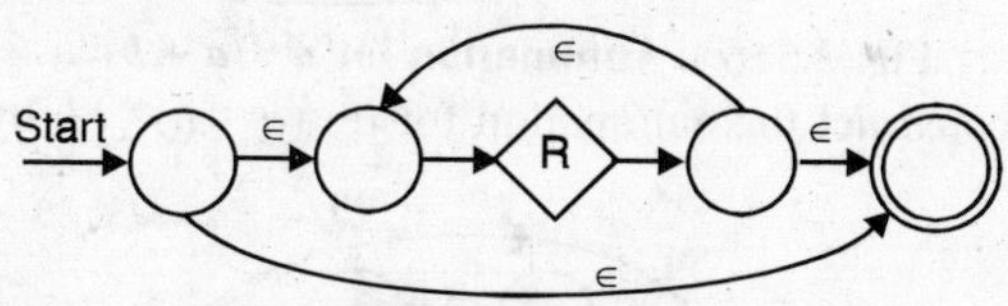

Fig. 2.50 (*c*).

(*i*) Directly from the start state to the accepting state along a path labelled ∈.

(*ii*) To the start state of automation for *r*, through that automation one or more times, and then to accepting state. This set of paths allow us to accept strings in *L*(*r*), *L*(*r*), *L*(*r*), *L*(*r*), *L*(*r*), *L*(*r*), and so on, thus covering all strings in $L(r^*)$ except perhaps ∈ which was covered by the direct arc to the accepting state.

where *R* is automation for *r*.

4. The expression is (*r*) for the smaller expression *r*. The automation for *r* also serves as the automation for (*r*), since the parentheses do not change the language defined by the expression.

Example 2.41. One the basis of above discussion find the automation for regular expression *a* . (*a* + *b*)* . *b* . *b*.

Solution. Let us solve this problem step by step:

Step 1. The basic regular expression involved are *a* and *b*, we start with automation for *a* and automation for *b*. Since brackets are evaluated first. We first constructed automation for (*a* + *b*), using automation for *a* and automation *b* shown in Fig. 2.51(*a*)

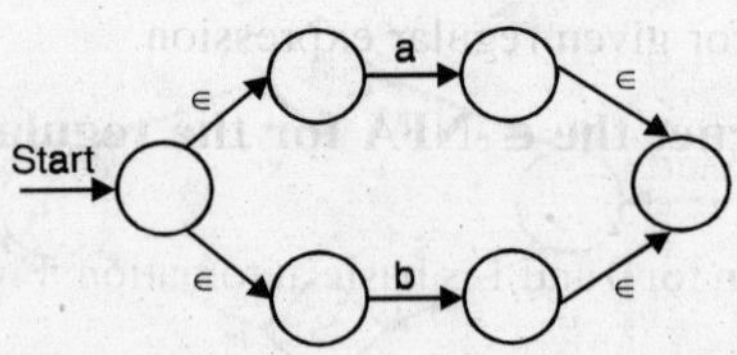

Fig. 2.51(*a*). Automation for (*a* + *b*).

Step 2. Since closure is required to take next, we construct automation for $(a + b)^*$ using automation for $(a + b)$ as shown in Fig. 2.51(*b*).

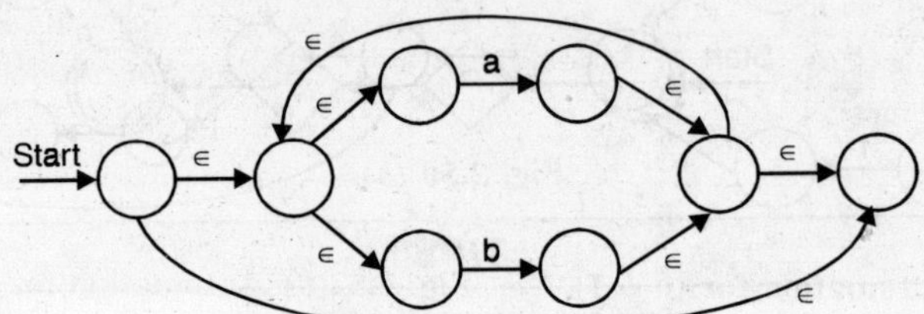

Fig. 2.51(*b*). Automation for (*a* + *b*)*

Step 3. Next we construct the automation for $a \,.\, (a + b)^*$ as shown in Fig. 2.51(*c*).

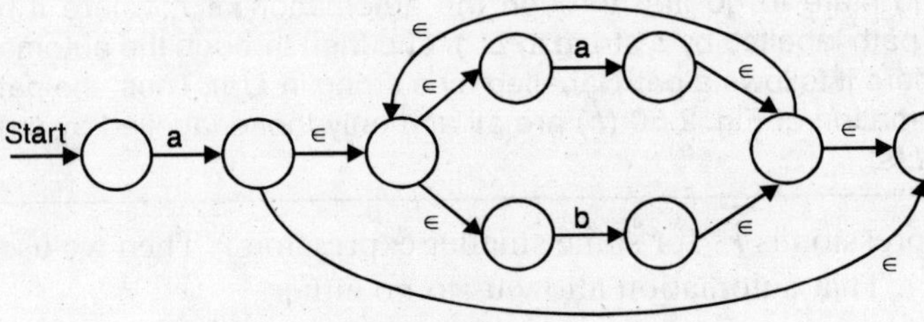

Fig. 2.51(*c*). Automation for *a* . (*a* + *b*)*.

Step 4. Next we construct the automation for $a \,.\, (a + b)^* \,.\, b$ by using automation for $a \,.\, (a + b)^*$.

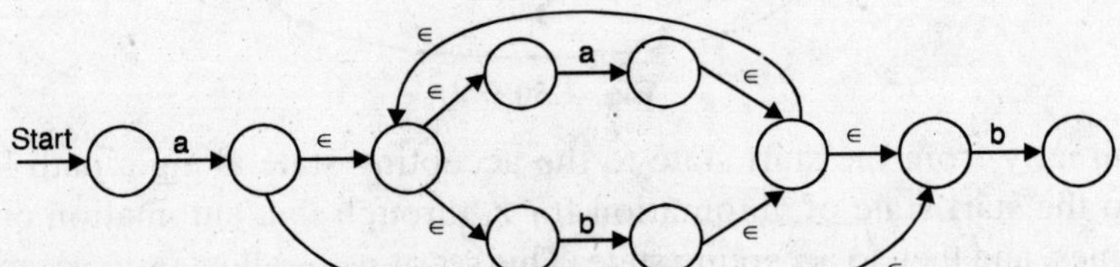

Fig. 2.51(*d*). Automation for *a* . (*a* + *b*)* . *b*.

Step 5. Now finally we can construct automation for $a \,.\, (a + b)^* \,.\, b \,.\, b$ by using automation for $a \,.\, (a + b)^* \,.\, b$.

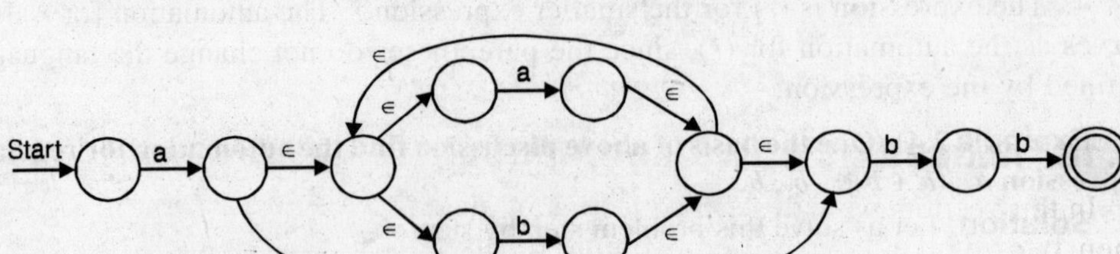

Fig. 2.51(*e*). Automation for *a* . (*a* + *b*)* . *b* . *b*.

It is required automation for given regular expression.

Example 2.42. Construct the ∈-NFA for the regular expression (0 + 1)* 1 (0 + 1).

Solution. Here automation for 0 and 1 is basic automation. First we will make automation for (0 + 1) as in Fig. 2.52(*a*).

Now we will convert it into automation for (0 + 1)* as shown in Fig. 2.52(*b*).

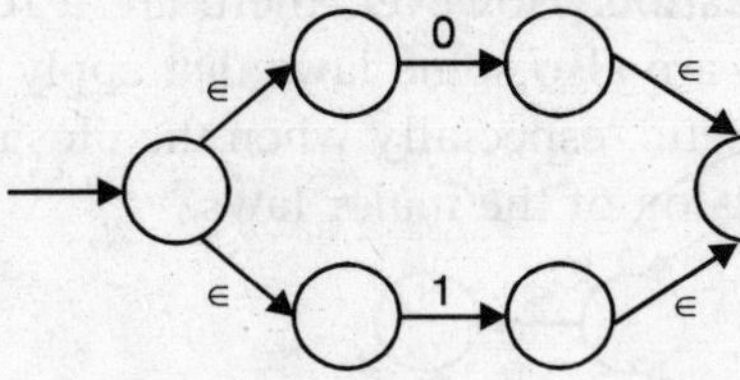

Fig. 2.52(*a*). Automation for (0 + 1).

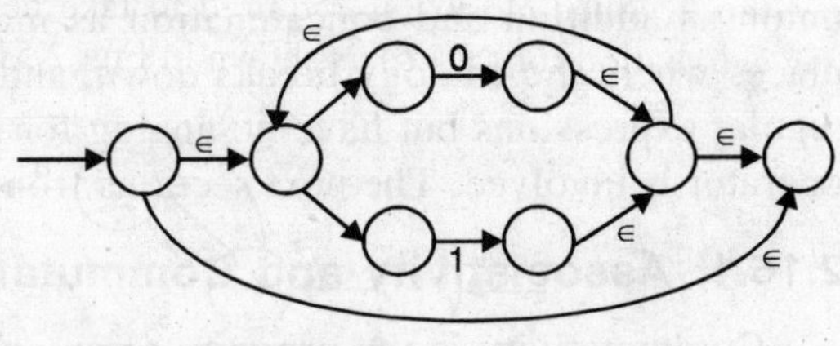

Fig. 2.52(*b*). Automation for (0 + 1)*.

Now we will construct automation for 1 . (0 + 1) as shown in Fig. 2.52(*c*).

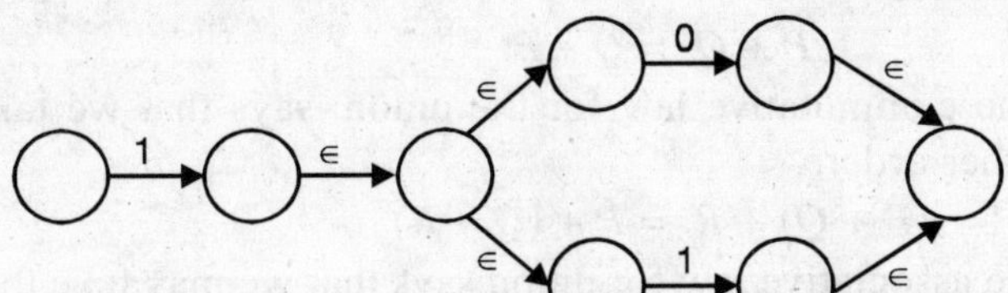

Fig. 2.52(*c*). Automation for 1 . (0 + 1).

Now we have to connect ∈-NFA of Fig. 2.52(*b*) and ∈-NFA of Fig. 2.52 by a ∈-transition to find out the ∈-NFA for given regular expression as shown in Fig. 2.52(*d*).

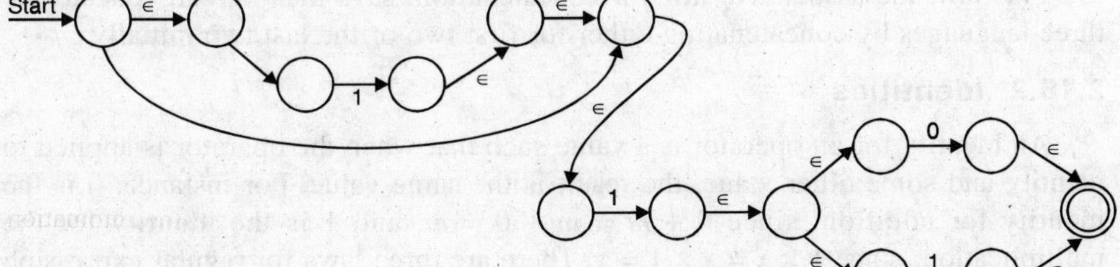

Fig. 2.52 (*d*). Automation for (0 + 1)* 1(0 + 1).

2.16. ALGEBRAIC LAWS FOR REGULAR EXPRESSIONS

In this section, we will see a collection of algebraic laws that bring the issue of when two regular laws are equivalent. Instead of examining specific regular expressions, we shall consider pairs of regular expressions with variables as arguments.

TIPS

Two expressions with variables are equivalent if whatever languages we substitute for the variables, the result of the two expressions are the same language.

Like arithmetic expressions, the regular expressions have a number of laws that work for them. Many of these are similar to the laws for arithmetic, if we think of union as addition and concatenation as multiplication. However, there are a few places where the analogy breaks down, and there are also some laws that apply to regular expressions but have no analog for arithmetic, especially when the closure operator is involved. The next sections from a catalog of the major laws.

2.16.1. Associativity and Commutativity

Commutativity is the property of an operator that says that we can switch the order of its operands and get the same result. For example, the associative law of multiplication is $(x \times y) \times z = x \times (y \times z)$. Here are three laws of these types that hold for regular expression:

$$P + Q = Q + P.$$

This law, the commutative law for the union, says that we take union of two languages in either order.

$$(P + Q) + R = P + (Q + R)$$

This law, the associative law for union says that we may take the union of three languages either by taking the union of the first two initially, or taking the union of the last two initially.

Note that, together with the commutative law for union, we conclude that we can take the union of any collection of languages with any order and grouping, and the result will be the same. Intuitively, a string is in $L_1 \cup L_2 \cup ... + L_K$ if and only if it is in one or more of the L_i's.

$$(PQ)R = P(QR)$$

This law, the associative law for concatenation, says that we can concatenate three languages by concatenating either for first two or the last two initially.

2.16.2. Identities

An identity for an operator is a value such that when the operator is applied to identity and some other value, the result is the same value. For instance, 0 is the identity for addition, since $0 + m = m + 0 = m$, and 1 is the identity for the multiplication, since $1 \times x = x \times 1 = x$. There are three laws for regular expression involving these concepts; we list them below.

- $\phi + L = L + \phi = L$. This law assets that ϕ is the identity for union.
- $\in L = L\in = L$. This law assets that $\in$ is the identity for concatenation.
- $\phi L = L\phi = \phi$. This law assets that ϕ is the annihilater for concatenation.

2.16.3. Distributed Law

A distributed law involves two operators, and assets that one operator can be pushed down to be applied to each argument of the operator individually. The most common example is the distributed law of multiplication over addition, that is

$$x \times (y + z) = x \times y + x \times z$$

However, there is an analogous law for regular expression, that we most state in two forms. Since concatenation is not commutative. These laws are:

- $P(Q + R) = PQ + PR$. This law, is the left distributed law of concatenation over union.
- $(Q + R)P = QP + RP$. This law, is the right distributed law of concatenation over union.

2.16.4. The Idempotent Law

An operator is said to be idempotent if the result of applying it to two of the same values as arguments is that value. The common arithmatic operators are not idempotent, $x + x \neq x$ in general and $x \times x \neq x$ in general (although there are some values of x for which the equality holds, such as $0 + 0 = 0$). However, union and intersection are common examples of idempotent operators. Thus for the regular expression, we may assert the following law:

- $L + L = L$. This law, the idempotent law for union, states if we take the union of two identical expression, we can replace them by one copy of the expression.

2.16.5. Laws for Closure

There are number of laws involving the closure operators. We shall list them here, and give some explanation for why they are true.

1. $(L^*)^* = L^*$. This law says that closing an expression that is already closed does not change the language. The language of $(L^*)^*$ is all strings created for concatenating strings in the language L^*. But those strings are themselves composed of strings from L. Thus, the string in $(L^*)^*$ is also a concatenation of strings from L and therefore in the language of L^*.
2. $\phi^* = \in$. The closure of ϕ contains only the string $\in$.
3. $\in^* = \in$. It is easy to check that only string that can be formed by concatenating any number of copies of the empty string is the empty string itself.
4. $L^+ = LL^* = L^*L$. Recall that positive closure of L, L^+ is defined to be $L + LL + LLL + \ldots$. Also, $L^* = \in + L + LL + LLL + \ldots$

   Thus $\quad LL^* = L + LL + LLL + LLLL + \ldots$

   That proves $\quad L^+ = LL^*$.
5. $L^* = L^+ + \in$. The proof is easy, since the expansion of L^+ includes every term in the expansion of L^* except $\in$. Note that if the language L contains the string $\in$ then the additional $+\in$ is not needed; that is $L^+ = L^*$ in this special case.
6. $L? = \in + L$. This rule is really the definition of the operator ?

2.16.6. Identities for Regular Expression

On the basis of above discussion we will see some identities for the regular expression. Let p, q, r are three regular expression. Two regular expression p and q are equivalent (we say $p = q$) if p and q represent the same set of strings.

Some identities are following:

$I^1 \quad \phi + r = r$

$I^2 \quad \phi r = r\phi = \phi$

I^3 $\quad \in r = r\in = r$
I^4 $\quad \in^* = \in$ and $\phi^* = \in$
I^5 $\quad r + r = r$
I^6 $\quad r^*r^* = r^*$
I^7 $\quad rr^* = r^*r = r^+$
I^8 $\quad (r^*)^* = r^*$
I^9 $\quad \in + rr^* = r^* = \in + r^*r$
I^{10} $\quad (pq)^*p = p(qp)^*$
I^{11} $\quad (p + q)^* = (p^*q^*)^* = (p^* + q^*)^*$
I^{12} $\quad (p + q)r = pr + qr$ and $r(p + q) = rp + rq$

Example 2.43. Prove that (1 + 00*1) + (1 + 00*1) (0 + 10*1)* (0 + 10*1) = 0*1(0 + 10*1)*

Solution. L.H.S. $= (1 + 00^*1)\ (\in + (0 + 10^*1^*(0 + 10^*1)$ using I^{12}
$= (1 + 00^*1)\ (0 + 10^*1)^*$ using I^9
$= (\in + 00^*)1\ (0 + 10^*1)^*$ using I^{12} for $(1 + 00^*1)$
$= 0^*1(0 + 10^*1)^*$ using I^9
= R.H.S.

Hence it is proved.

Example 2.44. Prove $t + RR^* = R^*$ where t is null string and R is a regular expression.

Solution. Two regular expressions P and Q are equivalent (we write $P = Q$). If P and Q are represent the same set of strings.

From Arden's Theorem, we know that, the solution of equation $R' = Q + R'P$ is $R' = QP^*$

Now $\quad t + RR^* = R^*$

L.H.S. $= t + RR^*$ Compare it with Arden's theorem
$\simeq Q + R'P$
$Q = t$
$R' = R$
$P = R^*$

So, the solution of this equation will be $R' = QP^*$

$R_1 = QP^*$
$= tR^*$
$= R^*$

So we can write that $t + RR^* = R^*$

Example 2.45. Prove the following identity:

$$(a^*ab + ba)^*\ a^* = (a + ab + ba)^*.$$

Solution. $(a^*ab + ba)^*a^* = (a + ab + ba)^*$

By taking L.H.S. $= (a * ab + ba)^*\ a^*$
$= (a\ (ab)^* + (ba)^*)^*a^*$
$= (ab^* + (ba)^*\ a^*)$

$$= (ab^* + (ba)^* a^*)$$
$$= (a + ab + ba)^*$$
$$= \text{R.H.S.}$$

Example 2.46. Write down regular expression corresponding to each of the following languages over {0, 1}. The langauge of all strings in which every 0 is followed immediately by 11.

Solution. Let us analyse strings, as follows:

1, 1, 111, 011, 011011, 0111, 1011 etc.

So regular expression will be

$$r = (1 + 011)^*$$

Example 2.47. Find a regular expression corresponding to each of the following subset [0, 1]:

(i) The language of all strings containing at least two 0's.

(ii) The language of all string containing at most two 0's.

(iii) The language of all strings ending with 1 and don't contain 00.

(iv) The language of all strings in which both the number of 0's and number of 1's are odd.

Solution. (*i*) $r = (0 + 1)^* \, 0 \, (0 + 1)^* \, 0 \, (0 + 1)^*$

(*ii*) $r = 1^* + 1^*01^* + 1^*01^*01^*$

It means string may contain all 1's, at most one 1 or atmost two 0's.

(*iii*) $r = (1 + 01)^* \, 1$

(*iv*) $r = (1(11)^* + 0(00)^*)$

Example 2.48. Find the regular expression representing the following sets:

(i) The set of all strings over [0, 1] having at most one pair of 0's or at most one pair of 1's.

(ii) The set of all strings over [*a*, *b*] in which there are at least two occurrence of b between any two occurrences of *a*.

Solution.

(*i*) Given t = [0, 1] at most means upto one pair of 0's and 1's.

$$0^* = \{\in, 0, 00, 000, 0000, ...\} \quad ...(i)$$
$$1^* = \{\in, 1, 11, 111, 1111, ...\} \quad ...(ii)$$
$$01^* = \{0, 01, 011, 0111, 01111, ...\} \quad ...(iii)$$
$$10^* = \{1, 10, 100, 1000, 10000, ...\} \quad ...(iv)$$

Now we can find that which regular expression does not have a pair of 0's *i.e.*, 00.

Hence the set 1* and 01* (*i.e.*, from 2 and 3) not containing pair of 0's.

Now we can write as:

$$1^*+01^*$$
$$(1 + 01)^* \quad ...(v)$$

Thus (1 + 01)* is regular expression that contain no pair of 0's. To find the string that contain exactly one pair of zero's, we simply write the string which have not a pair of 0's on both side of the pair of 0's *i.e.,* 00. Thus, the string which have exactly one pair 0's can be shown by

$$(1 + 01)^*\ 00\ (1 + 01)^* \qquad ...(vi)$$

Similarly, regular expression does not have pair of 1's *i.e.*, 0* and 10* (*i.e.*, from (*i*) and (*iv*)) we can write these two as

$$0^* + 10^*$$

$$(0 + 10)^* \qquad ...(vii)$$

Similarly, as above, the string that not have pair of 1's on the both side of the pair of 1's *i.e.*, 11. Thus this string which have exactly one pair of 1's:

$$(0 + 10)^*\ 11\ (0 + 10)^* \qquad ...(viii)$$

From (*v*), (*vi*), (*vii*) and (*viii*), we get a set of strings over {0, 1} having at most one pair of 0's or 1's *i.e.*,

$$r = (1 + 01)^* + (1 + 01)^*\ 00\ (1 + 01)^* + (0 + 10)^* + (0 + 10)^*\ 11\ (0 + 10)^*$$

(*ii*) When we say $(bb)^*\ a\ (bb)^*$ it means that one *a* between even *b*. Then if we do $(a(bb)^*\ a\ b(bb)^*\ a)^*$ then obviously there will be at least two *b* between two '*a*' hence corresponding regular expression (RE) will be:

$$r = (a(bb)^*\ a(bb)^*a)^*$$

Example 2.49. Let $\Sigma = \{a, b\}$ be an alphabet. Show that

(*i*) $a^* (a \cup b)^*$

(*ii*) $aa^* (a \cup b)^*\ a$

(*iii*) $(a \cup b)^* (a \cup \phi)$

are regular expression over Σ.

Solution.

(*i*) $\Sigma = \{a, b\}$ so a^* and a are regular expression over Σ. $a \cup b$ is also regular expression by the definition of regular expression, therefore $(a \cup b)^*$ is regular expression over Σ. Hence by concatenation of two regular expressions $a^* (a \cup b)^*$ is a regular expression.

(*ii*) *a*, *b* and a^* are regular expression and hence aa^* is r.e. (regular expression) by concatenation $aa^* (a \cup b)$ is regular expression. This regular expression concatenates with the regular expression a produces regular expression $aa^* (a \cup b)^*\ a$ on Σ.

(*iii*) Since $a, b \in \Sigma, \phi, a, b$ are regular expressions. Hence $a \cup \phi$ and therefore $(a \cup \phi)^*$ is a regular expression. Again $(a \cup b)$ and $(a \cup b)^*$ is a regular expression. By concatenation $(a \cup b)^* (a \cup \phi)$ is a regular expression over Σ. Note that ϕ represents the empty set, that is the set with no strings.

Example 2.50. Find regular expression '*r*' over $\Sigma = \{a, b\}$ such that $L(r) = \{a^m b^m : m > 0\}$.

Solution. Given language is $L = \{a^m b\, a^n : m > 0\ ;\ n > 0\}$

Regular expression, $r = aa^*\ baa^*$

Example 2.51. Exhibit the language *L* (a^*, $(a + b)$) in set notaion.

Solution. $L(a^* . (a + b)) = L(a^*) . L(a + b)$

$$= (L(a))^* . (L(a) + L(b))$$

$$= \{\in, a, aa, aaa, ...\} . \{a + b\}$$
$$= \{a, aa, aaa, ..., b, ab, aab, ...\}$$

Example 2.52. Find a regular expression for the language $L = \{w \in (0, 1)^* : w$ has no pair of consecutive zero's}.

Solution. If language is set of strings which does not have any consecutive zero's as substring we can analyse that L as the repetition of string 1 and 01 only. So as per this logic regular expression is:

$$r = (1 + 01)^* \underline{(0 + \in)}$$

String may end in 0 or 1.

Example 2.53. Let $L = \{uvu: u, v \in (a, b)^*, | w | = 2\}$, find regular expression for the same.

Solution. We can have following case for $| w | = 2$.

(*i*) $u = aa$ (*ii*) $u = ab$

(*iii*) $u = ba$ (*iv*) $u = bb$

So regular expression is

$$r = aa(a + b)^* aa + ab(a + b)^* ab + ba(a + b)^* ba + bb(a + b)^* bb$$

EXERCISE

1. Build an FA that accepts the words *baa*, *ab* and *abb* and no other string longer or shorter.
2. For the $\Sigma = \{a, b\}$
   (*i*) Build an FA that accepts only those words that have more than four letters.
   (*ii*) Build an FA that accepts only those words that do not end with *ba*.
3. Describe the langauge accepted by following FA's.

(*i*)

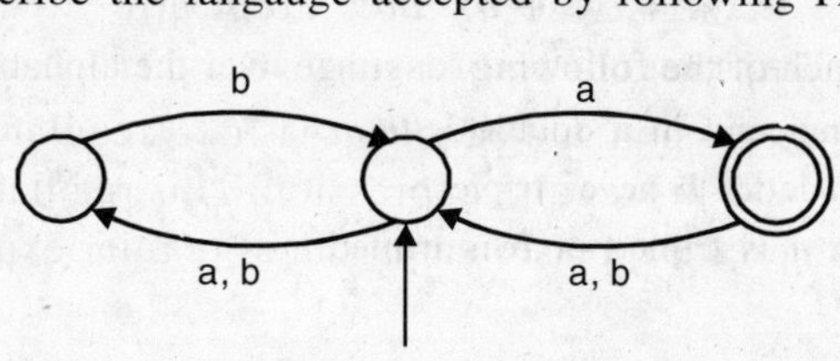

(*ii*)

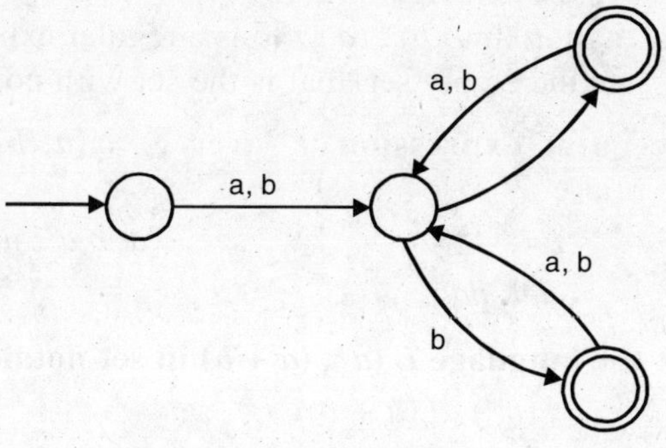

4. Convert each of the following NFA's into FA's.

(*i*)

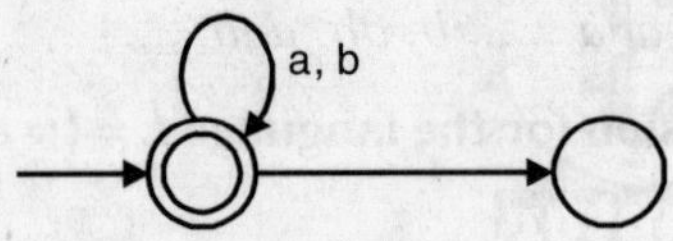

(*ii*)

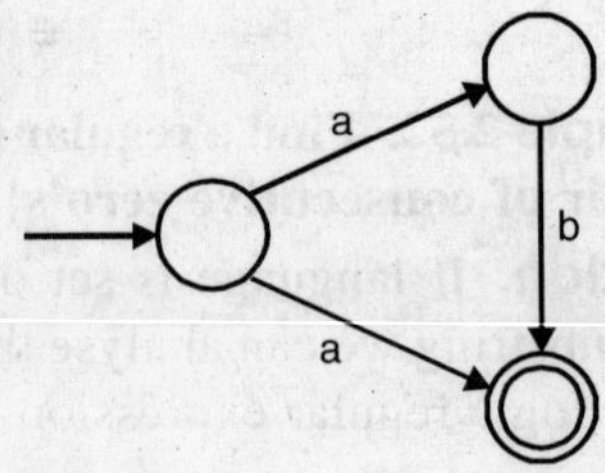

5. Design a NFA with no more than five states for the set $\{abab^n: n \geq 0\} \cup \{aba^n : n \geq 0\}$.

6. Find the minimal DFA's for the following given languages:

(*i*) $L = \{a^n b^m : n \geq 2, m \geq 1\}$

(*ii*) $L = \{a^n : n \geq 0, n \neq 3\}$

(*iii*) $L = \{a^n : n \neq 2 \text{ and } n \neq 4\}$

7. Find the regular expression for the following:

(*a*) $L = \{a^n b^m, n \geq 1, m \geq 1, nm \geq 3\}$

(*b*) $L = \{ab^n w : n \geq 3, w \in (a, b)^+\}$

(*c*) $L = \{a^n : n \geq 0, n \neq 3\}$

8. Give the English description for the following regular expression:

(*a*) $r = a(a + b) * ab$ (*b*) $r = a * b + b * a$

(*c*) $r = (aa + b)* (bb + a)*$

9. Construct the transition system for the following regular expression:

(*a*) $r = 1 + 00 + 010 * 1$ (*b*) $r = (ab + c*) * b$

10. Construct a finite automata for the regular expression

$$r = (a + b)^* abb.$$

11. Find the R.E. defining each of the following language over the alphabet $\Sigma = \{a, b\}$.

(*a*) All strings that do not end in a double letter.

(*b*) All strings in which letter *b* never triples.

(*c*) All strings in which *a* is tripled or *b* is tripled.

3

Chapter

LEXICAL ANALYSIS

INSIDE THIS CHAPTER

3.1. Introduction **3.2.** Reasons for Decouple the Scanner From the Parser **3.3.** The String Table **3.4.** Strategies in Lexical Analyzer Construction.

3.1. INTRODUCTION [MDU 2007]

We have already seen brief idea of lexical analysis in chapter 1. Now we can understand that analysis of a source program during compilation is often complex. To make construction of compiler easy we can separate analysis of source program in two parts.

(*a*) In the first part we can identify the low-level language constructs (tokens) such as variable names, key words, labels and operators.

(*b*) In the second part we can determine the syntactic organization of the program.

By studying chapter two we are proficient enough to understand the process of lexical analysis. There are two main aspects of analyser (scanner):

(*i*) Representation of the tokens of the language for which are constructing the compiler and that representation is done by regular expression or regular grammar. Both description methods are equivalent in the sense that both describe the set of regular languages.

(*ii*) The second aspect of analyser deals with the recognition of tokens. Finite-state machines that are well suited to this recognition task primarily because they can be specified pictorially by using transition diagrams.

Definition So now it is clear that lexical analyser is a software (part of compiler) which works as an interface between the source program and the parser. The lexical analyser, through a character-by-character examination of the input text, separates the source program into pieces called tokens, which represent the variable names, operators, labels, and so on that comprise the source program.

So now we can summarise the job of lexical analyser: [MDU 2007, 2008, 2009]

(*i*) Convert source file characters into token stream such as keywords, identifiers, constants, string literals, operators and punctuators in C.

(*ii*) Remove the content-free characters such as comment and white space.

(*iii*) Insert the line number during analysis.

(*iv*) Evaluating constants.

(*v*) Detect lexical errors such as

(*a*) Numeric literals are too long.

(*b*) Identifiers that are too long (often a warning is given)

(*c*) Ill-formed numeric literals.

(*d*) Input characters that are not in the source language.

(*vi*) Output of lexical analysis is input to syntax analysis (parser).

3.2. REASONS FOR DECOUPLE THE SCANNER FROM THE PARSER

We can follow two approaches during the construction of lexical analyser. We can construct it as a separate module or it can be a part of scanner. If scanner is the part of parse then in this approach parser usually generates a syntax tree of the source program as defined by a grammar. The leaves of the tree are the terminal symbols of the grammar. These terminal symbols or tokens pass to the parser. But generally we choose to handle analysis separately for better modularity and portability, and to allow make syntax analysis easier, reasons for decoupling the scanner from parser can be understood as follows:

(*i*) Techniques for lexical analysis can be simpler than those required for syntax analysis (DFA vrs. PDA). Separation also simplifies the syntax analyser.

(*ii*) Separation into different modules makes it easier to perform simplifications and optimizations unique to the different paradigms.

(*iii*) Due to input/output character set variations, lexical analysers are not always machine independent.

(*iv*) A symbol like an identifier is lexically equivalent to a "reserved word", the distinction may sensibly be made as soon as the basic token has been synthesized.

(*v*) A character set may vary from machine to machine, a variation easily isolated in this phase.

(*vi*) The semantic analysis of a numeric literal constant (deriving the internal representation of its value from the characters is easily performed in parallel with lexical analysis.

(*vii*) The scanner can be made responsible for screening out separators, like blanks and comments, which are rarely of interest in the formulation of the higher level grammar.

The difficulty even can be understood by the block diagram as follows:

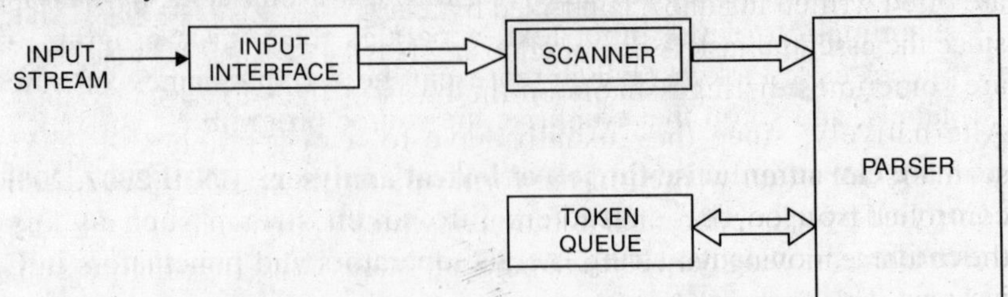

Fig. 3.1. The scanner fits in between the input interface and the parser.

TIPS

So it can be seen that decoupling of parser from lexical analyser increases the modularity.

3.3. THE STRING TABLE

Definition

The string table is a data structure for unique lexemes.

The string table can be queried for the lexeme by a numeric code.

Lexemes are generally inserted in the string table by the lexical analyser.

The string table can be used for:

(*a*) Error messages

(*b*) Memory map listings

(*c*) Intermodule linkage.

Index	String	Token
0	Fptr	[ident]
1	Number	[ident]
2	5-1	[f – lit]
3		

Fig. 3.2. String table.

It is not recommended to use the string table for reserved words. Use a separate structure to determine if an identifier is a reserved word, and then assign a token value accordingly. If it is not a reserved word insert it in the string table if necessary.

TIPS

Do not confuse the String Table with Symbol Table. The string table is for unique spelling of lexemes, allowing index comparisons instead of string comparison during syntax analysis and code generation. The symbol table is a dynamic data structure used differently than the string table.

3.4. STRATEGIES IN LEXICAL ANALYZER CONSTRUCTION [MDU 2007]

There are two main strategies that are employed in scanner construction:

(*i*) Rather than being decomposed into a set of recursive routines, simple scanners are often written in adhoc, controlled by a large CASE or SWITCH statement, since the essential task is one of choosing between a number of tokens, which are sometimes distinguishable on the basis of their initial characters.

(*ii*) Alternatively, since they usually have to read a number of characters, scanners are often written in the form of a **finite state automation (FSA)** controlled by a loop, on each iteration of which a single character is absorbed, the machine moving between a number of "states", determined by the character just read. This approach has the advantage that the construction can be formalized in terms of an extensively developed automata theory, leading to algorithm from which scanner generators can be constructed automatically.

3.4.1. Issues in Lexical Analyzer **[MDU 2007]**

Following issues must be addressed by any lexical analyser:

1. The lexical analyzer has to recognize the longest possible string.

   Example: identifier newval — n ne new newv newva newval
2. What is the end of a token? Is there any character which marks the end of a token?
   - It is normally not defined.
   - If the number of characters in a token is fixed, in that case no problem : + –
   - But < → < or <> (in Pascal)
   - The end of an identifier: the characters cannot be in an identifier can mark the end of token.
   - We may need a lookhead . Let us following example:

     In Prolog: *p*:- *X* is 1. *p*:- *X* is 1.5.

     The dot followed by a white space character can mark the end of a number. But if that is not the case, the dot must be treated as a part of the number.
3. Skipping comments
   - Normally we don't return a comment as a token.
   - We skip a comment, and return the next token (which is not a comment) to the parser.
   - So, the comments are only processed by the lexical analyzer, and the don't complicate the syntax of the language.
4. Symbol table interface
   - Symbol table holds information about tokens (at least lexeme of identifiers)
   - How to implement the symbol table, and what kind of operations.

     (*a*) hash table – open addressing, chaining

     (*b*) putting into the hash table, finding the position of a token from its lexeme.
5. Positions of the tokens in the file (for the error handling).

Example 3.1. Write the regular expression for identifier, draw FSA and finally write the code to recognise the symbol.

Solution. Let regular expression be r such that every string of r starts with a letter and followed by any combination of letter and digit.

So r = letter (letter + digit)*

Let FSA for r be $M = (Q, \Sigma, \delta, q_0, F)$

$Q = \{q, f\}$

Σ = {letter, digit}

= {(*a*, *b*, ..., *z*), (0, 1, ..., 9)}

q is initial state

$F = \{f\}$

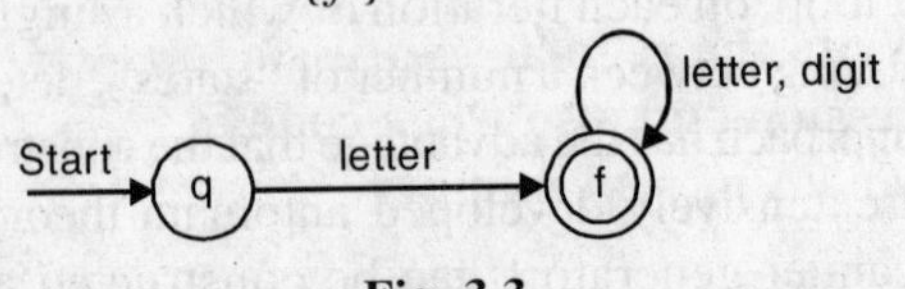

Fig. 3.3.

Now let us write the code for recognising the symbol.

```
# include <stdio . h>
# include <ctype . h>
void main ( )
  {
  char Ch ;
  ch = get char ( ) ;
  if (isalpha (ch))
  ch = getchar ( ) ;
  else error ( ) ;
  while (isalpha (ch) || is digit (ch))
  ch = getchar ( ) ;
  }
```

Here function getchar ( ) is to read next character, is alpha (Ch) is for checking that input character is letter, is digit (Ch) is function for checking that input character is digit and finally error ( ) is a function which is capable to report error and it depends on programmer.

FA given in Fig. 3.3 can be used to produce a recogniser. For example following code implements a recogniser for identifier.

```
/* program for a recogniser of identifier */
# include <stdio . h>
# include <ctype . h>
char main  ( )
  {
case 1
char state ;
char ch ;
  state = 'q0'
  ch = getchar ( ) ;
While (isalpha (ch) || isdigit (ch))
  {
switch (state)
  {
  case 'q' if (isalpha (ch))
     state = 'f'
     else
     error ( ) ;
     break ;
  case 'f': state = 'f'
     break ;
  }
  ch = getchar ( ) ;
}
  return (state = = 'f') ;
}
```

Example 3.2. Write the regular expression for real number, draw the FA for that regular expression and also write code in 'C' Language for recognising a real number and for FA of real number.

Solution. The regular expression for real number is r

$$r = (+ \mid -) \text{ digit }^* . \text{ digit digit}^*$$

The FA is as follows:

$$M = (Q, \Sigma, \delta, q_0, F)$$

where $Q = \{q, p, r, f\}$

$$\Sigma = \{+, -, \text{digit}, \text{point}\}$$

(where digit = {0, 1, ..., 9} and point = { . })

$$q_0 = \{q\}$$

$$F = \{f\}$$

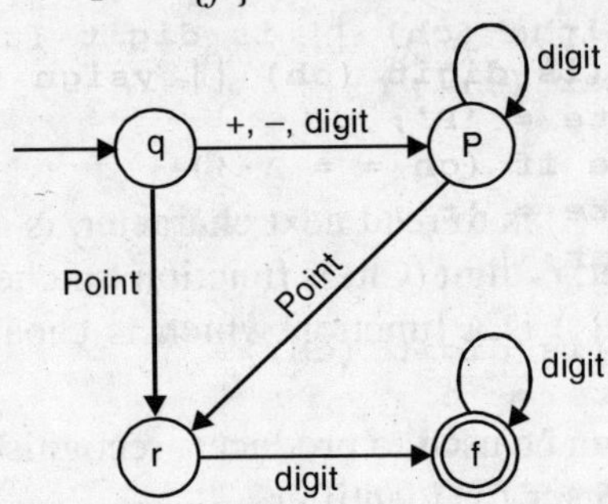

Fig. 3.4.

```
/* Code for recognising a real number */
# include <stdio . h>
# include <ctype.h>
void main ( )
{
    char ch ;
    ch = getchar ( ) ;
    if (ch = '+' || ch = = '-')
    ch = getchar ( ) ;
    while (isdigit (ch))
    ch = getchar ( ) ;
    if (ch = = ' • ' )
    ch = getchar ( ) ;
    else error ( );
    if (is digit (ch))
    ch = getchar ( ) ;
    else error ( ) ;
    while (isdigit (ch))
    ch = getchar ( ) ;
    Print ( "OK\h") ;
}
```

The program is self explanatory.

```
/* Program for FA of real number */
# include <stdio.h>
# include <ctype.h>
/* ysign is function for recognising sign */
int ysign (char sign)
    {
    return (sign = = '+' || sign = = '-') ;
    }
```

```
/* main function */
 char main ( )
 {
 char state;
 char ch;
 state = 'q';
 ch = getchar ();
while (is digit (ch) || ysign (ch) || ch = = '.')
 {
 switch (state)
 {
 case 'q':
       if (is digit (ch) || ysign (ch))
       state = 'P';
       else if (ch = = '.')
       state = 'r';
       break;
 case 'p':
       if (is digit (ch))
       state = 'p';
       else if (ch = = '.')
       state = 'r' ;
       else error ( );
       break ;
 case 'r':
       if (is digit (ch))
       state = 'f';
       else error ();
       break;
 case 'f':
       if (is digit (ch))
       state = 'f';
       else error () ;
       break ;
 }
 ch = get char ();
 }
 return (state = = 'f');
 }
```

Here error ( ) function is like trap state of FA and ysign( ) has the value true when its parameter is '+' or '–', and false otherwise.

Example 3.3. Draw the FA for a language which consists of the tokens :

<, <=, =, >=, >, < >, (,), +, –, *, /,:=, ;, identifier, keywords, constants and literals (which are enclosed in apostrophes). Comments, which begin with /* and end with a */, and blanks are ignored by the scanner (that is blanks are treated as blank separator only).

Solution. Let us take assumption that "not" preceding a character indicates that all input characters other than one listed can cause the indicated state transition. The arc labelled "error" indicates that an invalid character has been found in the source text.

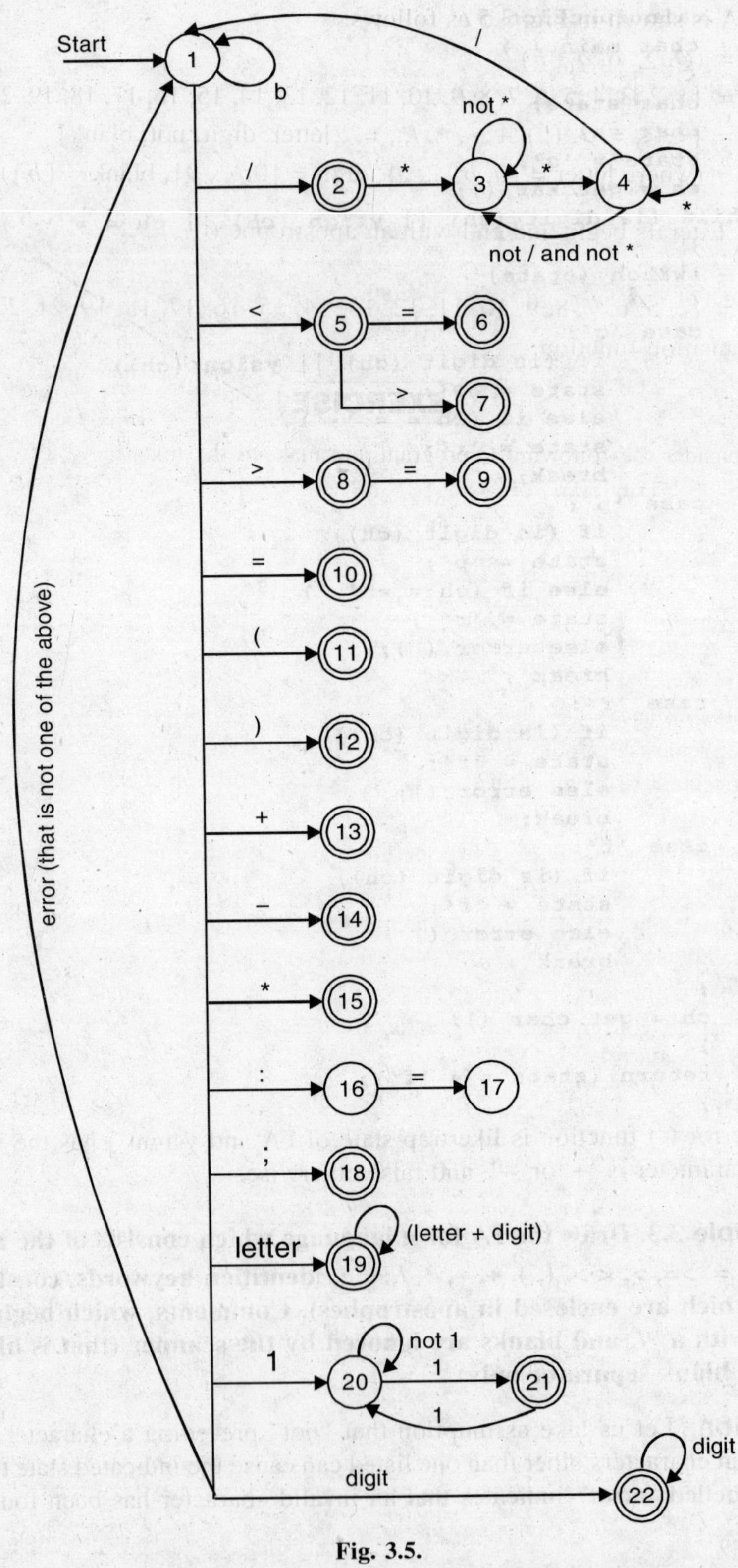
Start
b
/
not *
*
*
*
not / and not *
error (that is not one of the above)
<
=
>
=
(
)
+
–
*
:
=
;
letter
(letter + digit)
not 1
1
1
1
digit
digit

Fig. 3.5.

The FA is shown in Fig. 3.5 as follows:

$M = (Q, \Sigma, \delta, q_0, F)$

$Q = \{1, 2, 3, 4, 5, 6, 7, 8, 9, 10, 11, 12, 13, 14, 15, 16, 17, 18, 19, 20, 21, 22\}$

$\Sigma = \{<, <, =, >, (,), +, -, *, /, :, =, ;$ letter, digit, not, blank$\}$

(where letter = $\{a, b, ..., z\}$, digit = $\{0, ..., 9\}$, blank = $\{b\}$)

Note: Literals begin and end with an apostrophe (').

$q_0 = \{1\}$

$F = \{2, 5, 6, 7, 8, 9, 10, 11, 12, 13, 14, 15, 16, 17, 18, 19, 21, 22\}$

δ is transition function.

EXERCISE

1. Consider the following C program and makeup the tokens:

```
int min (i, j)
{
    int i, j;
    return i > j ? i: j;
}
```

2. Design the FA for the following:
   (*i*) identifier
   (*ii*) constant
   (*iii*) relops
   Consider the language C.
3. Specify the lexical form of numeric constant in the language of C.
4. Specify the lexical form of identifiers and keywords in the language C.
5. Write a C program for the FA of constant.
6. What is the use of determinate finite automata in lexical analysis ? Also give example.
7. What is the use of regular expression in lexical analysis ? Also give example.
8. What do you understand by automatic lexical generator ?

❑❑❑

SECTION-B

- Syntax Analysis
- Top-Down Parsing Techniques
- Bottom-Up Parsing Techniques

SYLLABUS

Syntax Analysis : Role of Parsers, Context Free Grammars, Definition of Parsing.

Parsing Technique : Shift-Reduce Parsing, Operator Precedence Parsing, Top Down Parsing, Predictive Parsing.

LR Parsers, SLR, LALR and Canonical LR Parser.

4

Chapter

SYNTAX ANALYSIS

INSIDE THIS CHAPTER

4.1. Grammars **4.2.** Chomsky Hierarchy **4.3.** Left Most and Right Most Derivations **4.4.** Parse Tree **4.5.** Phrase Structure and Lexical Structure **4.6.** Syntax Diagrams **4.7.** Reduction of Context Free Grammar **4.8.** Ambiguity in Grammars and Languages **4.9.** Syntax Analysis **4.10.** Role of Parsers.

4.1. GRAMMARS

The third chapter showed that lexical structure of tokens could be specified by regular expressions. We have also discussed that valid tokens can be recognised by the Finite automata, in second chapter. Now we will study the syntax analysis and for syntactic and semantic specification of any program, we must understand the context-free grammars, since CFG notations are used to syntactic specification of any program. So before studying the syntax analysis, it becomes necessary to study the grammars of the Programming Languages in detail and in this chapter, we will do that and main focus will be on context-free grammars.

Initially Linguists were trying to define precisely valid sentences and to give structural description for these sentences. They tried to define rules for natural languages like Hindi, English etc. Noam Chomsky gave a mathematical model for the grammars in 1956. Although it was useless to describe natural languages but it becomes very useful for computer languages.

The original motivation for grammars was the description of natural languages. We can write rules for the grammars of natural languages as follows:

<sentence> → <noun phrase> <verb phrase>

<noun phrase> → <article> <noun>

According to above set of rules "The girl eats" is valid sentence.

It can be easily observed that, some words in the grammar works as terminator for the sentence, these symbols are called terminal symbol and rest of the symbols of the vocabulary are called non-terminals. So it is clear that terminals and non-terminals forms vocabulary for any language.

Let us now formalise the idea of a grammar and how it is used. There are four important components in a grammatical description of a language:

1. There is a finite set of symbols that form the strings of the language being defined. We call these alphabets terminal symbols. Represented by V_t.

2. There is a finite set of variables, also called sometimes non-terminals or syntactic categories. Each variable represents a language; that is, a set of strings represented by V_n.
3. One of the variable represents the language being defined; it is called the start symbol. Generally it is denoted by *S*.
4. There is a finite set of rules or productions that represents the recursive definition of a langauge. Each production consists of:
   (*a*) A variable that is being (partially) defined by the production. This variable is often called the head of the production.
   (*b*) The production symbol $\rightarrow$
   (*c*) A string of zero or more terminals and variables. This string, called the body of the production, represents one way to form strings in the language of the variable of the head. So, we leave terminals unchanged and substitute for each variable of the body, with any string that is known to be in the language of that variable.

After this discussion, we are able to define a mathematical model for the grammar

$$G = (V_n, V_t, P, S)$$

G : is the notation for the grammar.

V_n : A finite set of non-terminals, generally represented by capital letters, *A*, *B*, *C*, *D*, ...

V_t : A finite set of terminals, generally represented by small letters, like, *a*, *b*, *c*, *d*, *e*, *f*, ...

S : Starting non-terminal, called start symbol of the grammar. *S* belongs to V_n.

P : Set of rules or productions in CFG.

All the productions in *P* have the form:

$$\alpha \rightarrow \beta$$

where $\alpha \in (V_n \cup V_t)^+$

and $\beta \in (V_n \cup V_t)^*$

These kind of grammars are known as unrestricted grammar, or phase-structure grammar.

4.1.1. Derivations and Their Variations

- Now we will define the notations to represent a derivation. First we define two notations $\underset{G}{\Rightarrow}$ and $\overset{*}{\underset{G}{\Rightarrow}}$. If $\alpha \rightarrow \beta$ is a production of *P* in CFG and *a* and *b* are strings in $(V_n \cup V_t)^*$, then

$$a\,\alpha\,b \underset{G}{\Rightarrow} a\,\beta\,b.$$

We say that the production $\alpha \rightarrow \beta$ is applied to the string $a\,\alpha\,b$ to obtain $a\,\beta\,b$ or we say that $a\,\alpha\,b$ directly drives $a\,\beta\,b$.

- Now suppose $\alpha_1, \alpha_2, \alpha_3, ..., \alpha_m$ are string in $(V_n \cup V_t)^*$, $m \geq 1$ and $\alpha_1 \underset{G}{\Rightarrow} \alpha_2, \alpha_2 \underset{G}{\Rightarrow} \alpha_3, \alpha_3 \underset{G}{\Rightarrow} \alpha_4, ..., \alpha_{m-1} \underset{G}{\Rightarrow} \alpha_m$

Then we say that $\alpha_1 \overset{*}{\underset{G}{\Rightarrow}} \alpha_m$, *i.e.*, we say α_1, drives α_m in finite number of steps. If α drives β by exactly *i* steps, we say $\alpha \overset{i}{\underset{G}{\Rightarrow}} \beta$.

➤ In terms of notation, a **sentential form** is the goal or start symbol, or any string that can be derived from it, that is, any string *w* such that $s \overset{*}{\Rightarrow} w$ where $w \in (V_n \cup V_t)^*$

➤ A **grammar is called recursive** if it permits derivations of the form $A \overset{*}{\Rightarrow} w_1 A w_2$ (where $A \in V_n$, and $w_1, w_2 \in (V_n \cup V_t)^*$). More specifically, **it is called left recursive** if $A \overset{*}{\Rightarrow} Aw$ and right recursive if $A \overset{*}{\Rightarrow} wA$.

➤ A **grammar is self-embedding** if it permits derivations of the form $A \overset{*}{\Rightarrow} w_1 A w_2$ (where $A \in V_n$, and where $w_1, w_2 \in (V_n \cup V_t)^*$, but where w_1 or w_2 contains at least one terminal (that is $(w_1 \cap V_t) \cup (w_2 \cap V_t) \neq \phi$).

➤ From all we can define a language *L*(*G*) produced by a grammar *G* by the relation

$$L(G) = \{w/w \in V_t^*;\ s \overset{*}{\Rightarrow} w\}.$$

4.1.2. Classic BNF Notation for Productions

As we have remarked, a production is a rule relating to a pair of strings, say α and β, specifying how one may be transformed into the other. This may be denoted $\alpha \rightarrow \beta$, and for simple theoretical grammars use is often made of this notation, using the conventions about the use of upper case letters for non-terminals and lower case ones for terminals. For more realistic grammars, such as those used to specify programming languages, the most common way of specifying productions for many years to use alternative notation invented by Bcaus, and called Backus-Naur-Form. Backus and Naur were largely responsible for the Algol 60 report (Naur, 1960 and 1963), which was the first major attempt to specify the syntax of a programming language using this notation. Regardless of what the acronym really stands for, the notation is now universely known as BNF.

In BNF, productions have the form

$$\text{left side} \rightarrow \text{definition}$$

where leftside $\in (V_n \cup V_t)^+$ and definition $\in (V_n \cup V_t)^*$ although we must be more restrictive than that, for leftside contain one nonterminal, so that we must also have

$$\text{left side} \cap V_n = \phi$$

Frequently we find several productions with the same leftside, and these are often abbreviated by lising the definitions as a set of one or more alternatives, separated by a vertical bar symbol "/".

Let production for any grammar are:

$$S \rightarrow aSa$$
$$S \rightarrow bSb$$
$$S \rightarrow c$$

then in BNF we can represents, above productions as follows:

$$S \rightarrow aSa/bSb/c$$

4.2. CHOMSKY HIERARCHY [MDU 2009]

We can exhibit the relationship between grammars by the Chomsky Hierarchy. Noum Chomsky, a founder of formal language theory, provided an initial classification in to four languages types:

Type – 0 (*Unrestricted grammar*)
Type – 1 (*Context sensitive grammar*)
Type – 2 (*Context free grammar*)
Type – 3 (*Regular grammar*)

Type 0 languages are those generated by unrestricted grammars, that is, the recursively enumerable languages. Type 1 consist of the context-sensitive languages, Type 2 consists of the context-free languages and Type 3 consists of the regular languages. Each language family of type k is a proper subset of the family of type $k - 1$. Following diagram shows the original Chomsky Hierarchy.

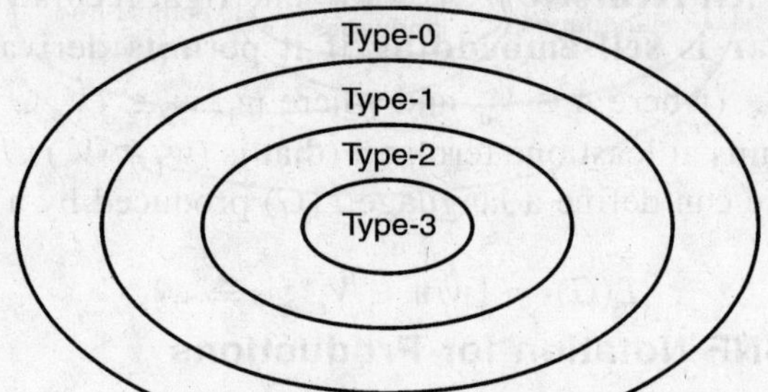

Fig. 4.1. The Original Chomsky Hierarchy.

We have also met several other language families that can be fitted in to this picture. Including the families of deterministic context-free languages (L_{DCF}), and recursive languages (L_{REC}). The modified Chomsky Hierarchy can be seen in Fig. 4.2.

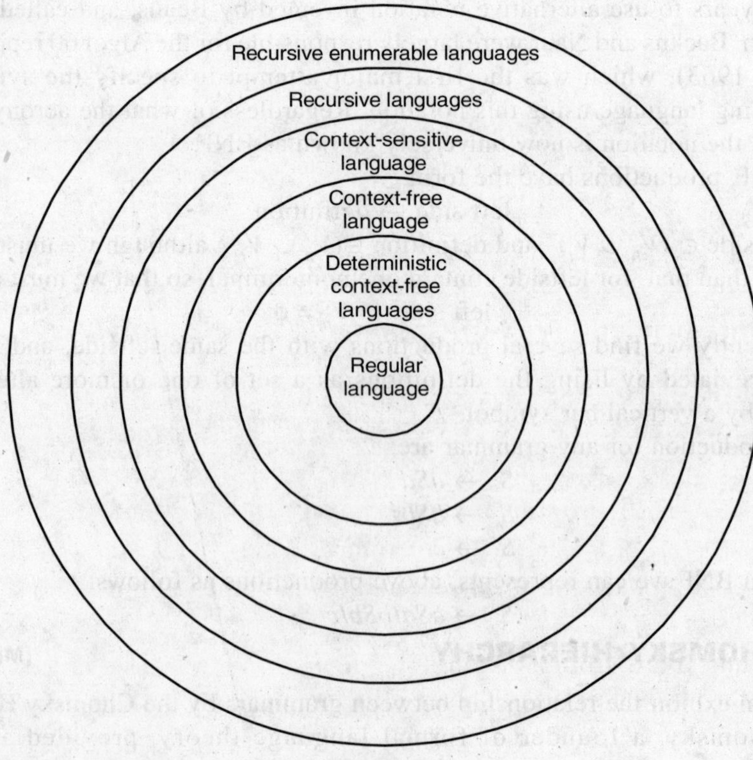

Fig. 4.2.

We know that $L = \{w: n_a(w) = n_b(w)\}$ is deterministic, but not linear. On the other hand, the language

$$L = \{a^n b^n\} \cup \{a^n b^{2n}\}$$

is linear, but not deterministic. The relationship between, linear, deterministic context-free and non-deterministic context-free language is shown in Fig. 4.3.

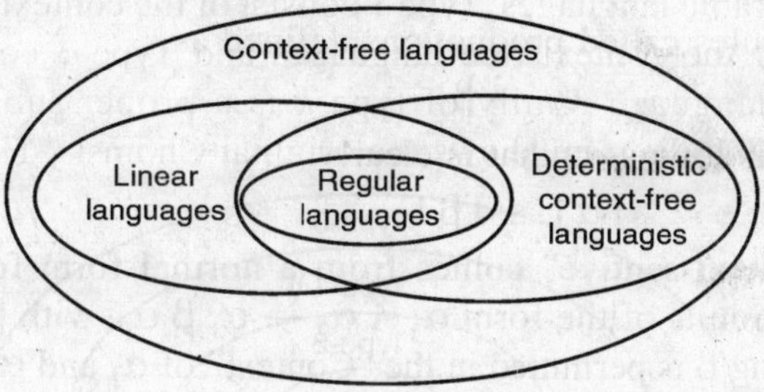

Fig. 4.3.

In the last we can summarise our discussion as follows:

Type	Name of languages generated	Production Restrictions $A \to B$	Acceptor
0	Unrestricted (reversively-enumerable)	A = any string with non-terminal B = any string	Turing machine
1	Context-sentive	A = any string with non-terminals B = any string as long as or longer than A	Linear bound automata
2	Context-free	A = one non-terminal B = any string	Pushdown automata
3	Regular	A = one non-terminal $B = aX$ or $B = a$, where a is a terminal and X is a non-terminal.	Finite automata

4.2.1. Unrestricted (Type-0) Grammars

Definition The unrestricted grammar is defined as

$$G = (V_n, V_t, P, S)$$

where V_n = a finite set of non-terminals

V_t = a finite set of terminals

S = starting non-terminal, $S \in V_n$

and P is set of productions of the following form

$$\alpha \to \beta$$

where α and β are arbitrary string of grammar symbols with $\alpha \neq \in$. These grammars are known as type-0, phase-structure or unrestricted grammars.

4.2.2. Context Sensitive Grammar

Let $G = (V_n, V_t, P, S)$ be context sentive grammar

where V_n = finite set of non-terminals

V_t = finite set of terminals

S = starting non-terminals $S \in V_n$

and P is the set of rules called productions defined as

$$\alpha \rightarrow \beta$$

where β is at least as long as α that is clearly

$$|\alpha| \leq |\beta|$$

The term "Context-sentive" comes from a normal form for these grammars, where each production is of the form $\alpha_1 A \alpha_2 \rightarrow \alpha_1 \beta \alpha_2$, with $\beta \neq \in$. Replacement of variable A by string β is permitted in the "Context" of α_1 and α_2.

For example, language $L = \{a^n b^n c^n : n \geq 0\}$ is context sensitive. We can show this exhibiting a context sensitive grammar for the language L_1 grammar as

$$S \rightarrow abc/aAbc$$
$$Ab \rightarrow bA$$
$$Ac \rightarrow Bbcc$$
$$bB \rightarrow Bb$$
$$aB \rightarrow aa/aaA$$

"A type 0 grammar is called length-increasing if for all rules $\alpha \rightarrow \beta$ in P we have $|\alpha| \leq |\beta|$" (same as context sensitive grammar).

So we can say that $\in$ free type-0 grammar is said to be length increasing grammar.

4.2.3. Context-Free Grammars

Now let us formalize the concept of context free grammar, as we discussed intuitive notions for context free grammar in previous section.

Definition

Mathematically context-free grammar is defined as follows:

"A grammar $G = (V_n, V_t, P, S)$ is said to be context-free"

where V_n: A finite set of non-terminals, generally represented by capital letters, $A, B, C, D, \ldots$

V_t: A finite set of terminals, generally represented by small letters, like, $a, b, c, d, e, f, \ldots$

S: Starting non-terminal, called start symbol of the grammar. S belongs to V_n.

P: Set of rules or productions in CFG.

G is context-free and all productions in P have the form

$$\alpha \rightarrow \beta$$

where $\alpha \in V_n$ and $\beta \in (V_t \cup V_n)^*$

Every regular grammar is context-free, so a regular language is also a context-free one. It is already proved by pumping lemma that language $\{a^n b^n / n > 0\}$ is not regular, but it is possible to design a context-free-grammar for these languages (we will design it in next section). **So now it is very clear that "the family of regular language is a proper subset of the family of context-free language".**

TIPS

Context-free grammars derive their name from the fact that the substitution of the variable on the left of a production can be made any time, such a variable appears in the sentential form. It does not depend on the symbols in the rest of the sentential form (the context). This feature is the consequence of allowing only a single variable on the left side of the production.

4.2.4. Regular Grammars

Definition

A regular grammar may be left linear or right linear.

"If all production of a CFG are of the form $A \to wB$ or $A \to w$, where A and B are variables and $w \in V_t^*$, then we say that grammar is right liner".

"If all the production of a CFG are of the form $A \to Bw$ or $A \to w$, we call it left linear."

A right or left linear grammar is called a regular grammar.

Example 4.1. Write a CFG, which generates string of balanced parenthesis.

Solution. This grammar will accept the balanced right and left parenthesis. For example ( ) ( ) is acceptable, ((( ))) is also accepted.

Let us design the CFG for this

Let CFG be $G = (V_n, V_t, P, S)$ where

V_n = set of non-terminals = $\{S\}$

V_t = set of terminals = $\{(,)\}$

and set of production P is given by

$$S \to SS$$
$$S \to (S)$$
$$S \to \in$$

Now we see what this grammar generates:

$$\begin{aligned} S &\Rightarrow SS \\ &\Rightarrow (S)\,S \\ &\Rightarrow (S)\,SS \\ &\Rightarrow (S)\,(S)\,S \\ &\Rightarrow (S)\,(S)\,(S) \\ &\Rightarrow (\,)\,(S)\,(S) \\ &\Rightarrow (\,)\,(\,)\,(S) \\ &\Rightarrow (\,)\,(\,)\,(\,) \end{aligned}$$

Thus the above grammar always generates balanced pairs of parenthesis.

4.3. LEFT MOST AND RIGHT MOST DERIVATIONS

In order to restrict the number of choices of replacement of variables, if at each step we replace the "left most" variable by one of its production bodies such a derivation is called a "left most derivation", and we indicate that a derivation is left most by using relation $\underset{lm}{\Rightarrow}$ and $\overset{*}{\underset{lm}{\Rightarrow}}$ for one or many steps, respectively.

Similarly, it is possible that at each step the "right most" variable is replaced by one of its bodies. If so, we call the derivation "right most" and use symbols $\underset{rm}{\Rightarrow}$ and $\overset{*}{\underset{rm}{\Rightarrow}}$ to indicate one or many right most derivation steps, respectively.

4.4. PARSE TREE

There is a tree representation for derivation that has proved extremely useful. It is the second way of showing derivations, independent of the order in which productions are used, is also called "derivation tree". "A parse tree is an ordered tree in which nodes are labeled with the left sides of productions and in which the children of a node represent its corresponding right sides".

4.4.1. Definition

Let $G = (V_n, V_t, P, S)$ be a context-free grammar. An ordered tree for this CFG, G is a derivation tree if and only if it has the following properties.

(*a*) The root is labeled by the starting non-terminal of the CFG that is S.

(*b*) Every leaf of the ordered tree has a label from $V_t \cup \{\in\}$.

(*c*) Every Interior node of ordered tree has a label from V_n.

(*d*) Let us assume that a vertex has label $X \in V_n$, and its children are labeled (from left to right) $y_1, y_1, y_3, ..., y_n$, then production must contain a production of the form

$$X \rightarrow y_1, y_2, ..., y_n.$$

(*e*) A leaf labeled $\in$ has no siblings, that is, vertex with a child labeled $\in$ can have no other children. Clearly if the leaf is labeled $\in$, then it must be the only child of its parent.

4.4.2. The Yield of Parse Tree

If we look at the leaves of any parse tree and can concate them from the left, we get a string, called the yield of the tree, which is always a string that is derived from the root will be proved shortly of special importance are those parse trees such that :

(*a*) The yield is a terminal string. That is all leaves are labeled either with a terminal or with $\in$.

(*b*) The root is labeled by the start symbol.

Now let us see some examples based on the parse tree.

Example 4.2. Consider the CFG

$$S \rightarrow XX$$

$$X \rightarrow XXX/bX/Xb/a$$

Find the parse tree for the string *bbaaaab*.

Solution. Given CFG is

$$S \rightarrow XX$$
$$S \rightarrow XXX/bX/Xb/a$$

and string is $w = bbaaaab$

We begin with S and apply the production $S \rightarrow XX$.

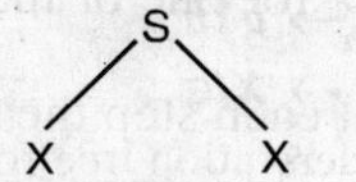

Fig. 4.4.

To the left hand X, let us apply the production $X \rightarrow bX$. To the right hand X, let us apply $X \rightarrow XXX$.

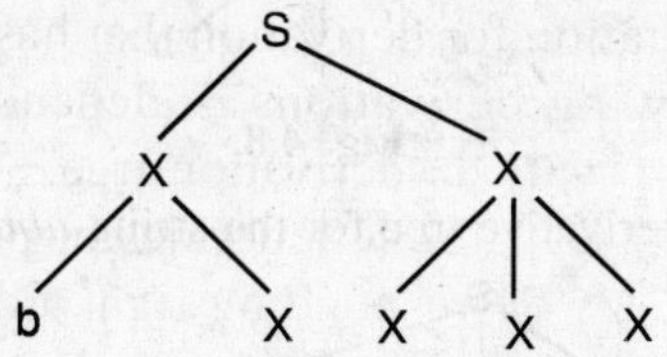

Fig. 4.5.

The b that we have on the bottom line is a terminal, so it does not descend further. In the terminology of trees, it is called a "terminal node". Let the four A's left to right, undergo the production $A \rightarrow bA$, $A \rightarrow a$, $A \rightarrow a$, and $A \rightarrow Ab$, respectively. We now have

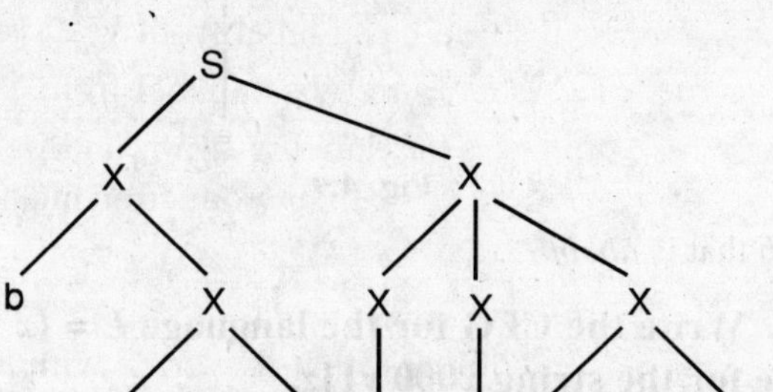

Fig. 4.6.

Let us finish off the generation of a word with the production $X \rightarrow a$ and $X \rightarrow a$:

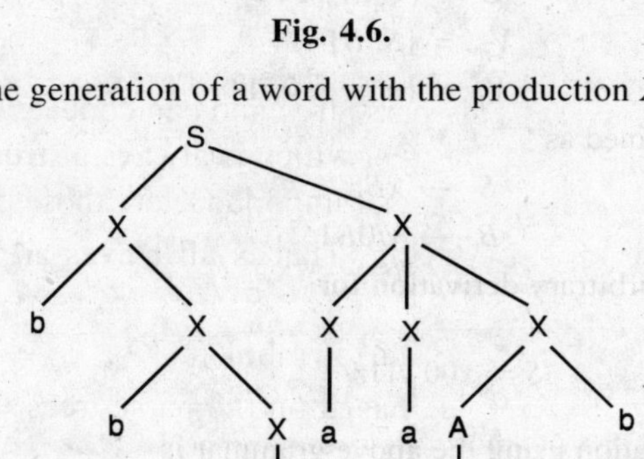

Fig. 4.7.

Reading from left to right, we see the word we have produced is *bbaaaab*.

These tree diagrams are called "syntax trees", "parse trees", "generation" trees, "production trees" or derivation trees.

Example 4.3. Consider the grammar G, with production

$$S \to aXY$$
$$X \to bYb$$
$$Y \to X/\in.$$

Solution. Let us see a partial derivation tree for G, this tree gives *abbbb* as yield.

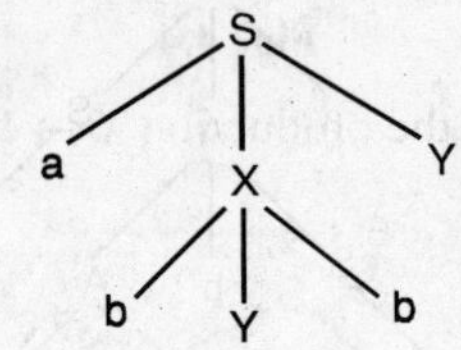

Fig. 4.8.

Now extend this partial derivative tree for the string *abbbb*.

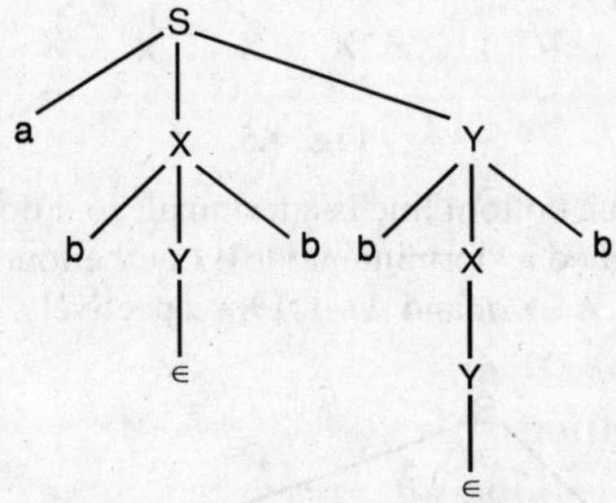

Fig. 4.9.

the yield is $ab\in bb\in b$ that is *abbbb*.

Example 4.4. Write the CFG for the language $L = \{x\ 0^n y 1^n z / n >= 0\}$, and give the parse tree for the string $x000\ y11z$.

Solution. Let the CFG for the language L is

$$G = (V_n, V_t, P, S)$$

where
$$V_n = \{S, B\}$$
$$V_t = \{x.\ y,\ z,\ 0,\ 1\}$$

and production P is defined as

$$S \to xBz$$
$$B \to y/0B1.$$

Now construct an arbitrary derivation for

$$S \underset{G}{\overset{*}{\Rightarrow}} x00\ y11z.$$

One possible derivation using the above grammar is

$$S \Rightarrow x\underline{B}z \Rightarrow x0B1z \Rightarrow x00\underline{B}11z \Rightarrow x00y11z$$

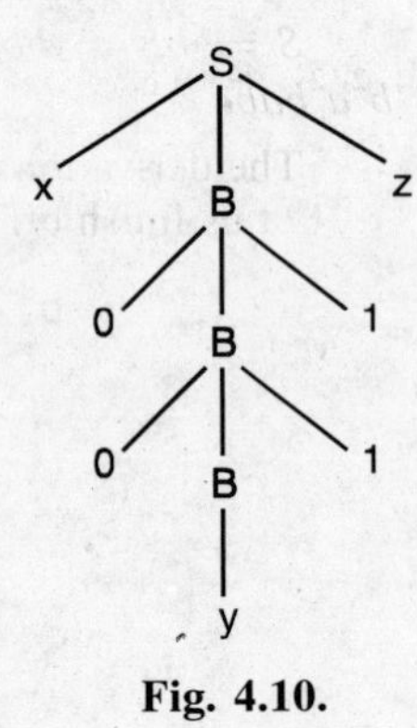

Fig. 4.10.

Now we are able to design the parse tree for the string $x\ 00y11z$ from the above CFG. Parse tree is shown in Fig. 4.10.

Example 4.5. Consider the CFG 'G' whose productions are

$$S \rightarrow aAS/a$$
$$A \rightarrow SbA/SS/ba.$$

show that $S \overset{*}{\Rightarrow}$ *aabbaa* and construct a derivation tree whose yield is *aabbaa*.

Solution. $S \Rightarrow a\underline{A}S \Rightarrow a\underline{S}bAS \Rightarrow aab\underline{A}S \Rightarrow aabbaS \Rightarrow a^2b^2a\underline{S} \Rightarrow a^2b^2a^2$

The derivation tree is given in Fig. 4.11.

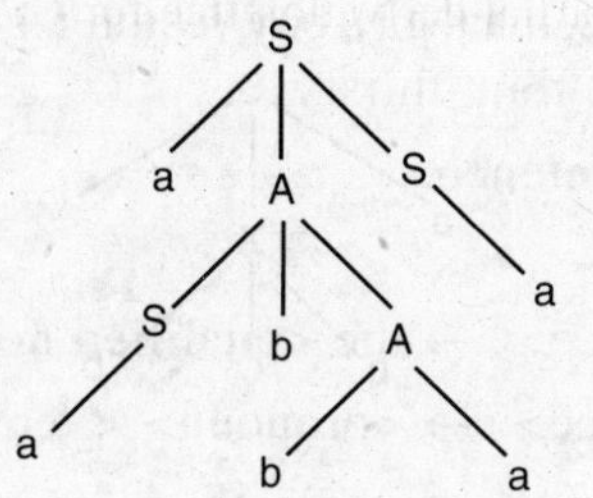

Fig. 4.11.

Derivation tree with yield *aabbaa*.

Example 4.6. Let *G* be CFG

$$S \rightarrow bB/aA,$$
$$A \rightarrow b/bS/aAA$$
$$B \rightarrow a/aS/bBB.$$

For the string *bbaababa* find

(*i*) left most derivation

(*ii*) right most derivation and

(*iii*) parse tree.

Solution. (*i*) Left most derivation for string $w = bbaababa$ is

$S \Rightarrow b\underline{B} \Rightarrow bb\underline{B}B \Rightarrow bba\underline{B} \Rightarrow bbaa\underline{S} \Rightarrow b^2a^2b\underline{B} \Rightarrow b^2a^2ba\underline{S} \Rightarrow b^2a^2bab\underline{B} \Rightarrow \underline{b}^2a^2baba$

(*ii*) The right most derivation is

$S \Rightarrow b\underline{B} \Rightarrow bbB\underline{B} \Rightarrow bbBa\underline{S} \Rightarrow bbBab\underline{B} \Rightarrow b^2Baba\underline{S} \Rightarrow b^2Babab\underline{B} \Rightarrow b^2\underline{B}ababa \Rightarrow b^2a^2baba$

The derivation tree is following in Fig. 4.12.

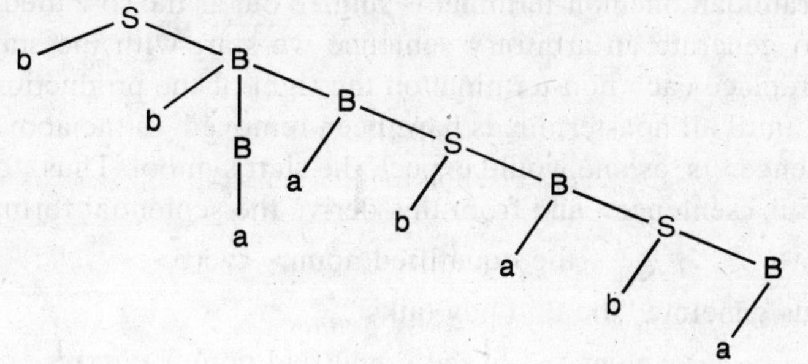

Fig. 4.12.

Yield is *bbaababa*.

Example 4.7. Write a context-free grammar which is able to generate "the thin boy talks" English sentence, also give the parse tree.

Solution. We are carefully reads the problems then it is clear that we have to write a **CFG for a tiny subset of English itself.**

Let us assume that CFG be $G = (V_n, V_t, P, S)$

where V_n = {<sentence>, <qualified noun>, <noun>, <pronoun>, <verb>, <adjective>}.

V_t = {the, man, girl, boy, lecturer, he, she, drinks, sleeps, mystifies, tall, thin, thirsty}

S = <sentence>

P = {

<sentence> → the <qualified noun> <verb> ...(*i*)

<sentence> → <pronoun> <verb> ...(*ii*)

<qualified noun> → <adjective> <noun> ...(*iii*)

<pronoun> → man/girl/boy/lecturer ...(*iv*)

<verb> → he/she ...(*v*)

<verb> → talks/listens/mystifies ...(*vi*)

<adjective> → fall/thin/sleepy ...(*vii*)

The set of productions defines the non-terminal <sentence> as consisting of either the terminal "the" followed by a <qualified noun> followed by a <verb>, or as a <pronoun> followed by a <verb>. A <qualified noun> is an <adjective> followed by a <noun>, and a <noun> is one of the terminal symbols "man" or "girl" or "boy" or "lecturer".

A <pronoun> is either of the terminals "he" or "she", while a <verb> is either "talks" or "listens" or "mystifies". Here <sentence>, <noun>, <qualified noun>, <pronoun> <adjective> and <verb> are nonterminals. These do not appear in any sentence of the language, which includes such majestic prose as

he talks

the thin lecturer mystifies

the sleepy boy listens

From a grammar, one non-terminal is singled out as the so-called **start symbol.** If we want to generate an arbitrary sentence we start with the start symbol and successively replace each non-terminal on the right of the production defining that non-terminal, until all non-terminals have been removed. In the above example, the symbol <sentence> is, as one would expect, the start symbol. Thus, for example, we could start with <sentence> and from this derive the sentential form

the <qualified noun> <verb>

Now let us generate "the thin boy talks".

<sentence> → the <qualified noun> <verb>

→ the <adjective> <noun> <verb>

→ the thin <noun> <verb>

→ the thin boy < verb>

→ the thin boy talks.

Parse tree for above derivations is as follows:

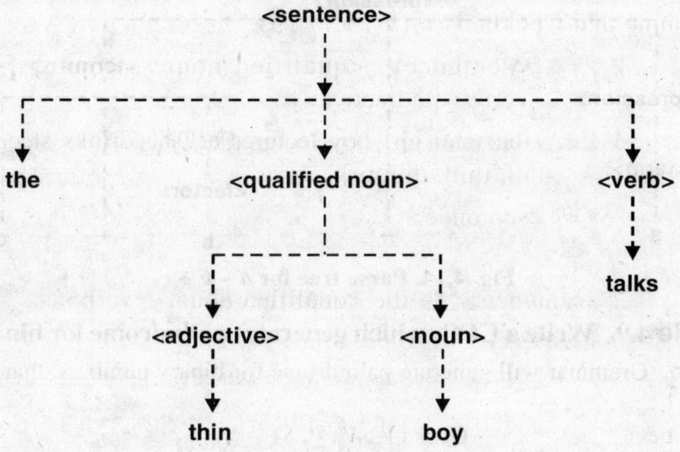

Fig. 4.13.

Example 4.8. Design a context-free grammar which generates the mathematical expressions for '+' and '–' operators.

Solution. Let CFG be $G = (V_n, V_t, S, P)$

where V_n = {<goal>, <expression>, <term>, <factor>}

$V_t = \{a, b, c, -, +\}$

S = <goal>

and productions are defined as follows:

<goal> → <expression>

<expression> → <term>/<expression> – <term>

<term> → <factor>/<term> + <factor>

<factor> → a/b/c

Let us apply above grammar on algebraic expression $a - b + c$.

<goal> → <expression>

→ <expression> – <term>

→ <term> – <term>

→ <factor> – <term>

→ a – <term>

→ a – <term> + <factor>

→ a – <factor> + <factor>

→ $a - b$ + <factor>

→ $a - b + c$

The parse tree for above string is as follows:

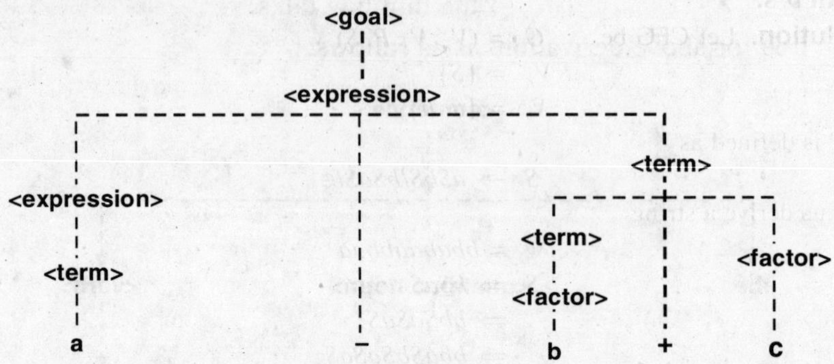

Fig. 4.14. Parse tree for $a - b + c$.

Example 4.9. Write a CFG, which generates palindrome for binary numbers.

Solution. Grammar will generate palindrome for binary numbers, that is 00, 010, 11, 101, 11100111 ...

Let CFG be $G = (V_n, V_t, P, S)$

where V_n = set of non-terminal = $\{S\}$

V_t = set of terminals = $\{0, 1\}$

and production rule P is defined as

$$S \rightarrow 0S0/1S1$$
$$S \rightarrow 0/1/\in$$

Obviously this grammar generates palindrome for binary numbers, it can be seen by the following derivation

$$\begin{aligned} S &\Rightarrow 0S0 \\ &\Rightarrow 01S10 \\ &\Rightarrow 010S010 \\ &\Rightarrow 0101010 \end{aligned}$$

which is a palindrome.

Example 4.10. Write a CFG for the regular expression

$$r = 0^*\, 1(0 + 1)^*$$

Solution. Let us analyse regular expression

$$r = 0^*\, 1\, (0 + 1)^*$$

Clearly regular expression is the set of strings which starts with any number of 0's, followed by a one and end with any combination of 0's and 1's.

Let CFG be $G = (V_n, V_t, P, S)$

where $V_n = \{S, A, B\}$

$V_t = \{0, 1\}$

and productions P are defined as

$$S \rightarrow A1B$$
$$A \rightarrow 0A/\in$$
$$B \rightarrow 0B/1B/\in$$

Let us see the derivation of the string 00101

$$\begin{aligned} S &\Rightarrow A1B \\ &\Rightarrow 0A10B \\ &\Rightarrow 00A101B \\ &\Rightarrow 00101 \end{aligned}$$

So clearly G is CFG for regular expression r.

Example 4.11. Write a CFG which generates strings having equal number of *a*'s an *b*'s.

Solution. Let CFG be $G = (V_n, V_t, P, S)$

$$V_n = \{S\}$$

$$V_t = \{a. b\}$$

where *P* is defined as

$$S \rightarrow aSbS/bSaS/\in$$

Let us derive a string

$$w = bbabaabbaa$$

$$S \Rightarrow b\underline{S}aS$$
$$\Rightarrow bb\underline{S}aSaS$$
$$\Rightarrow bbaSb\underline{S}aSaS$$
$$\Rightarrow bbaSba\underline{S}bSaSaS$$
$$\Rightarrow bba\underline{S}baaSbSbSaSaS$$
$$\Rightarrow bbabaa\underline{S}bSbSaSaS$$
$$\Rightarrow bbabaaba\underline{S}bSaSaS$$
$$\Rightarrow bbabaabb\underline{S}aSaS$$
$$\Rightarrow bbabaabba\underline{S}aS$$
$$\Rightarrow bbabaabbaa\underline{S}$$
$$\Rightarrow bbabaabbaa$$

Example 4.12. Design a CFG, which can generate string, having any combination of *a*'s and *b*'s, except null string.

Solution. Let CFG be $G = (V_n, V_t, P, S)$

$$V_n = \{S\}$$

$$V_t = \{a, b\}$$

Productions *P* are defined as

(P_1, say) $S \rightarrow aS$

(P_2, say) $S \rightarrow bS$

(P_3, say) $S \rightarrow a$

(P_4, say) $S \rightarrow b$

We can produce the word *baab* as follows:

$S \Rightarrow bS$ (by P_2)

$\Rightarrow baS$ (by P_1)

$\Rightarrow baaS$ (by P_1)

$\Rightarrow baab$ (by P_4)

The language generated by this CFG is the set of all possible strings of the letters *a* and *b* except for the null string, which we cannot generate.

We can generate any word by the following algorithm:

At the beginning, the working string is the start symbol *S* select a word to be generated. Read the letters of the desired word from left to right, one at a time. If an *a* is read that is not the last letter of the word, apply P_1 to the working string. If '*b*' is read that is not the last letter of the word, apply P_2 to the working string.

If the last letter is read and it is an *a*, apply P_3 to the working string. If the last letter is read and it is *b*, apply P_4 to the working string.

Productions 3 and 4 can be used only once and one of them can be used. For example, to generated *babb*, we apply in order productions P_2, P_1, P_2 and P_4 as below

$$S \Rightarrow bS \Rightarrow baS \Rightarrow babS \Rightarrow babb.$$

Example 4.13. Design CFG for regular expression

$$r = (a + b)^* \, aa \, (a + b)^*.$$

Solution. Let CFG be

$$G = (V_n, V_t, P, S)$$
$$V_n = \{S, T\}$$
$$V_t = \{a, b\}$$

Productions *P* are defined as

$S \rightarrow TaaT$ (say P_1)

$T \rightarrow aT$ (say P_2)

$T \rightarrow bT$ (say P_3)

$T \rightarrow \in$ (say P_4)

Last three productions, *i.e.*, P_2, P_3 and P_4 allows us to generate any word we want from terminal *T*. If the non-terminal *T* appears in any working string, we can apply productions to turn in to any string we want. Therefore, the words generated from *S* have the form any substring *aa* any substring

or $(a + b)^* \, aa \, (a + b)^*$

which is the language of the all words with a double '*a*' in them somewhere.

For example to generate *baabaab*, we can proceed as follows:

$$\begin{aligned} S &\Rightarrow TaaT \\ &\Rightarrow bTaaT \\ &\Rightarrow baTaaT \\ &\Rightarrow baaTaaT \\ &\Rightarrow baabTaaT \\ &\Rightarrow baab\in aaT \\ &\Rightarrow baabaaT \\ &\Rightarrow baabaabT \\ &\Rightarrow baabaab\in \\ &\Rightarrow baabaab \end{aligned}$$

There are other sequences that can also derive the word *baabaab*.

Example 4.14. Design a CFG for the regular expression $r = (a + b)^*$.

Solution. Let *G* be CFG

$$G = (V_n, V_t, P, S)$$
$$V_n = \{S\}$$
$$V_t = \{a, b\}$$

P are defined as

$$\begin{aligned} S &\rightarrow aS \\ S &\rightarrow bS \\ S &\rightarrow a \\ S &\rightarrow b \\ S &\rightarrow \in \end{aligned}$$

The word *ab* can be generated by the derivation

$$\begin{aligned} S &\Rightarrow aS \\ &\Rightarrow abS \\ &\Rightarrow ab\in \\ &\Rightarrow ab \end{aligned}$$

or by the derivation

$$\begin{aligned} S &\Rightarrow aS \\ &\Rightarrow ab. \end{aligned}$$

Example 4.15. The difference between even 'palindrome' and 'odd palindrome' (whose definitions is obvious) is that when we are finally ready to get rid of *S* in the even palindrome working string, we must replace it with ∈. If we were force to replace it with an *a* or *b* instead, we would create a central letter and the result would be a grammar for odd palindrome as follows :

$$S \to aSa$$
$$S \to bSb$$
$$S \to a$$
$$S \to b$$

Solution. If we allow the option of turning the central *S* into either ∈ or a letter, we would have a grammar for the entire language palindrome:

$$S \to aSa$$
$$S \to bSb$$
$$S \to a$$
$$S \to b$$
$$S \to \in$$

The language $\{a^n b^n\}$ and palindrome are amazingly similar in grammartical structure while the first is nearly a regular expression and the other is for from it.

Example 4.16. Write a CFG for the language $L(G) = \{ww^r : w \in \{0, 1\}^*\}$.

Solution. Let grammar be *G*.

$$G = (V_n, V_t, P, S)$$
$$V_n = \{S\}$$
$$V_t = \{0, 1\}$$

Productions are defined as

$$S \to 0S0 \qquad (P_1)$$
$$S \to 1S1 \qquad (P_2)$$
$$S \to \in \qquad (P_3)$$

Let us derive a string $\underset{w}{\underline{001}}\ \underset{w^r}{\underline{100}}$ by above CFG.

$$S \Rightarrow 0S0 \qquad (\text{by } P_1)$$
$$\Rightarrow 00S00 \qquad (\text{by } P_1)$$
$$\Rightarrow 001S100 \qquad (\text{by } P_2)$$
$$\Rightarrow 001100 \qquad (\text{by } P_3)$$

Example 4.17. Design a CFG for the language

$$L(G) = \{ab\,(bbaa)^n\,bba\,(ba)^n : n \geq 0\}.$$

Solution. Let CFG be *G*

$$G = (V_n, V_t, P, S)$$
$$V_n = (S, X, Y)$$
$$V_t = \{a, b\}$$

Productions are defined as

$$S \to abX$$
$$X \to bbYa$$
$$Y \to aaXb$$
$$Y \to \in$$

or we can also defined production as:

$$S \to abX$$

$$X \rightarrow bbaaXba$$
$$X \rightarrow bba$$

Let us derive a string $ab(bbaa)^2\, bba(ba)^2$

$$\begin{aligned} S &\Rightarrow ab\underline{X} \\ &\Rightarrow abbb\underline{Y}a \\ &\Rightarrow abbbaa\underline{X}ba \\ &\Rightarrow abbbaabb\underline{Y}aba \\ &\Rightarrow abbbaabbaa\underline{X}baba \\ &\Rightarrow ab(bbaa)^2 bbYa(ba)^2 \\ &\Rightarrow ab(bbaa)^2 bb\in a(ba)^2 \\ &\Rightarrow ab(bbaa)^2 bba(ba)^2 \end{aligned}$$

or

$$\begin{aligned} S &\Rightarrow abX \\ &\Rightarrow abbbaaXba \\ &\Rightarrow abbbaabbaaXbaba \\ &\Rightarrow ab(bbaa)^2 bba(ba)^2. \end{aligned}$$

Example 4.18. Design a CFG for the language

$$L = \{a^n\, b^m : n \neq m\}.$$

Solution. If $n \neq m$ then there are only two cases are possible.

Case 1. $n > m$

Let us say langauge L on condition $n > m$ is

L_1 and $L_1 = \{a^n\, b^m : n > m\}$.

Let us say G_1 be the CFG for the language L_1,

$$\begin{aligned} G_1 &= (V_n^1, V_t^1, P^1, S^1) \\ V_n^1 &= \{S_1, A, S^1\} \\ V_t^1 &= \{a, b\} \end{aligned}$$

then productions P^1 are defined as

$$\begin{aligned} S^1 &\rightarrow AS_1 \\ S_1 &\rightarrow aS_1b/\in \\ A &\rightarrow aA/a \end{aligned}$$

Case 2. $n < m$

Let us say L on condition $n < m$ is L_2

and $L_2 = \{0^n 1^m : n < m\}$

Let CFG for G_2 be G_2

$$\begin{aligned} G_2 &= (V_n^2, V_t^2, P^2, S^2) \\ V_n^2 &= \{S^2, S_2, B\} \\ V_t^2 &= \{a, b\} \end{aligned}$$

P^2 are defined as follows:

$$\begin{aligned} S^2 &\rightarrow S_2B \\ S_2 &\rightarrow aS_2b/\in \\ B &\rightarrow bB/b \end{aligned}$$

by the help (by the combining G_1 and G_2) of G_1 and G_2 we can write the CFG for L as

$$S \rightarrow S^1/S^2$$

where S is start symbol of CFG for L.

Example 4.19. Design a CFG for the language

$$L = \{0^n\ 1^n/n \geq 0) \cup (1^n\ 0^n/n \geq 0)\}.$$

Solution. We can assure L as

$$L = L_1 \cup L_2$$
$$L_1 = \{0^n\ 1^n/n \geq 0\}$$

Let say G_1 be CFG for the language L_1

$$G_1 = (V_n, V_t, P, S)$$
$$V_n = \{S_1\}$$
$$S = \{S_1\}$$
$$V_t = \{0, 1\}$$

and P is defined as

$$S_1 \rightarrow 0S_1 1/\in$$

and $$L_2 = \{1^n 0^1/n \geq 0\}$$

Let say G_2 be CFG for language L_2

$$G_2 = (V_n, V_t, P, S)$$
$$V_n = \{S_2\}$$
$$S = \{S_2\}$$
$$V_t = \{0, 1\}$$

P is defined as

$$S_2 \rightarrow 1S_2 0/\in$$

Since $L = L_1 \cup L_2$, suppose G be CFG for language L, with starting non-terminal S. Then productions in G will be

$$S \rightarrow S_1/S_2$$
$$S_1 \rightarrow 0S_1 1/\in$$
$$S_2 \rightarrow 1S_2 0/\in.$$

Example 4.20. Give the simple description of the language generated by the grammar with production

$$S \rightarrow aA$$
$$A \rightarrow bS$$
$$S \rightarrow \in.$$

Solution. Given CFG is $S \rightarrow aA$ (say P_1)

$A \rightarrow bS$ (say P_1)

$S \rightarrow \in$ (say P_3)

From the production P_3, CFG can generate null string. Now let us see production P_1 (*i.e.,* $S \rightarrow aA$), once we use it then we have to $A \rightarrow bS$. Now again we have to use either production P_1 ($S \rightarrow aA$) or production P_3 ($S \rightarrow \in$), so clearly the langauge of above CFG will be as follows :

$$L = \{(ab)^n/n \geq 0\}$$

Clearly language L is the set of strings. Starts with *ab* followed by any number of *ab*'s. Null string ($\in$) also included in language L.

4.5. PHRASE STRUCTURE AND LEXICAL STRUCTURE

It should not take much to see that a set of production for a real programming language grammar will usually divided into two distinct groups. In such language we can distinguish between the productions that specify the **phrase structure**—the way in which the words or tokens of the language are combined to form components of programs—and the production that specify the **lexical structure** or **lexicon**—the way in which individual characters are combined to form such words or tokens. Some tokens are easily specified as simple constant strings for themselves. Others are more generic–lexical tokens such as Identifier, literal constants, and strings are themselves specifier by means of productions (or, in many cases, by regular expression).

As I have already hinted, the recognition of tokens for a real programming language is usually done by a scannar (lexical analyser) that returns tokens to the parser (syntax analyser) on demand.

Sometimes the distinction between "lexical" and "syntactic" elements is taken to great lengths. For example we might find:

Const Declaration $\rightarrow$ const sym

Const Identifier equals const expression semicolon

{const identifier equals const expression semicolon}.

With production like

Const sym $\rightarrow$ "CONST"

Semicolon $\rightarrow$ ";"

equals $\rightarrow$ "="

and so on. This may seem rather long-winded, but there are occasional advantages, for example in allowing alternative for limited character set machines, as in

left bracket $\rightarrow$ "[" / "("

Pointer sym $\rightarrow$ "^" / "β"

as in used in some Pascal systems.

4.6. SYNTAX DIAGRAMS

An entirely different method of syntax definition is by means of the graphic representation known as syntax diagram, syntax charts, or define the syntax of Pascal, modula-2 and Fortran 77. The rules take the form of flow diagrams, the possible path representing the possible sequence of symbols. One stats at the left of a diagram, and traces a path which may incorporate terminals, or incorporate transfer to other diagrams if a word is reached that corresponds to a non-terminal. For example, and identifier might be defined by

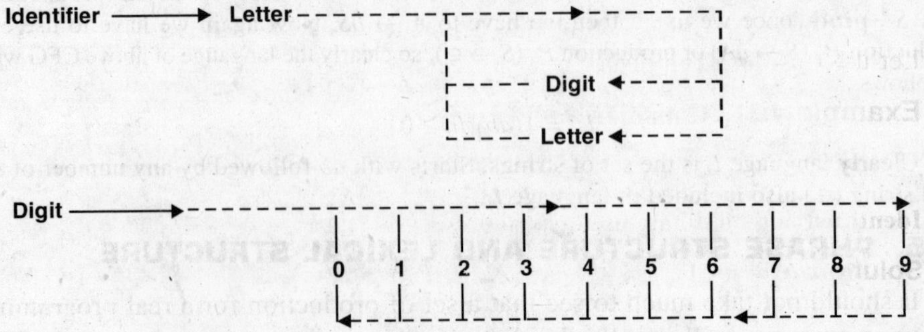

Fig. 4.15.

With a similar diagram applying to letter.

4.7. REDUCTION OF CONTEXT FREE GRAMMAR

There are several ways in which one can restrict the format of context free grammar without reducing the language generation power of context free grammar. Let L be a non-empty context free language, then it can be generated by a context-free grammar G with the following properties:

(*a*) We must eliminate useless symbols, those variables or terminals that do not appears in any derivation of a terminal string from the start symbol.

(*b*) We must eliminate $\in$-productions, those of the form $X \to \in$ for some variable X.

(*c*) We must eliminate unit production, those of the form $X \to Y$ for variables X and Y.

4.7.1. Eliminating Use Less Symbols

Here we are going to identified those symbols, which do not play any role in the derivation of any string w in $L(G)$ (these symbols are called use less symbols) and then eliminate the identified production, which contains useless symbols, from the context free grammar.

A symbols Y in a context-free grammar is useful it and only if:

(*a*) $Y \overset{*}{\Rightarrow} w$, where $w \in L(G)$ and w in V_t*, that is Y leads to a string of terminals. Here Y is said to be "generating".

(*b*) If there is a derivation $S \overset{*}{\Rightarrow} \alpha Y \beta \overset{*}{\Rightarrow} w$, $w \in L(G)$, for same α and β, then Y is said to be reachable.

Surely a symbol that is useful will be both generating and reachable. If we eliminate the symbols that are not generating first, and then eliminate from the remaining grammar those symbols that are not reachable. Then, after this process, context free grammar will have only useful symbols.

There for reduction of a given grammar G, involves following steps:

(*a*) Identified non-generating symbols in given CFG and eliminate those productions which contains non-generating symbols.

(*b*) Identified non-reachable symbols in grammar and eliminate those productions which contains non-reachable symbols.

Let us consider examples:

Example 4.21. Consider a CFG

$$S \to AB/a$$
$$A \to b$$

Identified and eliminate useless symbols.

Solution. Given CFG is

$$S \to AB/a$$
$$A \to b$$

Here by observing the given CFG, it becomes very clear that B is non-generating symbol. Since A derives b, S derives a but B is not deriving any string w in the V_t^*.

So we can eliminate $S \to AB$ from the context free grammar, now CFG becomes

$$S \to a$$
$$A \to b$$

Here A is non-reachable symbol, since it can not be reached by starting non-terminal S.

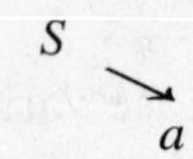

So we can eliminate $A \to b$, now grammar is

$$S \to a$$

which is reduced grammar.

Example 4.22. Remove the useless symbol from the given context free grammar :

$$S \to aB/bX$$
$$A \to BAd/bSX/a$$
$$B \to aSB/bBX$$
$$X \to SBD/aBx/ad.$$

Solution. First we choose those non-terminals which are deriving to the strings of terminals. The non-terminals A and X directly deriving to the strings of terminal a and ad respectively. Hence they are useful symbols. Since $S \to bX$ and X is useful symbol, hence S is also useful symbol. But B does not derive any string w in V_t^* so clearly B is a non-generating symbol, so eliminate those productions which contains B. Now grammar becomes:

$$S \to bX$$
$$A \to bSX/a$$
$$X \to ad$$

Again here A is non-reachable, since A can not be reached from S (starting non-terminal), it can be understand by following parse tree in Fig. 4.16.

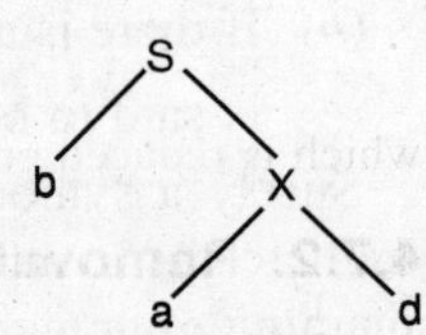

Fig. 4.16.

So eliminate $A \to bSX/a$, then reduced CFG is

$$S \to bX$$
$$X \to ad.$$

Example 4.23. Consider the following grammar and obtain an equivalent grammar containing no useless grammar symbol.

$$A \to xyz/Xyzz$$
$$X \to Xz/xYz$$
$$Y \to yYy/Xz$$
$$Z \to Zy/z.$$

Solution. Since $A \to xyz$ and $Z \to z$, hence A and Z are directly deriving to the string of terminals. Hence they are useful symbols.

Hence X and Y do not lead to a string of terminals, that is, X and Y are useless symbols. Therefore by eliminating these productions we get:

$$A \to xyz$$
$$Z \to Zy/z$$

Now since A is the starting non-terminal and right side of A does not contain Z, it means Z is not reachable. Hence after eliminating $Z \rightarrow Zy/z$, grammar becomes

$$A \rightarrow xyz.$$

which is a grammar containing no useless symbols.

Example 4.24. Find the reduced grammar that is equivalent to the CFG given below:

$$S \rightarrow aC/SB$$
$$A \rightarrow bSCa$$
$$B \rightarrow aSB/bBC$$
$$C \rightarrow aBC/ad.$$

Solution. Since $C \rightarrow ad$, therefore, C is generating symbol. Since $S \rightarrow aC$, therefore S is also a useful symbol according to production $A \rightarrow bSCa$, A is also generating. Right side of $B \rightarrow aSB$ and $B \rightarrow bBC$ contains B, and B is not terminating so B is not a generating symbol so we can eliminate those productions and grammar becomes

$$S \rightarrow aC$$
$$A \rightarrow bSCa$$
$$C \rightarrow ad$$

Now since S is the start symbol and right side of S does not contain A, hence A is not reachable as:

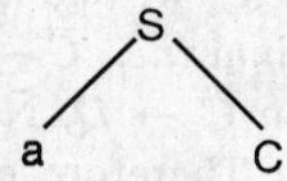

Fig. 4.17.

So by eliminating A we get

$$S \rightarrow aC$$
$$C \rightarrow ad$$

which is reduced grammar equivalent to the given grammar, containing no useless symbol.

4.7.2. Removal of Unit Production

Let us first define unit production:

A production of the form

Non-terminal $\rightarrow$ One non-terminal

that is a production of the form $A \rightarrow B$ (where A and B, both are non-terminals) is called unit production. Unit production increase the cost of derivation in a grammar. Following algorithm can be used to eliminate the unit production.

Algorithm: Removal of unit productions $\Rightarrow$

While (there exist a unit production, $A \rightarrow B$)

{

select a unit production $A \rightarrow B$, such that there exist a production $B \rightarrow \alpha$, where α is a terminal.

For (every non-unit production, $B \rightarrow \alpha$)

Add production $A \rightarrow \alpha$ to the grammar

Eliminate $A \rightarrow B$ from the grammar.

}.

Now let us apply this algorithm on the given CFG's.

Example 4.25. Consider the context free grammar G.

$$S \rightarrow AB$$
$$A \rightarrow a$$
$$B \rightarrow C/b$$
$$C \rightarrow D$$
$$D \rightarrow E$$
$$E \rightarrow a$$

remove the unit production.

Solution. Given CFG

$$S \rightarrow AB$$
$$A \rightarrow a$$
$$B \rightarrow C/b$$
$$C \rightarrow D$$
$$D \rightarrow E$$
$$E \rightarrow a$$

Contain three unit productions

$$B \rightarrow C$$
$$C \rightarrow D$$
$$D \rightarrow E$$

Now to remove unit production $B \rightarrow C$, we see if there exists a production whose left side has C and right side contains a terminal (*i.e.*, $C \rightarrow a$), but there is no such productions in G. Similar things holds for production $C \rightarrow D$. Now we try to remove unit production $D \rightarrow E$, because there is a production $E \rightarrow a$. Therefore, eliminate $D \rightarrow E$ and introduce $D \rightarrow a$, grammar becomes

$$S \rightarrow AB$$
$$A \rightarrow a$$
$$B \rightarrow C/b$$
$$C \rightarrow D$$
$$D \rightarrow a$$
$$E \rightarrow a$$

Now we can remove $C \rightarrow D$ by using $D \rightarrow a$, we get

$$S \rightarrow AB$$
$$A \rightarrow a$$
$$B \rightarrow C/b$$
$$C \rightarrow a$$
$$D \rightarrow a$$
$$E \rightarrow a$$

Similarly, we can remove $B \rightarrow C$ by using $C \rightarrow a$, we obtain

$$S \rightarrow AB$$
$$A \rightarrow a$$
$$B \rightarrow a/b$$
$$C \rightarrow a$$
$$D \rightarrow a$$
$$E \rightarrow a$$

Now it can be easily seen that productions $C \rightarrow a$, $D \rightarrow a$ $E \rightarrow a$ are useless because if we start deriving from S, these productions will never be used. Hence eliminating them gives,

$$S \rightarrow AB$$
$$A \rightarrow a$$
$$B \rightarrow a/b$$

which is completely reduced grammar.

Example 4.26. Consider the following unambiguous expression grammar

$$I \rightarrow a/b/Ia/Ib/I0/I1$$
$$F \rightarrow I/(E)$$
$$T \rightarrow F/T{*}F$$
$$E \rightarrow T/E + T$$

Identified unit production and remove them.

Solution. In the given grammar

$$F \rightarrow I$$
$$T \rightarrow F$$
$$E \rightarrow T$$

are three unit production.

Unit production $F \rightarrow I$ can be removed by the help of

$$I \rightarrow a/b/Ia/Ib/I0/I1$$

and grammar becomes

$$I \rightarrow a/b/Ia/Ib/I0/I1$$
$$F \rightarrow (E)/a/b/Ia/Ib/I0/I1$$
$$T \rightarrow F/T{*}F$$
$$E \rightarrow T/E + T$$

Now we can eliminate $T \rightarrow F$ by the help of

$$F \rightarrow (E)/a/b/Ia/Ib/I0/I1$$

and now grammar becomes

$$I \rightarrow a/b/Ia/Ib/I0/I1$$
$$F \rightarrow (E)/a/b/Ia/Ib/I0/I1$$
$$T \rightarrow T{*}F/(E)/a/b/Ia/Ib/I0/I1$$
$$E \rightarrow T/E + T$$

Now let us remove $E \rightarrow T$ by the help

$$T \rightarrow T{*}F/(E)/a/b/Ia/Ib/I0/I1$$

and finally, unit production free grammar is

$$I \rightarrow a/b/Ia/Ib/I0/I1$$
$$F \rightarrow (E)/a/b/Ia/Ib/I0/I1$$
$$T \rightarrow T{*}F/(E)/a/b/Ia/Ib/I0/I1$$
$$E \rightarrow E + T/T * F/(E)/a/b/Ia/Ib/I0/I1$$

Example 4.27. Identified and remove the unit productions from following grammar

$$S \rightarrow A/bb$$
$$A \rightarrow B/b$$
$$B \rightarrow S/a$$

Solution. Here unit productions are

$$S \to A$$
$$A \to B$$
$$B \to S$$

We list all unit productions and sequences of unit productions. Then we will handle one unit production or sequence at a time and try to prove generating and reachable for each non-terminal involve in the unit productions, as follows:

$S \to A$	generate $S \to b$
$S \to A \to B$	generate $S \to a$
$A \to B$	generate $B \to a$
$A \to B \to S$	generate $A \to bb$
$B \to S$	generate $B \to bb$
$B \to S \to A$	generate $B \to b$

The new CFG becomes

$$S \to bb/b/a$$
$$A \to b/a/bb$$
$$B \to a/bb/b$$

which hold no unit productions.

4.7.3. Removal of ∈-productions and Nullable Non-terminal

Any context-free language in which ∈ is a word, must have some ∈-productions in its grammar since otherwise we could never derive the word ∈ from S. The statement is obvious, but it should be given some justification, mathematically, this is easy; we observe that ∈-productions are only productions that shorten the working string. If we begin with the string S and apply only non-∈-productions, we never develop a word of length zero.

We now form our attention to the eliminate the productions of the form $A \to \in$, which are called ∈-productions. Surely if ∈ is in $L(G)$, we can not eliminate all ∈-productions from G, but if ∈ is not in $L(G)$, we can eliminate all ∈-productions from G.

"In a given CFG, we call a non-terminal N Nullable if

There is a production $N \to \in$ or

There is a derivation that starts at N and leads to ∈:

$$N \Rightarrow \ldots \Rightarrow \in$$"

To eliminate ∈-productions from a grammar G we use the following technique:

"If $A \to \in$ is a production to be eliminated then we look for all productions, whose right side contains A, and replace each occurrence of A in each of these productions to obtain the non ∈-productions. Now these resultant non ∈-production must be added to the grammar to keep the language generated, the same."

Let us apply this concept:

Example 4.28. Consider the following grammar

$$S \to aA$$
$$A \to b/\in$$

remove the Nullable non-terminal.

Solution. Given CFG is

$$S \rightarrow aA$$
$$A \rightarrow b/\in$$

Here $A \rightarrow \in$ is $\in$-production, so let us apply above described concept, put $\in$ in place of A, at the right side of productions and add the resulted productions to the grammar.

Here is only one productions $S \rightarrow aA$ whose right side contains A. So, replacing A by $\in$ in this production, we get $S \rightarrow a$; now add this new productions **to keep the language generated by this grammar same.** Therefore, the $\in$-free grammar, equivalent to above grammar is

$$S \rightarrow aA$$
$$S \rightarrow a$$
$$A \rightarrow b$$

or

$$S \rightarrow aA/a$$
$$A \rightarrow b$$

there is no $\in$-production in the above context-free grammar.

Example 4.29. Consider the following grammar *G*

$$S \rightarrow ABAC$$
$$A \rightarrow aA/\in$$
$$B \rightarrow bB/\in$$
$$C \rightarrow c$$

remove the $\in$-production from the above grammar.

Solution. Here we have two $\in$-production to remove $A \rightarrow \in$ and $B \rightarrow \in$. To eliminate $A \rightarrow \in$ from the grammar, the non $\in$-productions to be added are obtained as follows:

List of productions whose right side contain A are:

$$S \rightarrow ABAC$$
$$A \rightarrow aA$$

So on replacing each occurrence of A by $\in$ (one by one), we get four new productions to be added to the grammar:

$$S \rightarrow ABC/BAC/BC$$
$$A \rightarrow a$$

add these productions to the grammar and eliminate $A \rightarrow \in$, to obtain the following grammar

$$S \rightarrow ABAC/ABC/BAC/BC$$
$$A \rightarrow aA/a$$
$$B \rightarrow bB/\in$$
$$C \rightarrow c$$

Now to eliminate $B \rightarrow \in$, list all productions whose right side contains B, these are :

$$S \rightarrow ABAC/BAC/ABC/BC$$
$$B \rightarrow bB$$

Replace occurrence of B in each of these productions to obtain non $\in$-productions to be added to the grammar, we get following new productions.

$$S \rightarrow AAC/AC/C$$
$$B \rightarrow b$$

Add these productions to the grammar and eliminate $B \rightarrow \in$ from the grammar, to obtain

$$S \rightarrow ABAC/BAC/ABC/BC/AAC/AC/C$$
$$A \rightarrow aA/a$$
$$B \rightarrow bB/b$$
$$C \rightarrow c$$

which is the grammar without any $\in$-production.

Example 4.30. Consider the following grammar and remove ∈-productions

$$S \to aSa$$
$$S \to bSb/\in$$

Solution. Here $S \to \in$ is the ∈-production, so let us replace occurrence of S by ∈ in the following productions

$$S \to aSa$$
$$S \to bSb$$

We will get following new productions

$$S \to aa/bb$$

add these productions in the grammar and remove $S \to \in$ from the grammar

$$S \to aSa/aa/bb/bSb$$

It is ∈-production free grammar.

Example 4.31. Consider the following grammar

$$S \to a/Xb/aYa$$
$$X \to Y/\in$$
$$Y \to b/X$$

eliminate ∈-productions.

Solution. Here $X \to \in$ is ∈-productions. So this problem can also solved in similar fashion and grammar without ∈-productions will be

$$S \to a/Xb/aYa/b/aa$$
$$X \to Y$$
$$Y \to b/X.$$

Example 4.32. Design a CFG for regular expression $r = (a + b)^*\ bb\ (a + b)^*$, which is free from ∈-productions.

Solution. Let CFG G be required context free grammar with ∈-productions

$$S \to XY$$
$$X \to Zb$$
$$Y \to bW$$
$$Z \to AB$$
$$W \to Z$$
$$A \to aA/bA/\in$$
$$B \to Ba/Bb/\in$$

Finally here $A \to \in$ and $B \to \in$ are ∈-productions and A, B are nullable non-terminals.

The CFG without ∈-production can be achieved from G in similar fashion, as above discussed. Let it be G', then G' is

$$S \to XY$$
$$X \to Zb/b$$
$$Y \to bW/b$$
$$Z \to AB/A/B$$
$$W \to Z$$
$$A \to aA/bA/a/b$$
$$B \to Ba/Bb/a/b$$

Because ∈ was not generated by G, the new CFG G' generates exactly the same language.

4.7.4. Cycle-free Grammar

A production in which the right side consists of a single non-terminal

$$A \to B \text{ (where } A, B \in V_n)$$

is termed as **single production.** Fairly a single production of the form $A \rightarrow A$ serves no useful purpose, and should never be present. It could be argued that it causes no harm, for it presumably would be an alternative which was never used (so being useless, in a sense not quite discussed above). Let us consider an example as follows :

$$A \rightarrow B$$
$$B \rightarrow C$$
$$C \rightarrow A$$

Non only is the useless in this new sense, it is highly undesirable from the point of obtaining a unique parse, and so all parsing technique require a grammar to be cycle-free, it should not permit a derivation of the form $A \stackrel{+}{\rightarrow} A$.

4.8. AMBIGUITY IN GRAMMARS AND LANGUAGES

[MDU 2007, 2008, 2009, 2010]

Grammars can be used to put structure on programs and documents. The assumption was that a grammar uniquely determines a structure for each string in its language. However not every grammar does provide unique structure.

When a grammar fails to provide unique structures, it is sometimes possible to redesign the grammar to make the structure unique for each string in the language. Unfortunately some times we cannot do so. That is, there are some CFL's that are "inherently ambigous"; every grammar for the language puts more than one structure for some strings in the language.

Definition "A CFG is called ambigous if for at least one word in the language that it generates, there are two possible derivations of the word that correspond to different syntax trees. If a CFG is not ambiguous, it is called unambiguous.

Example 4.33. The language of all non null strings of *a*'s can be defined by a CFG as follows:

$$S \rightarrow aS/Sa/a$$

Solution. In this case, the word a^3 can be generated by four different tree's :

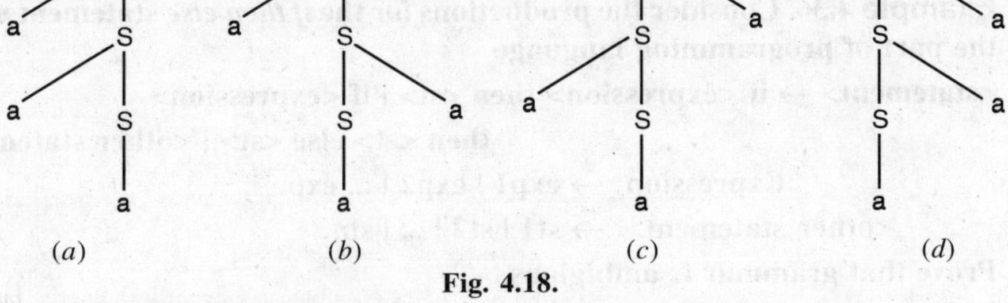

Fig. 4.18.

The CFG is therefor ambigious.

However, the same language can also be defined by the CFG

$$S \rightarrow aS/a$$

for which a^3 has only one production:

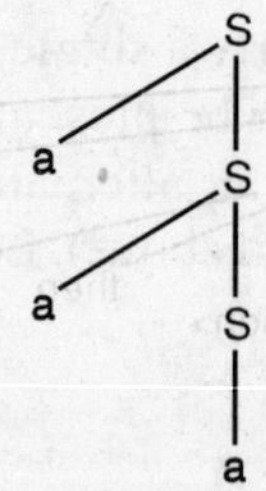

Fig. 4.19.

Example 4.34. The CFG $S \to aSb/SS/\in$ is ambiguous. The sentence *aabb* has the two derivation trees shown in Fig. 4.20 (*a*) and 4.20 (*b*).

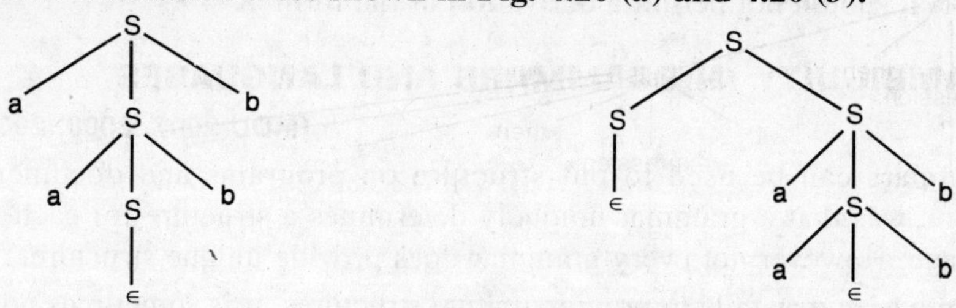

(*a*) (*b*)

Fig. 4.20.

Example 4.35. Using the same grammar as in Example 4.34, we find that string $a + b \times b$ has many different derivation trees, as follows.

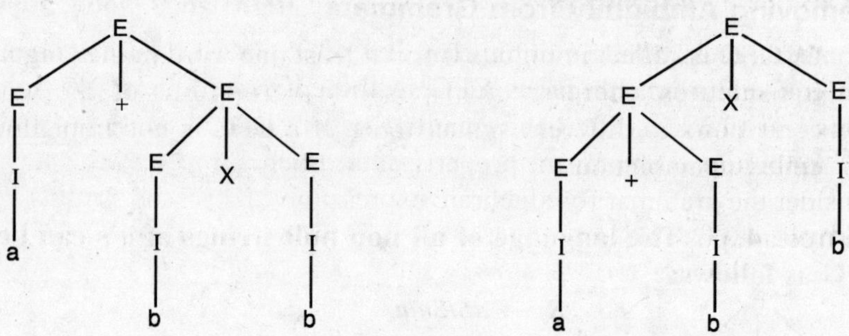

(*a*) (*b*)

Fig. 4.21.

Example 4.36. Consider the productions for the *if then else* statement which are the part of programming language

<statement> → if <expression> then <st> | if <expression> then <st> else <st> | <other statement>

Expression → exp1 | exp2 | ... exp.

<other statement> → st1 | st2 | ... | st*n*.

Prove that grammar is ambigious.

Solution. Consider the statement

if exp1 then if exp2 then st1 else st2

There are two different parse trees as shown in Fig. 4.22 and Fig. 4.23.

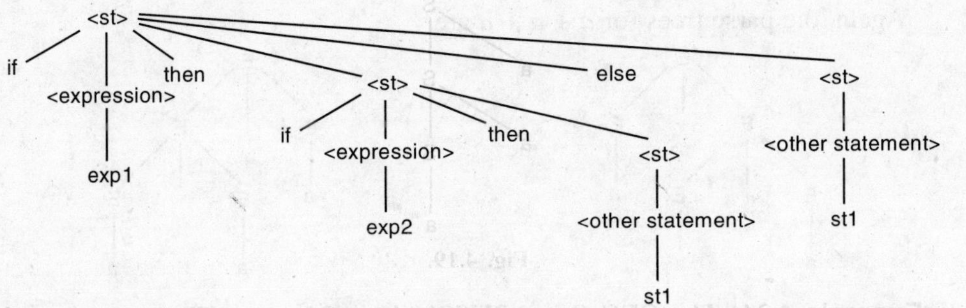

Fig. 4.22.

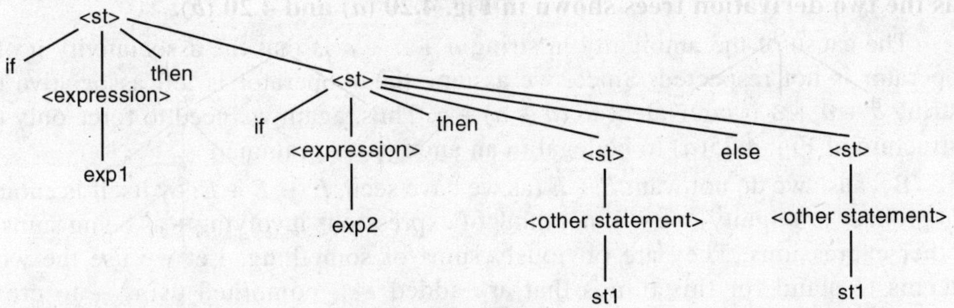

Fig. 4.23.

Hence grammar is ambiguous.

4.8.1. Removing Ambiguity from Grammars [MDU 2007, 2008, 2009, 2010]

If a context free grammar is ambiguous, it is often possible and usually desirable to find an equivalent unambiguous CFG. Although some CFGs are **"inherently ambiguous"** in the sense that they cannot be produced except by ambiguous grammars. Ambiguity is usually the property of the grammar rather than the language. Let us consider the grammar for algebraic expressions and for the string $a + a*a$ the two parse tree are possible.

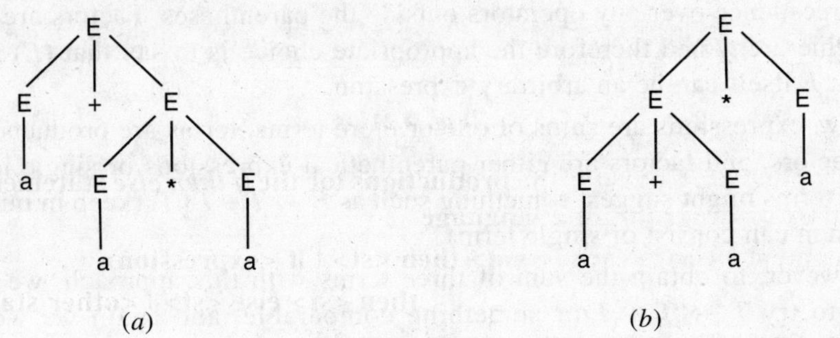

Fig. 4.24.

The first cause of ambiguity in the grammar is that the precedence of the operator + and * is not respected. Both parsing trees in Fig. 4.24 are valid. We need to force only the structure of Fig. 4.24(*a*) to be legal in an ambiguous grammar in which the * operator has higher precedence then + because $a + a*a$ is equivalent to $a + (a*a)$ and $a*a + a$ is equivalent to $(a*a) + a$.

Again, the parse trees for $a + a + a$ are

Fig. 4.25.

The cause of the ambiguity in string $a + a + a$ is that the associativity it of + operator is not respected. Since, we assume the + operator is left associative the string $a + a + a$ is equivalent to $(a + a) + a$. Thus, again we need to force only the structure of Fig. 4.25(*a*) to be legal in an ambiguous grammar.

Because we do not want $E + E$ (as we have seen, $E \to E + E$, by itself is enough to produce ambiguity) we will not think of expressions involving + as being sums of other expressions. They are obviously sums of something: Let we use the word **terms** to stand for this things that are added *i.e.,* combined using + to create expressions. The corresponding variable in the grammar will be T. Expressions can also be products; because the two expressions $a + b * c$ and $a * b + c$ are both sums, it is more appropriate to say that terms can be products. Let us say that "factors" are the things that are multiplied (combined using *) to produce terms. The corresponding variable in the grammar will be F.

Since the multiplication have higher precedence over addition. We can say that Expressions are sums of one or more terms, and terms are products of one or more factors.

Now, we must deal with parentheses. We might say that (E) could be an expression, a term or a factor. However, evaluation of a parenthetical expression takes precedence over any operators outside the parentheses. Factors are evaluated first in hierarchy and therefore the appropriate choice is to say that (E) is a factor and that E itself can be an arbitrary expression.

Now, expressions are sums of one or more terms, terms are products of one or more factors, and factors are either parenthetical expressions or single identifiers. Sum of terms might suggest something such as $E \to T + T \mid T$. (keep in mind that an expression can consist of single term).

However, to obtain the sum of three terms with this approach, we would be forced to try $T \to T + T$ or something comparable, and again we would have ambiguity. What we say instead is that an expression is either a single term or the sum of a term and another expression. The only question is whether we want $E \to E + T$ or $E \to T + E$. Since + operator is left associative, we would probably choose $E \to E + T$ as more appropriate. Similarly, we choose the production $T \to T * F$ rather than $T \to F * T$.

The resulting unambiguous grammar is

$$E \to E + T \mid T$$

$$T \rightarrow T * F \mid F$$
$$F \rightarrow (E) \mid a$$

In the similar fashion we can find the unambiguous grammar for algebraic expressions having other operators.

Example 4.37. Eliminate the ambiguity from the following grammar

<statement> → if <expression> then <st>
| if <expression> then <st>
else <st> | <other statement>
<other statement> → $S_1 \mid S_2 \mid S_3 \mid \ldots S_n$
<expression> → $E_1 \mid E_2 \mid E_3 \ldots \mid E_n$

Solution. The Grammar is ambiguous since the string "if E_1 then if E_2 then S_1 else S_2" has two parse trees.

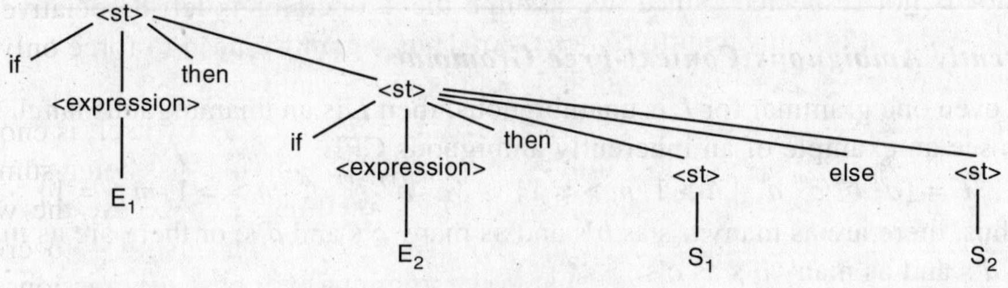

Fig. 4.26.

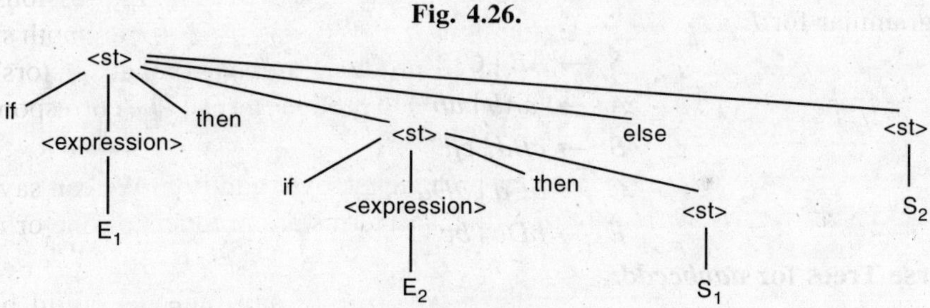

Fig. 4.27.

In all programming languages with conditional statements of this form, the first parse tree is preferred. The general rule is "Match each else with the closest previous unmatched then". Then unambiguous grammar is as follows :

<st> → <matched-statement> | <unmatched-statement>
<matched-statement> → if <expression> then <matched-statement>
else <matched statement> | <other statement>
<unmatched-statement> → if <expression> then <st>
| if <expressions> then <matched-statement>
else <unmatched-statement>
<expression> → $E_1 \mid E_2 \mid \ldots E_n$
<unmatched statement> → $S_1 \mid S_2 \ldots S_n$
<matched statement> → $S_1 \mid S_2 \ldots \mid S_n$

The idea in this grammar is that a statement appearing between a *then* and an *else* must be "matched" *i.e.,* it must not end with an unmatched *then* followed by *any statement*, for the *else* would then be forced to match this unmatched *then.* A matched statement is either an if then else statement containing no unmatched statements or it is any other kind of unconditional statement.

This grammar generates the same set of strings as previous one, but it allows only one parsing.

4.8.2. Inherent Ambiguity

TIPS

"If L is a context-free language for which there exists an unambiguous grammar, then L is said to be unambiguous. If every grammar that generates L is ambiguous, then the language is said to be "inherently ambiguous".

Inherently Ambiguous Context-Free Grammar

If even one grammar for L is unambiguous, then L is an unambiguous language. Let us see an example of an inherently ambiguous CFL.

$$L = \{a^n\ b^n\ c^m\ d^m \mid n > 1, m >= 1\} \cup \{a^n\ b^m\ c^m\ d^n \mid n >= 1, m >= 1\}$$

Thus, there are as many a's as b's and as many c's and d's, or there are as many a's as d's and as many b's as c's.

A grammar for L

$$\begin{aligned} S &\rightarrow AB \mid C \\ A &\rightarrow aAb \mid ab \\ B &\rightarrow cBd \mid cd \\ C &\rightarrow aCd \mid aDd \\ E &\rightarrow bDc \mid bc \end{aligned}$$

Parse Trees for *aabbccdd:*

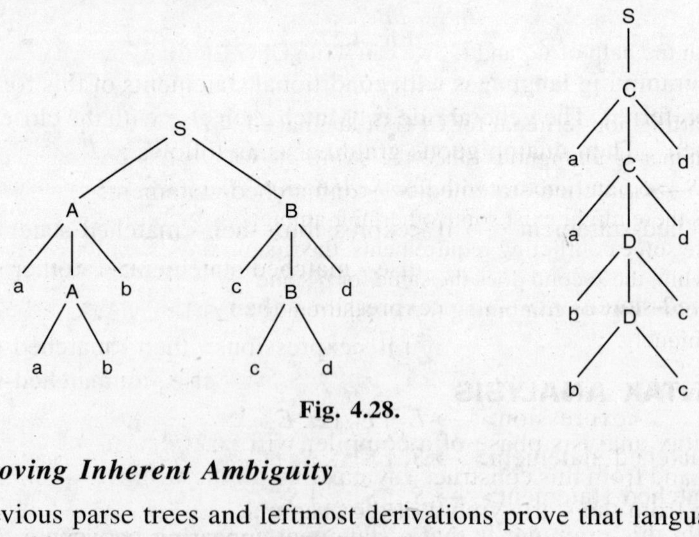

Fig. 4.28.

4.8.2.1. Proving Inherent Ambiguity

The previous parse trees and leftmost derivations prove that language L has at least some ambiguous grammars. This doesn't prove that all of L's grammars are ambiguous, so we don't know if L is ambiguous.

Proof Idea: We argue that all but a finite number of the strings, whose counts of the four symbols *a*, *b*, *c*, *d* are all equal, must be generated in two different ways :

(*i*) *a*'s and *b*'s are equal and *c*'s and *d*'s are equal.

(*ii*) *a*'s and *d*'s are equal and *b*'s and *c*'s are equal.

(*iii*) The only way to generate strings where *a*'s and *b*'s are equal is with a variable like *A*.

(*iv*) Variations can occur, but we must follow the mechanism for generating *a*'s in a way that matches the *b*'s.

(*v*) This idea carries for the variables *B*, *C*, and *D*.

(*vi*) This argument will show that no matter what modifications we make to the basic grammar, it will generate at least some of the strings of the form $a^n b^n c^n d^n$.

(*vii*) Following this proof idea will show that *L* is inherently ambiguous, because no matter how the grammar is changed, it will remain ambiguous.

Let us consider an example for it:

Example 4.38. Show that

$$L = \{a^n\ b^n\ c^m\} \cup \{a^n\ b^m\ c^m\}$$

with *n* and *m* non-negative, is an inherently ambiguous context-free language.

Solution. Let us say $L = L_1 \cup L_2$

where $L_1 = \{a^n\ b^n\ c^m\}$

and $L_2 = \{a^n\ b^m\ c^m\}$

Let us write CFG for L_1, let it is G_1, as follows:

$$S_1 \to S_1\ c/A$$
$$A \to aAb/\in$$

Similarly we can write CFG for L_2, Let it is G_2, as follows:

$$S_2 \to aS_2/B$$
$$B \to bBc/\in$$

Now with the help of G_1 and G_2 we can write CFG for the language *L*, as follows

$$S \to S_1/S_2$$

where *S* is starting non-terminal for CFG of language *L*.

The grammar is ambiguous since the string $a^n\ b^n\ c^m$ has two distinct derivation, one starting with $S \Rightarrow S_1$, and another with $S \Rightarrow S_2$. It does of course not follows that *L* is inherently ambiguous as there might exist some other non-ambiguous grammar for it. But in some way L_1 and L_2 have some conflicting requirements, the first putting a restriction on the number of *a*'s and *b*'s, while the second does the same for *b*'s and *c*'s. A few tries will quickly convince us of the impossibility of combining these requirements in a single set of rules that cover the case $n = m$ uniquely.

4.9. SYNTAX ANALYSIS

The syntax analysis phase of a compiler will take a string of tokens produced by the lexer, and from this construct a syntax tree for the string by finding a derivation of the string from the start symbol of the grammar.

This can be done by guessing derivations until the right one is found, but random guessing is hardly an effective method. Even so, some parsing techniques are based on "guessing" derivations. However, these make sure, by looking at the string, that they will always guess right. These are called predictive parsing methods. Predictive parsers always build the syntax tree from the root down to the leaves and are hence also called (deterministic) top-down parsers.

Other parsers go the other way: They search for parts of the input string that matches right-hand sides of productions and rewrite these to the left-hand nonterminals, at the same time building pieces of the syntax tree. The syntax tree is eventually completed when the string has been rewritten (by inverse derivation) to the start symbol. Also here, we wish to make sure that we always pick the "right" rewrites, so we get deterministic parsing. Such methods are called bottom-up parsing methods.

Let us see various definition of parsing as follows:

"A grammatical exercise that involves breaking down a text into its component parts of speech with an explanation of the form, function, and syntactical relationship of each part".

"Breaking a data block into smaller chunks by following a set of rules, so that it can be more easily interpreted, managed, or transmitted by a computer. Spreadsheet programs, for example, parse a data to fit it into a cell of certain size".

"Parsing splits a sequence of characters or values into smaller parts. It can be used for recognizing characters or values that occur in a specific order. In addition to providing a powerful, readable, and maintainable approach to regular expression pattern matching, parsing enables us to create our own custom languages for specific purposes".

The **parse** function has the general form:

```
parse series rules
```

The series argument is the input that is parsed and can be a string or a block. If the argument is a string, it is parsed by character. If the argument is a block, it is parsed by value.

The rules argument specifies how the series argument is parsed. The rules argument can be a string for simple types of parsing or a block for sophisticated parsing.

The **parse** function also accepts two refinements: **/all** and **/case**. The **/all** refinement parses all the characters within a string, including all delimiters, such as space, tab, newline, comma, and semicolon. The **/case** refinement parses a string based on case. When **/case** is not specified, upper and lower cases are treated the same.

4.10. ROLE OF PARSERS [MDU 2009]

Definition A parser for any grammar is a program (and parsing is a technique) that takes as input string *w* and produces as output either a parse tree for *w*, if *w* is a valid sentence of grammar, or an error message indicating that *w* is not a valid sentence of given grammar.

The **goal of parsing** is to determine the syntactic validity of a source string. If the string is valid, a tree is built for use by subsequent phases of the computer.

The tree reflects the sequence of derivations or reduction used during the parse. Hence it is called parse tree. If a string is invalid, the parser has to issue diagnostic message identifying the nature and cause of the errors in the string. Every elementary subtree in the parse tree corresponds to a production of the grammar. There are two ways of identifying an elementary subtree:

(*a*) By deriving a string from a non-terminal, or

(*b*) By reducing a string of symbols to an non-terminal.

Based on these, two fundamental approaches to parsing, namely **top down parsing** and **bottom up parsing**, are defined.

There are universal parsing methods that will parse any grammar but they are too inefficient to use in compilers. Almost all programming languages have such simple grammar that an efficient top-down or bottom-up parser can parse a source program with a single left-to-right so as of the input.

For example. The production rules of grammar G is:

list $\rightarrow$ list + digit | list-digit | digit

digit $\rightarrow$ 0 | 1 | ... | 9

Given expression is 9 – 5 + 2

Parse tree is:

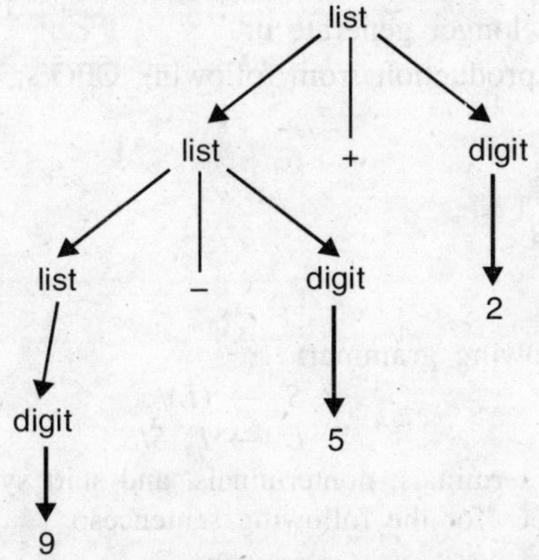

Fig. 4.29.

Each node in the parse tree is labeled by a grammar symbol. The interior node corresponds to the left side of the production. The children of the interior node correspond to the right side of production.

The language defined by above parse tree is lists of digits separated by plus and minus signs.

TIPS

For any language that can be described by CFG, the parsing requires $O(n^3)$ time to parse string of n token. However, most programming languages are so simple that a parser requires just $O(n)$ time with a single left-to-right scan over the input string of n tokens.

EXERCISE

1. Find the regular grammar for the following languages:

   (*i*) $L = \{aa^*(ab + a)^*\}$

   (*ii*) $L = \{a^n b^m \,/\, n \geq 2, m \geq 3\}$

   (*iii*) $L = \{w \in (a, b)^+ \,/\, w$ contains no more than two a's$]$

2. Convert the following CFG's to CFF:

   (*i*) $S \rightarrow SS/a$

   (*ii*) $S \rightarrow aSa/SSa/a$

   (*iii*) $S \rightarrow aXX$
   $X \rightarrow aS/bS/a$

   (*iv*) $E \rightarrow E + E$
   $E \rightarrow E * E$
   $E \rightarrow (E)$
   $E \rightarrow \rceil$
   the terminal are +, *, (,), ⌉.

   (*v*) $S \rightarrow ABABAB$
   $A \rightarrow a/\in$
   $B \rightarrow b/\in$

   Note that ∈ is the word in this language, but when converted into CNF, the grammar will not longer generate it.

3. Remove the unit production from following CFG's:

   (*i*) $S \rightarrow AB$
   $A \rightarrow B$
   $B \rightarrow aB/Bb/\in$

   (*ii*) $S \rightarrow ABABAB$
   $A \rightarrow a/\in$
   $B \rightarrow b/\in$

4. Consider the following grammar:

   $$S \rightarrow (L)/a$$
   $$L \rightarrow L, S/S$$

   (*i*) What are the terminals, nonterminals, and start symbol ?

   (*ii*) Find parse trees for the following sentences:

   (*a*) (*a*, *a*)

   (*b*) (*a*, (*a*, *a*))

   (*c*) (*a*, ((*a*, *a*), (*a*, *a*))

   (*iii*) Construct the leftmost derivation for each of the sentence in (b) part.

   (*iv*) Construct a rightmost derivation for each of the sentence in (b) part.

5. Consider the following grammar:

   bexpr → bexpr **or** bterm/bterm
   bterm → bterm **and** bfactor/bfactor
   bfactor → **not** bfactor/(bexpr)/true/false

   (*i*) Construct a parse tree for the sentence **not (true or false)**.

6. Consider the grammar

   $$S \rightarrow aSbS/bSaS/\in$$

   (*i*) Show that grammar is ambigious by constructing two different leftmost derivations for the sentence *abab*.

   (*ii*) Construct the corresponding rightmost derivation for *abab*.

   (*iii*) Construct the corresponding parse trees for *abab*.

❑❑❑

5

Chapter

TOP-DOWN PARSING TECHNIQUES

INSIDE THIS CHAPTER

5.1. Top-Down Parsing **5.2.** Issues of CFG for the Programming Languages **5.3.** Recursive-Decent Parsing **5.4.** Top-down Predictive Parsers **5.5.** LL (1) Grammars.

5.1. TOP-DOWN PARSING

For a given input string, top-down-parsing attempt to derive a string identical to it by successive application of grammar rules to the grammar's distinguished symbol. When such a string is obtained, a tree representing its derivation would be the syntax tree for the input string. Thus if *w* is the input string, a top-down parser determines a derivation sequence.

$$S \Rightarrow .... \Rightarrow .... \Rightarrow w$$

where *S* is starting non-terminal in given grammar.

TIPS

Basically top-down parsing attempts to find the left-most derivations for the input string *w*, since string *w* is scanned by the parser left to right, one symbol/token at a time, and the left-most derivations generate the leaves of the parse tree in left-to-right order, which matches the input scan order.

Example 5.1. Consider the following grammar.

$$V_N = \{\text{expr, term, rest}\}$$
$$V_t = \{+, -, 0, 1, 2, 3, 4, 5, 6, 7, 8, 9, \in\}$$
$$\text{expr} \rightarrow \text{term rest}$$
$$\text{rest} \rightarrow + \text{term rest} \mid - \text{term rest} \mid \in$$
$$\text{term} \rightarrow 0 \mid 1 \mid 2 \mid 3 \mid 4 \mid 5 \mid 6 \mid 7 \mid 8 \mid 9$$

and show the construction of the parse tree for the input string: 9 – 5 + 2.

Solution. Initialization: The root of the parse tree must be starting symbol (Non-terminal) of the grammar, expr.

↓

expr

Step 1. The only production for expr is expr → term rest, so the root node must have a **term node** and a **rest node** as children

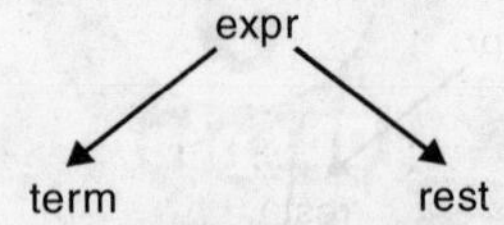

Fig. 5.1.

Step 2. The first token in the input is 9 and the only production in the grammar containing a 9 is: term → 9 so 9 must be a leaf with **term** node as parent.

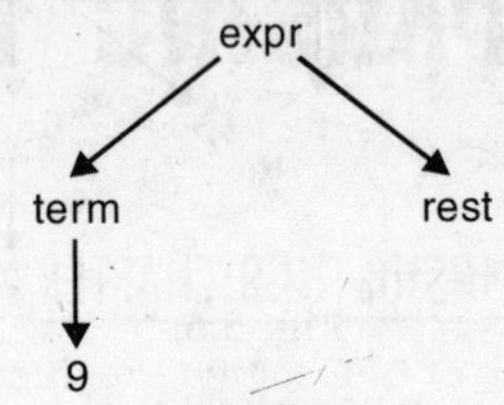

Fig 5.2.

Step 3. The next token in the input is the minus-sign and the only production in the grammar containing a minus-sign is: **rest → – term rest**. The rest node must have a minus-sign leaf, a **term** node and a **rest** node as children.

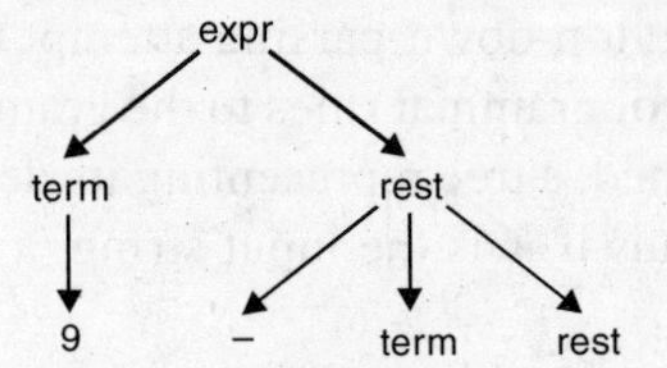

Fig. 5.3.

Step 4. The next token in the input is 5 and the only production in the grammar containing a 5 is: term → 5 so 5 must be a leaf with a **term** node as a parent.

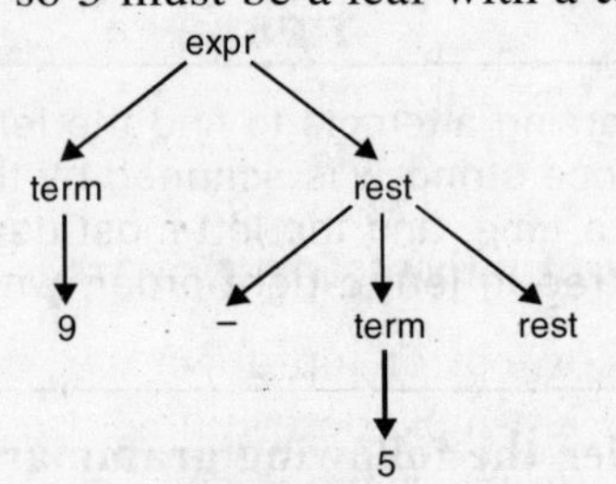

Fig. 5.4.

Step 5. The next token in the input is the plus-sign and the only production in the grammar containing a plus-sign is: **rest → + term rest.** A **rest** node must have a plus-sign leaf, a **term** node and a rest node as children.

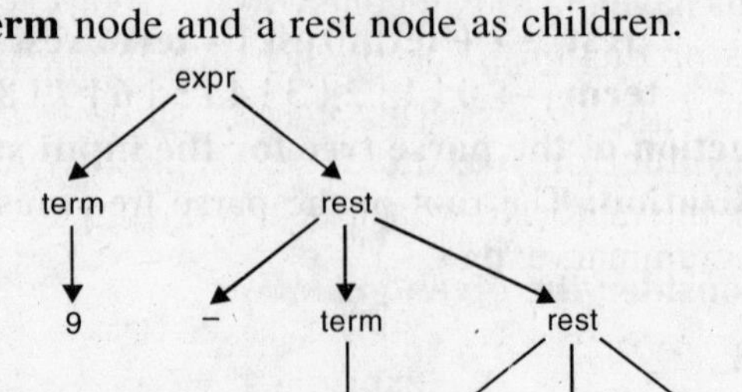

Fig. 5.5.

Step 6. The next token, in the input is 2 and the only production in the grammar containing a 2 is: term → 2 so 2 must be a leaf with a term node as a parent.

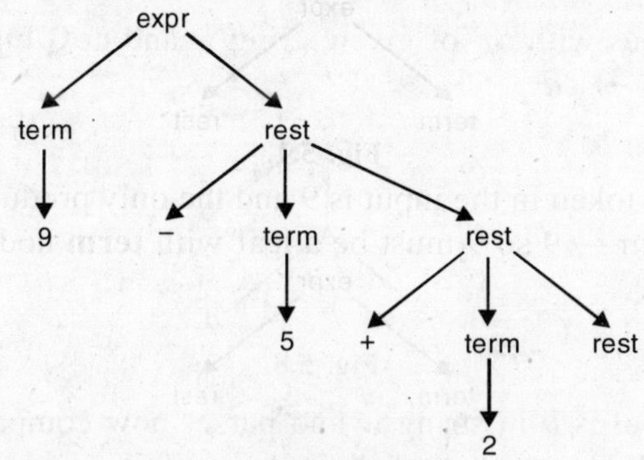

Fig. 5.6.

Step 7. The whole input has been absorbed but the parse tree still has a **rest** node with no children. The **rest → ∈ production** must now be used to give the **rest** node the empty string as a child.

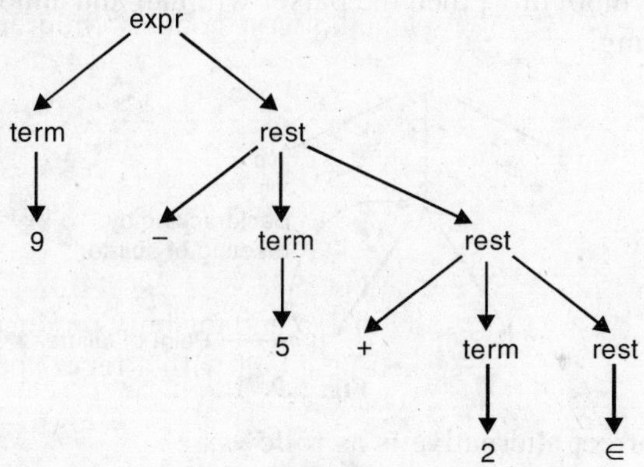

Fig. 5.7.

5.1.1. Backtracking a Problem of Top-down Parsing

Now by the study of concept of top-down parsing, it is clear that top-down parsing attempts to find the left-most derivations for input string. Basically in top-down mechanism, every terminal symbol generated by some production of the grammar (which is predicted) is matched with the input string symbol pointed to by the string marker. If the match is successful, the parse can continue. If a mismatch occurs, then predictions have obviously gone wrong. At this stage, it is necessary to reject (some) previous predictions. The prediction which led to the mismatching terminal symbol is rejected, and the string marker (pointer) is reset to its position when the rejected production was made. This is known as **backtracking**. Backtracking is one of the major drawback of top-down parser.

Example 5.2. Consider the given grammar

$$S \rightarrow aAb$$

$$A \rightarrow cd/c$$

and show the backtracking for string $W = acb$.

Solution. Let us start with

$$S \rightarrow aAb$$

'a' of aAb matches with 'a' of given string w and next input symbol is 'c', so use the production. $A \rightarrow cd$

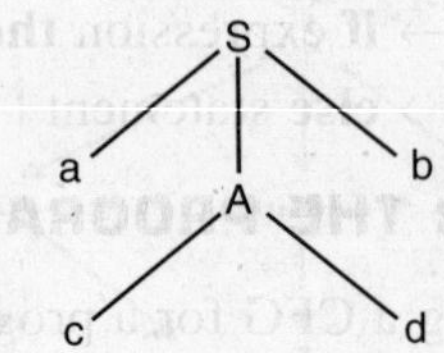

Fig. 5.8.

Next input symbol is b in string w and parser now compares b to the label of next leaf. If the label does not match d, it reports failure and goes back to A. The parser will also reset the input pointer to the second input symbol—the position it had when the parser encountered A, and it will try a second alternative for A in order to obtain the tree. If the leaf 'c' matches the second symbol, and if the next leaf b matches the third symbol of w, then the parser will halt and announce the successful completion of parsing.

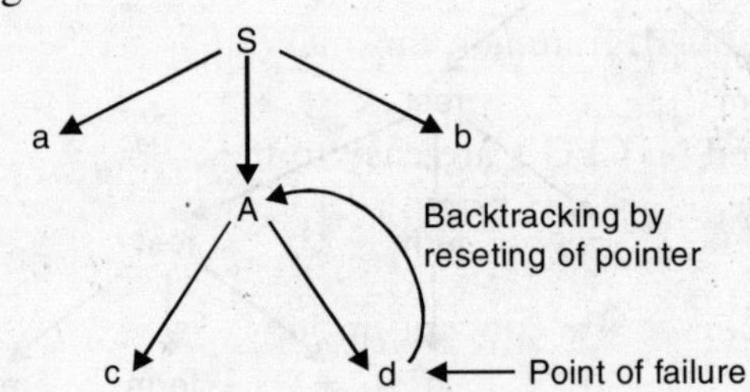

Fig. 5.9.

Now finally correct alternative is as follows:

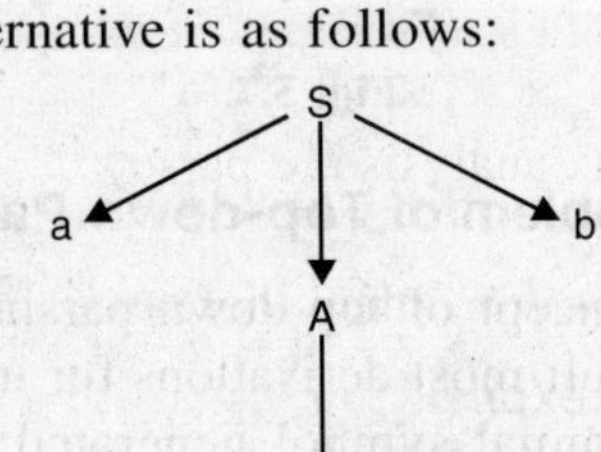

Fig. 5.10.

Yield is acb which matches to $W = acb$.

Example 5.3. Consider the following grammar

Statement $\rightarrow$ if expression then statement else statement

Statement $\rightarrow$ if expression then statement

Now alternate the production so that it may free from the backtracking.

Solution. When the input token is an **if** token should a top-down parser use the first or second production ? The parser would have to guess which one to use, continue parsing, and later on, if the guess is wrong, go back to the **if** token and try the other production.

Usually one can modify the grammar so a predictive top-down parser can be used: the parser always pick the correct production in each step of the parse so it never has to backtrack. To allow the use of a predictive parser, one replaces the two productions of the given grammar with:

Statement $\rightarrow$ **if** expression **then** statement *A*

A $\rightarrow$ **else** statement | $\in$.

5.2. ISSUES OF CFG FOR THE PROGRAMMING LANGUAGES

Issues to resolve when writing a CFG for a programming language:

(*a*) Ambiguity

(*b*) Left Recursion

(*c*) Indirect Left Recursion

(*d*) Left Factoring.

CFG is very important in compiler design and it can be discussed as follows :

(*i*) CFG gives a precise and easy to understand syntactic specification of a programming language.

(*ii*) Efficient parsers can be automatically constructed from a CFG.

(*iii*) A carefully defined grammar can make the detection of errors and code generation easier.

(*iv*) Compilers based on CFG's are easy to modify.

5.2.1. Ambiguity [MDU 2008]

We have already studied the concept in previous chapter but let us refresh the concept again.

Consider the following grammar:

< expr > $\rightarrow$ < expr > + < expr > | < expr > * < expr > | (< expr >) | < *id* >

< *id* > $\rightarrow$ *a/b/c*

The general result of syntax analysis is a parse tree description of the statements of a source program.

Let given string be $w = a + b * c$ then there exist two parse trees as follows:

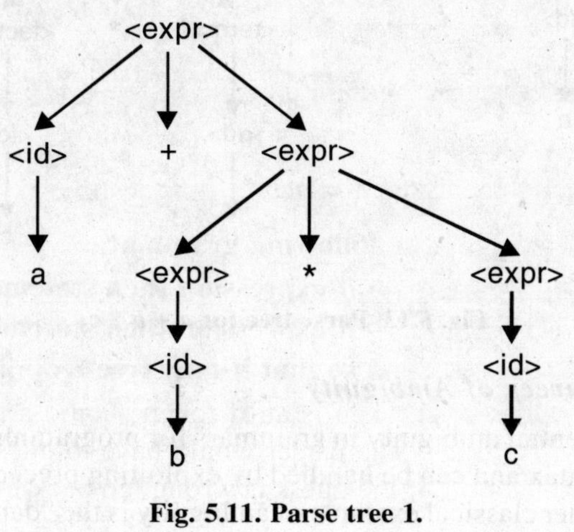

Fig. 5.11. Parse tree 1.

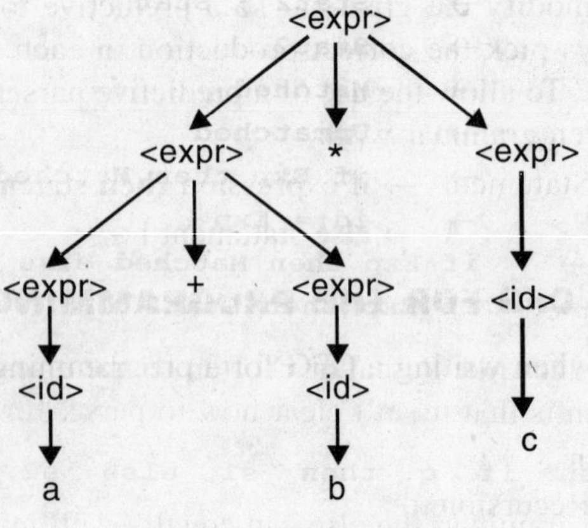

Fig. 5.12. Parse tree 2.

Yield of both the parse tree is $a + b * c$ so given grammar is ambiguous.

Now we can resolve the ambiguity by re-writing the grammar:

<expr> → <term> + <expr> | <term>

<term> → <term> * <factor> | <factor>

<factor> → (<expr>) | *<id>*

<id> → *a*/*b* | *c*

Now only one parse tree is possible for same string $a + b * c$.

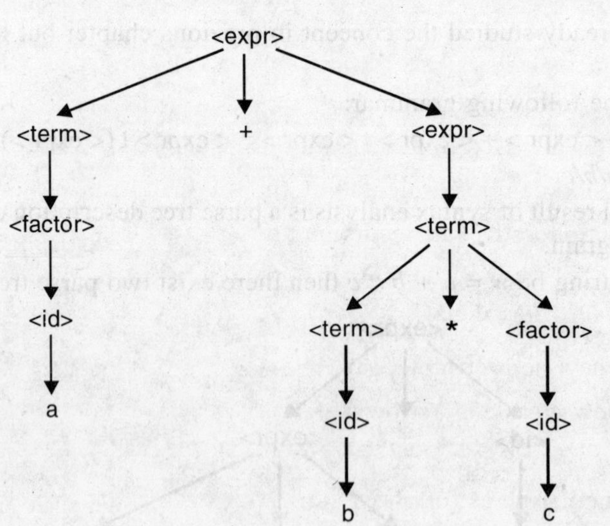

Fig. 5.13. Parse tree for $a + b * c$.

5.2.1.1. Other Sources of Ambiguity

Most of the potential ambiguity in grammars for programming languages comes from expression syntax and can be handled by exploiting precedence rules as shown in section 5.4. Another classical example of ambiguity is the "dangling else" problem.

```
Stat        →   Stat2 ; Stat
Stat        →   Stat2
Stat2       →   Matched
Stat2       →   Unmatched
Matched     →   if Exp then Matched else Matched
Matched     →   id:= Exp
Unmatched → if Exp then Matched else Unmatched
Unmatched → if Exp then Stat2
```

Imperative languages like Pascal or C often let the else-part of a conditional be optional. The problem is that it isn't clear how to parse, for example,

```
if p then if q then s1 else s2
```

According to the grammar, the else can equally well match either if. The usual convention is that an else matches the closest not previously matched if, which, in the example, will make the else match the second if.

5.2.2. Left Recursion and its Removal [MDU 2007, 2008]

A CFG containing rules of the form

$$A \rightarrow A\alpha/\beta$$

is called left recursive in A. The language generated by such rules is of the form $A \overset{*}{\Rightarrow} \beta\alpha^n$. If we replace the rules $A \rightarrow A\alpha/\beta$ by

$$A \rightarrow \beta A'/\beta$$

$$A' \rightarrow \alpha A'/\alpha$$

where A' is a new variable, then language generated by A is the same while no left-recursive A-rules are used in the derivation.

Left-recursion can be eliminated by the following scheme:

Let production under left-recursion are

$$A \rightarrow A\alpha_1/A\alpha_2 \ldots./A\alpha_m$$

$$A \rightarrow \beta_1/\beta_2/\ldots./\beta_n$$

Choose a new non-terminal, say A';

Add the new A'-rule as follows:

$$A' \rightarrow \alpha_i/\alpha_i A', \quad 1 \le i \le m$$

Replace the A-rule as follows:

$$A \rightarrow \beta_i/\beta_i A', \quad 1 \le i \le n$$

or, we can write as follows:

$$A \rightarrow \beta_1 A'/\beta_2 A'/\ldots./\beta_n A'/\beta_1/\beta_2/\ldots../\beta_n$$

$$A' \rightarrow \alpha_1 A'/\alpha_2 A'/\ldots./\alpha_m A'/\alpha_1/\alpha_2/\ldots./\alpha_m$$

or

$$A \rightarrow \beta_1 A'/\beta_2 A'/\ldots./\beta_n A'$$

$$A \rightarrow \beta_1/\beta_2/\ldots./\beta_n$$

$$A' \to \alpha_1 A'/\alpha_1 A'/..../\alpha_m A'$$
$$A' \to \alpha_1/\alpha_2/..../\alpha_m$$

Example 5.4. Consider the following CFG

$$E \to E + T/T$$
$$T \to T * F/F$$
$$F \to (E)/I$$
$$I \to a/b/c$$

and remove the left recursion.

Solution. Given CFG is

$$E \to E + T/T \quad : P_1$$
$$T \to T * F/F \quad : P_2$$
$$F \to (E)/I \quad : P_3$$
$$I \to a/b/c \quad : P_4$$

Only P_1 and P_2 are left recursive production so let us introduce another non-terminal A in P_1 and B in P_2 as follows:

$$E \to TA$$
$$A \to +TA/\in$$
$$T \to FB$$
$$B \to *FB/\in$$
$$F \to (E)/I$$
$$I \to a/b/c$$

Now all the productions are free from left recursion.

Example 5.5. Write an algorithm to remove the indirect left recursion.

Solution. *Removing Indirect Left-Recursion:*

Getting rid of **immediate left recursion** is not enough, one must get rid of indirect left recursion, where two or more non-terminals are **mutually left-recursive**. One can write any CFG to remove left recursion.

```
for i: = 1 to n do
    for j: = 1 to i - 1 do
        begin
        replace each Ai → Aj gamma with productions
        Ai → delta1 gamma/delta2 gamma
    end.
```

Eliminate immediate left recursion.

Example 5.6. Remove the left recursion from the following CFG.

expr → expr + term/expr – term/term

Solution. Given CFG

expr → expr + term/expr-term/term

can be re-written so it will be free from left recursion.

expr → term rest

rest → + term rest/–term rest/∈.

Example 5.7. Consider the following CFG.

$$S \rightarrow Bb/a$$
$$B \rightarrow Bc/Sd/e$$

Find out the left recursion and remove it.

Solution. Given CFG is

$$S \rightarrow Bb/a$$
$$B \rightarrow Bc/Sd/e$$

We can also write above CFG as follows:

$$S \rightarrow Bb/a$$
$$B \rightarrow Bc/Bbd/ad/e$$

Now we will remove immediate left recursion as follows:

$$S \rightarrow Bb/a$$
$$B \rightarrow adB'/eB'$$
$$B' \rightarrow cB'/bdB'/\in.$$

Now above CFG is free from left recursion.

5.2.3. Left-Factoring **[MDU 2008]**

Left factoring is useful for producing a grammar suitable for a predictive parser. Let us consider a production.

< statement > → **if** <expr> **then** <statement> **else** <statement>
| **if** <expr> **then** <statement>

Now we don't know which production rule to follow without knowing if there is an "**else**".

We can introduce another non-terminal *S*, as follows:

<statement> → **if** <expr> **then** <statement> *S*
S → **else** <statement>/∈.

So we can finally if situation of the productions are as follows:

$$A \rightarrow \alpha\beta_1/\alpha\beta_2$$

then it is difficult to decide whether to expand A to $\alpha\beta_1$ or to $\alpha\beta_2$. Now we will rewrite the production:

$$A \rightarrow \alpha A'$$
$$A' \rightarrow \beta_1/\beta_2$$

Now after seeing the input derived from α, we expand A' to β_1 or to β_2.

Left-Factoring–Algorithm **[MDU 2008]**

For each non-terminal A with two or more alternatives (production rules) with a common non-empty prefix, let say

$$A \rightarrow \alpha\beta_1 \mid ... \mid \alpha\beta_n \mid \gamma_1 \mid ... \mid \gamma_m$$

convert it into

$$A \rightarrow aA' \mid \gamma_1 \mid ... \mid \gamma_m$$
$$A' \rightarrow \beta_1 \mid ... \mid \beta_n$$

Example 5.8. Consider the following grammar and remove the left-factoring from it.

$$A \rightarrow xByA/xByAzA/a$$
$$B \rightarrow b$$

Solution. The given CFG is as follows:

$$A \rightarrow xByA/xByAzA/a$$
$$B \rightarrow b.$$

Left-factored, this grammar becomes

$$A \rightarrow xByAA'/a$$
$$A' \rightarrow zA/\in$$
$$B \rightarrow b$$

Example 5.9. Consider the following CFG.

Statement → variable ASSIGNOP expr | Procedure-call |
block | If expr THEN statement ELSE statement |
while expr do statement.

Three of the production for *statement* begin with non-terminals:

variable → *ID* | *ID* (expr)
Procedure call → ID | *ID* (expr-list)
block → Begin opt_statements End.

Now remove the left factoring from the given grammar.

Solution. In the productions for statement we replace non-terminals: variable, procedure-call, and block, by the right-sides of their production to obtain :

Statement → ID ASSIGNOP expr |
ID (expr) ASSIGNOP expr |
ID |
ID (expr_list) |
Begin opt_statement End |
if expr THEN statement ELSE statement |
while expr do statement

Now every production for Statement begins with a terminal but four of the productions begin with the same terminal, ID, so we add a new non-terminal, statement-rest, to the grammar and left factor ID out of these four productions to obtain :

Statement → ID statement_rest |
Begin opt_statements End |
If expr THEN statement ELSE statement |
while expr do statement

Statement_rest → ASSIGNOP expr |
(expr) ASSIGNOP expr |
(expr_list) | ∈

Note that all underlined tokens are terminals.

5.3. RECURSIVE-DECENT PARSING [MDU 2009]

Definition "A top-down parser that executes a set of recursive procedures to process the input without backtracking is called recursive-decent parser, and parsing is called recursive-decent parsing".

The recursive procedures can be easy to write and fairly efficient if written in a language that implements the procedure call efficiently.

There is a procedure for each non-terminal in the grammar. We assume a global variable, **lookhead**, holding the current input token and a procedure **match (Expected Token)** is the action of recognizing the next token in the parsing process and advancing the input stream pointer, such that **lookhead** points to the next token to be parsed. match ( ) is effectively a call to the lexical analyzer to get the next token.

For example input stream is: $a + b\$$

```
then        lookhead = = a
            match ( )
            lookhead = = +
            match ( )
            lookhead = = b
            _ _ _ _ _ _ _
            _ _ _ _ _ _ _
```

In this manner parsing can be done.

Consider the following grammar:

$$E \rightarrow TA$$
$$A \rightarrow +TA/\in$$
$$T \rightarrow FB$$
$$B \rightarrow *FB/\in$$
$$F \rightarrow (E)/id$$

Now let us write the algorithm of recursive-decent parser for given grammar.

```
E ( )
    {
    T ( ) ;
    A ( ) ;
    }.
T ( )
    {
    F ( ) ;
    B ( ) ;
    }.
B ( )
{
if (lookhead = = [ast])
    {
```

```
        match ( ) ;
        F ( ) ;
        B ( ) ;
        }.
    A ( )
        {
        if (lookhead = = [plus])
        {
        match ( ) ;
        T ( ) ;
        A ( ) ;
        }
    }
    F ( )
    {
        if (lookhead = = [id])
        {
        match ( ) ;
        }
    else if (lookhead = = [l paren] )
        {
        match ( ) ;
        E ( ) ;
        if ( lookhead = = [r paren ])
        {
        match ( ) ;
        }
        else ERROR
        }
    else ERROR
  } |* end of F ( ).*|
```

If there is any ∈-production for a non-terminal then the procedure for that non-terminal selects it whenever none of the other productions are suitable. If there is no production for a non-terminal and none of its productions are suitable then the procedure should report a syntax error.

TIPS

But one of the major drawback of recursive-decent parsing is that, it can be implemented only for those language which supports recursive procedure calls, and it suffers with the problem of left recursion.

Example 5.10. Write a code for the Recursive-decent parsing of the following grammar.

$$\begin{aligned} \textbf{expr} &\rightarrow \textbf{term rest} \\ \textbf{rest} &\rightarrow \textbf{+ term rest} \mid \textbf{– term rest} \mid \in \\ \textbf{term} &\rightarrow \textbf{0} \mid \textbf{1} \mid \ldots \mid \textbf{9} \end{aligned}$$

Solution. We know that in a recursive-decent parsing, we write code for each non-terminal of a grammar. In the case of above grammar, we should have three procedure, correspond to non-terminal expr, rest, and term.

Since there is only one production for non-terminal expr, the procedure <u>expr</u> is :

```
expr ( )
     {
     term ( ) ;
     rest ( ) ;
     return
     }
```

Since there are three productions for rest, procedure <u>rest</u> uses a global variable, <u>'lookhead'</u> to select the correct production or simply selects "no action" that is, ∈-production, indicating that lookhead variable is neither + nor – (neither plus nor minus).

```
rest ( )
{
If (lookhead = = '+')
     {
     match ( ) ;
     term ( ) ;
     rest ( ) ;
     return ( ) ;
     }
else if (lookhead = = ' - ')
     {
     match ( ) ;
     term ( ) ;
     rest ( ) ;
     return ;
     }
else
     {
     return ;
     }
}
```

The procedure <u>term</u> checks whether global variable is a digit.

```
term ( ) {
     if (is digit (lookhead))
     {
     match ( ) ;
     return ;
     }
    else
     {
     Error ( ) ;
     }
```

After loading first input token into variable 'lookhead' predictive parser is started by calling starting symbol, <u>expr</u>.

If the input is error free, the parser conducts a depth-first traversal of the parse tree and return to caller routine through expr.

5.4. TOP-DOWN PREDICTIVE PARSERS [MDU 2007]

Definition A predictive parser is an efficient way of implementing recursive-decent parsing since a stack is maintained in predictive parsing for handling the activation records. In top-down predictive parsers, the grammar will be able to predict the right alternative for the expansion of non-terminal during the parsing process; and hence, it need no backtrack.

Let $B \rightarrow \alpha_1/\alpha_2/ \ldots / \alpha_m$ are the production in the grammar, then during the parsing process parser has to decide that B is to be expanded or not. In predictive technique parser can exactly decide that which production of B should be used to expand B. Basically predictive parser looks at the next input symbol and find out which of the α_i derives to a string that start with the terminal symbol comes next in the input.

The predictive parser program maintains a stack of grammar symbols and uses a two-dimensional M-table created from the grammar. A special symbol, \$ marks the bottom of the stack and also the end of the input string. **The parser is initialized with the start symbol of grammar on the stack top and the input pointing to the first token.**

Pictures view of parser program can be viewed as follows:

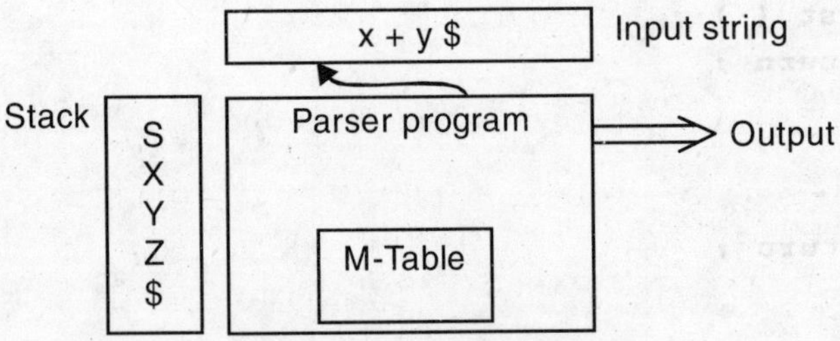

Fig. 5.14. Predictive parser.

The action of the parser depends on the grammar symbol on the top of the stack, Y and the current input token, b:

(*a*) If $Y = b = \$$ then parsing halts and parser announces successful completion of the parsing.

(*b*) $Y = b$ but does't equal \$ then parser pops Y off the stack and advances the input to the next token.

(*c*) If Y is a terminal not equal to b then parser announces an error.

(*d*) If Y is a non-terminal then parser consults entry $M[Y, b]$ in the M-table.

If the $M[Y, b]$ entry is a production for Y then the parser pops Y off the stack and pushes the symbols on the right-side of the production on to the stack (pushing the rightmost symbol of the right-side first and pushing the left most symbol on the right-side last).

If the $M[Y, b]$ has no entry then the parser announces the error and calls an error recovery routine.

Let us consider the following example to make the concept clear

$$E \rightarrow TE'$$
$$E' \rightarrow +TE' / \in$$
$$T \rightarrow FT'$$
$$T' \rightarrow *FT' / \in$$
$$F \rightarrow id \,/\, (E)$$

Generation of the *M*-table from the grammar will be described later. The table given below shows the moves made by the parser when the input is ***id + id * id*. In the table the stack is shown with the bottom on the left and the top on the right.**

Stack	Input string	Production used
E $	id + id * id $	
TE′ $	id + id * id $	E → TE′
FT′E′$	id + id * id $	T → FT′
idT′E′$	id + id * id $	F → id
T′E′$	+id * id $	
E′$	+id * id $	T′ → ∈
+TE′$	+id * id $	E′ → + TE′
TE′$	id * id $	
FT′E′ $	id * id $	T → FT′
id T′E′ $	id * id $	F → id
T′E′ $	* id $	
*FT′E′ $	* id $	T → * FT′
FT′E′ $	id $	
id T′E′ $	id $	F → id
T′E′ $	$	
E′ $	$	T′ → ∈
$	$	E′ → ∈

Fig. 5.15.

5.4.1. First and Follow

First and follow are two functions associated with a grammar that help us fill in the entries of an M-table.

First ( ) : is a function which gives the set of terminals that begin the strings derived from the production rule.

A symbol c is in FIRST(α) if and only if $\alpha \Rightarrow c\beta$ for some sequence β of grammar symbols.

Follow ( ) : is a function which gives the set of terminals that can appear immediately to the right of a given symbol.

A terminal symbol a is in FOLLOW(N) if and only if there is a derivation from the start symbol S of the grammar such that $S \Rightarrow \alpha N a \beta$, where α and β are (possibly empty) sequences of grammar symbols.

In other words, a terminal c is in FOLLOW(N) if c may follow N at some point in a derivation.

Benefits of First ( ) and Follow ( ):

1. Can be used to prove the LL(K) characteristic of a grammar.
2. Can be used to aid in the construction of predictive parsing tables.
3. Provides selection information for recursive descent parsers.

5.4.2. Following Algorithm can be used to Compute First (X) for a Grammar Symbol X

(*i*) If X is a terminal, First (X) is $\{X\}$.

(*ii*) If $X \rightarrow \in$ is a production, then add $\in$ to the set of first (X).

(*iii*) If X is a non-terminal

(*a*) If $X \rightarrow Y$, First (Y) is an element of the set of first (X).

(*b*) If first (Y) has $\in$ as an element and $X \rightarrow YZ$ then first (X) = First (Y) $- \{\in\} \cup$ First (Z)

First (Z) can be calculated in identical manner.

For example if given grammar is

$$E \rightarrow TA$$
$$A \rightarrow +TA/\in$$
$$T \rightarrow FB$$
$$B \rightarrow *FB/\in$$
$$F \rightarrow (E)/id$$

then let us calculate First (E), First (A), First (T), First (B) and First (F).

First (E) = First (TA) (since $E \rightarrow TA$)

= First (T) (since $T \rightarrow \in$ is not in the grammar)

= First (FB) (since $T \rightarrow FB$)

= First (F) (since $F \rightarrow \in$ is not there in the given grammar)

= {(, *id* } (since $F \rightarrow (E)/id$, so terminals are (and *id*)

First (A) = {+, $\in$} (since $A \rightarrow +TA/\in$ so terminals are + and $\in$)

First (T) = {(, *id* } (as above)

First (B) = { *, $\in$ } (since $B \rightarrow *\ FB/\in$, so terminals are + and $\in$)

First (F) = {(, *id* }

So for the given grammar First (E) = {(, *id* }, First (A) = {+, $\in$}

First (T) = { (, *id* }, First (B) = { *, $\in$ }

and First (F) = { (, *id* }.

5.4.3. Algorithm to Calculate the Follow (X) for a Non-terminal X

(*i*) \$ is an element of follow (S), where S is the start symbol and \$ indicates the end of the input.

(*ii*) If $A \rightarrow \alpha XY$ is a production, the set First (Y) is in the set follow (X), excluding $\in$.

(*iii*) If $A \to \alpha X$ or $A \to \alpha XY$ are productions, and First (Y) has an element $\in$, the set Follow (X) is in the set Follow (A).

Let us consider the same example as follows to calculate the grammar:

$E \to TA$ First (E) = { (, *id* }

$A \to +TA/\in$ First (A) = { +, $\in$ }

$T \to FB$ First (T) = { (, *id* }

$B \to *FB/\in$ First (B) = { *, $\in$ }

$F \to (E)/id$ First (F) = { (, *id* }.

Follow (E) = { \$,) } (Here \$ is in follow (E) since E is starting nonterminal and ')'is for production $F \to (E)/id$)

Follow (A) = { \$,) } (Since $E \to TA$ and $A \to \in$ so follow E is the follow (A))

Follow (T) = First (A) by rule 2 (since $E \to TA$)
∪ Follow (E) by rule 3 on production $E \to TA$
(since First (A) contains $\in$)
= { +, $\in$ } – $\in$ ∪ { \$,) }
= { +, \$,) }

Follow (B) = Follow (T) by the rule 3 on production $T \to FB$
= { +,), \$ }

Follow (F) = (First (B) by rule 2) ∪ (Follow (T) by rule 3 on $T \to FB$) ∪ (Follow (B) by rule 3 on $B \to FB$)
= { * } ∪ { +,), \$} ∪ (+,), \$ }
= { +,), *, \$ }.

5.4.4. Constructing M-table/Predictive Parsing Table

Many book follows various techniques to construct the *M*-table, here we will discuss another approach, by calculating a new function select ( ).

> **Definition** **Select ( ):** is a function which can be defined by the First ( ) and Follow ( ) functions and by using select ( ) function, we can easily construct the *M*-table.

Let $A \to W$ is any production then

Select ($A \to W$) = First (First (W) follow (A))

Let us continue with previous example:

$E \to TA$	First (E) = { (, *id* }	Follow (E) = { \$,) }
$A \to +TA/\in$	First (A) = { +, $\in$ }	Follow (A) = { \$,) }
$T \to FB$	First (T) = { (, *id* }	Follow (T) = { +,), \$ }
$B \to *FB/\in$	First (B) = { *, $\in$ }	Follow (B) = { +,), \$ }
$F \to (E)/id$	First (F) = { (, *id* }	Follow (F) = { +,), *, \$ }

$$\begin{aligned}
\text{Select } (E \to TA) &= \text{First (First } (TA) \times \text{Follow } (E)) \quad \text{*/ here } \times \text{ represents cross product */} \\
&= \text{First } (\{ (, id \} \times \{ \$,) \}) \\
&= \text{First } (\{ \underline{(}\ \$, \underline{(}\ ), \underline{id}\ \$, \underline{id}\ ) \}) \\
&= \{ (, id \} \text{ */ First symbol of each pair */} \\
\text{Select } (A \to +TA) &= \text{First (First } (+TA) \times \text{Follow } (A)) \\
&= \text{First } (\{ + \}, \times \{ \$,) \}) \\
&= \text{First } (\underline{+}\ \$, \underline{+}\ ),) \\
&= \{ + \} \\
\text{Select } (A \to \in) &= \text{First (First } (\in) \times \text{Follow } (A)) \\
&= \text{First } (\{ \in \} \times \{ \$,) \}) \\
&= \text{First } (\$,) \} \\
&= \{ \$,) \} \\
\text{Select } (T \to FB) &= \text{First (First } (FB) \times \text{Follow } (T)) \\
&= \text{First } (\{ (i, id \} \{ +,), \$, \}) \\
&= \text{First } (\underline{(}\ +, \underline{(}\ ), \underline{(}\ \$, \underline{id} +, \underline{id}\ ), \underline{id}\ \$) \\
&= \{ (, id \} \\
\text{Select } (B \to *FB) &= \text{First (First } (*FB) \times \text{Follow } (B)) \\
&= \text{First } (\{ * \}, \times \{ +,), \$ \}) \\
&= \{ * \} \\
\text{Select } (B \to \in) &= \text{First (First } (\in) \times \text{Follow } (B)) \\
&= \text{First } (\{ \in \} \times \{ +,), \$ \}) \\
&= \text{First } (\underline{+}, \underline{)}, \underline{\$}) \\
&= \{ +,), \$ \} \\
\text{Select } (F \to (E)) &= \text{First (First } (E) \times \text{Follow } (F)) \\
&= \text{First } (\{ (\} \times \{ +,), *, \$ \}) \\
&= \{ (\} \\
\text{Select } (F \to id) &= \text{First (First } (id) \times \text{Follow (F)}) \\
&= \text{First } (\{ id \} \{ +,), *, \$ \}) \\
&= \text{First } (\underline{id} +, \underline{id}\ ), id *, \underline{id}\ \$) \\
&= \{id\}
\end{aligned}$$

So for the predictive parsing table:

$E \to TA$	Select (E)	$= \{ (, id \}$
$A \to +TA$	Select (A)	$= \{ + \}$
$A \to \in$	Select (A)	$= \{ \$,) \}$
$T \to FB$	Select (T)	$= \{ (, id \}$
$B \to *FB$	Select (B)	$= \{ * \}$
$B \to \in$	Select (B)	$= \{ +,), \$ \}$

$F \rightarrow (E)$ Select $(F) = \{ (\}$

$F \rightarrow id$ Select $(F) = \{ id \}$

Finally with the help of these select values we can construct M-table as follows :

Look Ahead Tokens

Non-terminals	(	id	+	*	)	$
E	*TA*	*TA*				
A			*+TA*		∈	∈
T	*FB*	*FB*				
B			∈	**FB*	∈	∈
F	*(E)*	*id*				

Fig. 5.16.

Construction is clear now since we have the values of select ( ) for all non-terminals.

For example if production is $T \rightarrow FB$, select (T) is $\{ (, id \}$ then corresponding entry for $M [T, (]$ will be FB, and for $M [T, id]$ will be FB.

Similar for $A \rightarrow \in$ select is $\{), \$ \}$ so $M [A,)]$ and $M[A, \$]$ will be $\in$ and $\in$ respectively.

In similar fashion the complete parsing table is constructed.

5.5. LL (1) GRAMMARS [MDU 2007]

Now we have fair idea about predictive parsing technique and we know that in **some of the situation M-table (predictive parsing table) may have same entry for two or more different productions, it simply means that the grammar is ambiguous and/or left-recursive and/or not left factored.**

Definition **"A grammar is an LL(1)-grammar if and only if its M-table has no entries that are multiply defined."**

We can also define LL(1) grammar mathematically as follows:

"If each non-terminal of given grammar, which appears on the left hand side of more than one production have the LL(1) property then grammar is said to be LL(1)"

(If a non-terminal appears on the left side of more than one production and select ( ) for those production are disjoint then, it has LL(1) property).

For example if $A \rightarrow ai$ and $A \rightarrow aj$

If select $(A \rightarrow ai) \cap$ select $(A \rightarrow aj) = 0$ then

A has the LL(1) property.

Let us consider the same grammar

$$E \rightarrow TA$$
$$A \rightarrow +TA/\in$$
$$T \rightarrow FB$$
$$B \rightarrow *FB/\in$$
$$F \rightarrow (E)/id$$

Here only A and B and F are present in more than one production.

$$A \to +TA, \qquad A \to \in$$
$$B \to *FB, \qquad B \to \in$$
$$F \to (E), \qquad F \to id$$

Select ($E \to TA$) = {(, id}
Select ($A \to +TA$) = {+}
Select ($A \to \in$) = {\$,)}
Select ($T \to FB$) = {(, id}
Select ($B \to *FB$) = {*}
Select ($B \to \in$) = {+,), \$}.
Select ($F \to (E)$) = {(}
Select ($F \to id$) = {id}.

Now,

For $A \to +TA, A \to \in$, Select ($A \to +TA$) $\cap$ Select ($A \to \in$) = ϕ

For $B \to *FB, B \to \in$, Select ($B \to *FB$) $\cap$ Select ($B \to \in$) = ϕ

For $F \to (E), F \to id$, Select ($F \to (E)$) $\cap$ Select ($F \to id$) = ϕ

So A, B and F have LL(1) property.

So finally we can say that given grammar is LL(1).

TIPS

The first *L* indicates the reading direction (left-to-right), the second *L* indicates the derivation order (left) and the 1 indicates that there is a one-symbol lookahead. A grammar that can be parsed using LL(1) parsing is called an LL(1) grammar.

5.5.1. Advantages and Disadvantages of LL(1) Parsing

The reason why LL(1) parsing by recursive descent is appealing to many people is its naturalness, and the method is extensively taught as an introduction to, and a framework for, compiling programming language. In addition, it is easy to implement and to have confidence in its correctness. As well as performing the parsing process, the functions written may also contain actions to perform type checking and other forms of checking, as well as synthesis actions such as storage allocation and code generation.

However, we have also identified some disadvantages of recursive desent parsing, such as the inefficiency of function calls and the need to perform grammar transformations, without even knowing whether suitable transformations exist or not. The problem is not just finding the transformation but ensuring that they are performed correctly. There are, therefore, good reasons for using tools to perform the transformations rather than relying on a purely manual approach. **Other disadvantages of recursive descent parsing are**:

(*i*) The very large parsers often produced.

(*ii*) The tendency for actions which are part of different phases of compilation to appear in the same function bodies.

The following are the requirements for effective use of recursive descent:

(*i*) A good grammar transform that will usually be able to transform a grammar into LL(1) form.

(*ii*) The ability to represent the equivalent of the recursive descent parser in tabular form. This means that the parser, instead of entering and leaving functions as it checks the input text, will instead move about the tabular equivalent of the grammar, stacking return addresses as necessary.

Example 5.11. Consider the following grammar

$$S \rightarrow AaAb/BbBa$$
$$A \rightarrow \in$$
$$B \rightarrow \in$$

Test whether the grammar is LL(1) or not, and construct a predictive parsing table for it.

Solution. Given grammar is as follows:

$$S \rightarrow AaAb/BbBa$$
$$A \rightarrow \in$$
$$B \rightarrow \in$$

Let us calculate First and Follow:

$$\text{First }(S) = \text{First }(AaAb) \cup \text{First }(BbBa)$$
$$\text{First }(AaAb) = \text{First }(A) - \{\in\} \cup \text{First }(aAb) = \{a\}$$
$$\text{First }(BbBa) = \text{First }(B) - \{\in\} \cup \text{First }(bBa) = \{b\}$$

So $\text{First }(S) = \{a\} \cup \{b\} = \{a, b\}$

$$\text{First }(A) = \{\in\}$$
$$\text{First }(B) = \{\in\}$$

Follow (S) = \$

Follow $(A) = \{a, b\}$ (since in $S \rightarrow AaAb$, A is followed by both a and b)

Follow $(B) = \{b, a\}$ (since in $S \rightarrow BbBa$, B is followed by both b and a)

Now let us calculate select ( ) for different productions:

$$\text{Select }(S \rightarrow AaAb) = \text{First (First }(AaAb)\text{ Follow }(S))$$
$$= \text{First }(\{a\}, \{\$\}) = a$$
$$\text{Select }(S \rightarrow BbBa) = \text{First (First }(BbBa)\text{ Follow }(S))$$
$$= \text{First }(\{b\}\ \{\$\}) = \text{First }(b\$) = b$$
$$\text{Select }(A \rightarrow \in) = \text{First (First }(\in)\text{ Follow }(A))$$
$$= \text{First }(\{\in\}\ \{a, b\}) = \text{First }(\{a, b\}) = \{a, b\}$$
$$\text{Select }(B \rightarrow \in) = \{a, b\}$$

Here S (non-terminal) is in the left hand side of two productions as:

$$S \rightarrow AaAb/BbBa$$

So Select $(S \rightarrow AaAb) \cap (S \rightarrow BbBa)$

Now let us make

$$= a \cap b$$
$$= \phi$$

So we can say that S follows LL(1) properties, now finally we can say that given grammar is LL(1).

Now it is easy to construct the parse table by the help of select ( ).

	a	b	\$
S	$S \to AaAb$	$S \to BbBa$	
A	$A \to \in$	$A \to \in$	
B	$B \to \in$	$B \to \in$	

Fig. 5.17.

Since Select $(S \to AaAb) = a$

Select $(S \to BbBa) = b$

Select $(A \to \in) = \{a, b\}$

Select $(B \to \in) = \{a, b\}$

Example 5.12. Consider the following grammar:

$$S \to 1AB/\in$$
$$A \to 1AC/0C$$
$$B \to 0S$$
$$C \to 1$$

and test that whether the grammar is LL(1) or not.

Solution. Let us calculate First and Follow:

$$\text{First }(S) = \text{First }(1\text{ AB}) \cup \text{First }(\in) = \{1, \in\}$$
$$\text{First }(A) = \{1, 0\}$$
$$\text{First }(B) = \{0\}$$
$$\text{First }(C) = \{1\}$$
$$\text{Follow }(S) = \{\$\}$$
$$\text{Follow }(A) = \text{First }(A) \cup \text{First }(B) = \{1, 0\} \cup \{0\} = \{1, 0\}$$
$$\text{Follow }(B) = \text{Follow }(S) = \$$$

Now only S and A derives two productions as

$$S \to 1AB, \quad S \to \in$$

and

$$A \to 1AC, \quad A \to 0C$$

$$\text{Select }(S \to 1AB) = \text{First (First }(1AB)\text{ Follow }(S))$$
$$= \text{First }(\{1\}\ \{\$\}) = \{1\}$$
$$\text{Select }(S \to \in) = \text{First (First }(\in)\text{ Follow }(S))$$
$$= \text{First }(\in \$) = \{\$\}$$
$$\text{Select }(A \to 1AC) = \text{First (First }(1AC)\text{ Follow }(A))$$
$$= \text{First }(\{1\}, \{1, 0\})$$
$$= \text{First }(11, 10)$$
$$= \{1\}$$
$$\text{Select }(A \to 0C) = \text{First (First }(0C)\text{ Follow }(A))$$
$$= \text{First }(\{0\}\ \{1, 0\})$$
$$= \text{First }(01, 00) = \{0\}$$

Now following condition should hold for LL(1) grammar:

$$\text{Select } (S \to 1\ AB) \cap \text{Select } (S \to \in) = \{1\} \cap \{\$\} = \phi$$

$$\text{Select } (A \to 1\ AC) \cap \text{Select } (A \to 0C) = \{1\} \cap \{0\} = \phi$$

So here both S and A are following LL(1) property so we can say that grammar is LL(1).

Example 5.13. Calculate the First, Follow and Select for following given grammar:

$$S \to A$$
$$A \to aB/Ad$$
$$B \to bBC/f$$
$$C \to g$$

Solution. By calculating select for S, A, B and C we can design the predictive parsing table for the given grammar.

$$\text{First } (S) = \text{First } (A) = \{a\}$$
$$\text{First } (B) = \{b, f\}$$
$$\text{First } (C) = \{g\}$$
$$\text{Follow } (S) = \{\$\}$$
$$\text{Follow } (A) = \{d\}$$
$$\text{Follow } (B) = \text{First } (C) = \{g\}$$
$$\text{Follow } (C) = \text{Follow } (B) = \{g\}$$

So $\quad \text{Select } (S \to A) = \text{First (First } (A) \text{ Follow } (S))$

$$= \text{First } (\{a\} \times \{\$\}) = \{a\}$$
$$\text{Select } (A \to aB) = \{a\}$$
$$\text{Select } (A \to Ad) = \{a\}$$
$$\text{Select } (B \to bBC) = \{b\}$$
$$\text{Select } (B \to f) = \{f\}$$
$$\text{Select } (C \to g) = \{g\}$$

Example 5.14. Construct M-table for following grammar:

$$S \to iEtSS_1/a$$
$$S_1 \to eS/\in$$
$$E \to b$$

Solution. $\quad \text{First } (S) = \{i, a\}$

$$\text{First } (S_1) = \text{First } (eS) \cup \text{First } (\in) = \{e, \in\}$$
$$\text{First } (E) = \text{First } (b) = \{b\}$$
$$\text{Follow } (S) = \{\$\}$$

Now by $\quad S \to iEtSS_1, \text{Follow } (E) = \{t\}$

$$\text{Follow } (S) = \text{First } (S_1) - \{\in\} \cup \text{Follow } (S)$$

$= \{e, \in\} - \{\in\} \cup \{\$\}$

$= \{e, \$\}$

Follow (S_1) = Follow $(S) = \{e, \$\}$

By $S_1 \to eS$

Follow (S) = Follow $(S_1) = \{e, \$\}$

Now calculate select by the help of First and Follow:

Select $(S \to iEt\ SS_1)$ = First (First $(iEtSS_1)$ Follow(S))

= First $(\{i\}\ \{e, \$\})$ = First $(ie, i\$)$

$= \{i\}$

Select $(S \to a)$ = First (First (a) Follow (S)) = First $(ae, a\$)$

$= \{a\}$

Select $(S_1 \to eS)$ = First (First (eS) Follow (S_1))

= First $(\{e\}, \{e, \$\})$ = First $(e, e\$)$

$= \{e\}$

Select $(S_1 \to \in)$ = First (First $(\in)$ Follow (S_1) = First $(\{\in\}, \{e, \$\})$

= First $(e, \$) = \{e, \$\}$

Select $(E \to b)$ = First (First (b) Follow (E))

$= \{b\}$

Now entries of M-table can be very easily filled by the looking the select for every production as follows:

	i	***a***	***b***	***e***	***t***	**$**
S	$S \to iEtSS_1$	$S \to a$				
S_1				$S_1 \to eS$ $S_1 \to \in$		$S_1 \to \in$
E			$E \to b$			

Fig. 5.18.

$M(S, i) = S \to iEtSS_1$ (Since select for $S \to iEtSS_1$ is i)

$M(S, a) = S \to a$ (Since select $(S \to a) = \{a\}$

$M(S_1, e) = S_1 \to eS$ (Since select $(S_1 \to eS) = \{e\}$

$= S_1 \to \in$ and select $(S_1 \to \in) = \{e, \$\}$)

$M(S_1, \$) = S_1 \to \$$

$M(E, b) = E \to b$

So clearly once we calculate the select then it becomes very easy to design the M-table for given grammar.

Since M-table have double entries for $M[S_1, e]$ as $S_1 \to eS$ and $S_1 \to \in$ so given grammar is not LL(1).

Example 5.15. Construct the M-table for given grammar:

$$\begin{aligned} S &\to aBDh \\ B &\to cC \\ C &\to bC/\in \end{aligned}$$

$$D \to EF$$
$$E \to g/\in$$
$$F \to f/\in$$

Solution. First (S) = First $(aBDh) = \{a\}$

First $(B) = \{c\}$

First $(C) = \{b, \in\}$

First (D) = First (EF) = First $(E) - \{\in\} \cup$ First (F) ...(1)

First $(E) = \{g, \in\}$

First $(F) = \{f, \in\}$

So First $(D) = \{g, \in\} - \{\in\} \cup \{f, \in\}$ by (1)

$= \{g, f, \in\}$

Follow (S) = \$

(*a*) Using the production $S \to aBDh$ we can get

Follow (B) = First (Dh) = First $(D) - \{\in\} \cup$ First (h)

$= \{g, f, \in\} - \{\in\} \cup \{h\} = \{g, f, h\}$

Follow (D) = First $(h) = \{h\}$

(*b*) Using the production $B \to cC$, we will get

Follow (C) = Follow $(B) = \{g, f, h\}$

(*c*) By using the production $C \to bC$, we will get

Follow (C) = Follow $(C) = \{g, f, h\}$

(*d*) Now by using the production $D \to EF$ we will get

Follow (E) = First $(F) - \{\in\} \cup$ Follow (D)

$= \{f, \in\} - \{\in\} \cup \{h\} = \{f, h\}$

Follow (F) = Follow $(D) = \{h\}$

Now let us calculate select as follows:

Select $(S \to aBDh)$ = First (First $(aBDh)$ Follow (S))

= First $(\{a\}, \{\$\})$ = First $(a\$) = \{a\}$

Select $(B \to cC)$ = First (First (cC) Follow (B))

= First $(\{c\}\ \{g, f, h\}) = \{c\}$

Select $(C \to bC)$ = First (First (bc) Follow (C)

= First $(\{b\}\ \{g, f, h\}) = \{b\}$

Select $(C \to \in)$ = First (First $(\in)$ Follow (C))

= First $(\{\in\}\ \{g, f, h\} = \{g, f, h\}$

Select $(D \to$ EF$)$ = First (First (EF) Follow (D))

= First $(\{g, f, \in\}\ \{h\})$

= First $(gh, fh, \in h)$

$= \{g, f, h\}$

Similarly: Select $(E \to g) = \{g\}$

Select $(E \to \in) = \{f, h\}$

Select $(F \to f) = \{f\}$

Select $(F \to \in) = \{h\}$

Now we can design *M*-table as follows:

$M(S, a) = S \to aBDh$ (Since select $(S \to aBDh) = \{a\}$)

$$M(B, c) = B \to cC \quad \text{(Since select } (B \to cC) = \{c\})$$
$$M(C, b) = C \to bC \quad \text{(Since select } (C \to bC) = \{b\})$$
$$M(C, g/f/h) = C \to \in \quad \text{(Since select } (C \to \in) = \{g, f, h\})$$
$$M(D, g/f/h) = D \to EF \quad \text{(Since select } (D \to EF) = \{g, f, h\})$$
$$M(E, g) = E \to g \quad \text{(Since select } (E \to g) = \{g\})$$
$$M(E, f/h) = E \to \in \quad \text{(Since select } (E \to \in) = \{f, h\})$$
$$M(F, f) = F \to f \quad \text{(Since select } (F \to f) = \{f\})$$
$$M(F, h) = F \to \in \quad \text{(Since select } (F \to \in) = \{h\})$$

Now we can design M-table as follows:

	a	***b***	***c***	***g***	***f***	***h***	**$**
S	$S \to aBDh$						
B			$B \to cC$				
C		$C \to bC$		$C \to \in$	$C \to \in$	$C \to \in$	
D				$D \to EF$	$D \to EF$	$D \to EF$	
E				$E \to g$	$E \to \in$	$E \to \in$	
F					$F \to f$	$F \to \in$	

Fig. 5.19.

EXERCISE

1. Consider the following grammar:

$$S \to AaAbBbBa$$
$$A \to \in$$
$$B \to \in$$

Is the grammar LL(1) ?

2. Given a grammar:

$$E \to E + T/T$$
$$T \to T * F/F$$
$$F \to id$$

which is a set of valid items for a viable prefix it.

3. Compute the First and Follow set of A in following grammar:

$$A \to (A)\ A/\in$$

4. Is the following grammar LL(1)

$$S \to aSa/\in$$

5. (*a*) Eliminate the left recursion from the following grammar:

$$S \to (L)/a$$
$$L \to L,\ S/S$$

(*b*) Construct the predictive parser for given grammar in (a).

(*c*) Show the behaviour of parser on the string:

(*i*) (*a*, *a*)

(*ii*) (*a*, (*a*, *a*))

6. The following is an LL(1) grammar for regular expression over alphabet $\{a, b\}$.

$$E \to TE'$$
$$E' \to +E/\in$$
$$T \to FT'$$
$$T' \to T/\in$$
$$F \to PF'$$
$$F' \to *F'/\in$$
$$P \to (E)/a/b/\in$$

(*i*) Compute First and Follow for each nonterminal of the above grammar.
(*ii*) Show the grammar is LL(1).
(*iii*) Construct the predictive parsing table for the grammar.
(*iv*) Construct a recursive-descent parser for the grammar.

7. Consider the following grammar, which generates all even length strings of *a*'s except for empty string.

$$S \rightarrow aSa/aa$$

(*i*) By training through the steps of a top-down parser (with back track) which tries the alternate *aSa* before *aa*, show that *S* succeeds on 2, 4 or 8 *a*'s, but fails on 6 *a*'s.
(*ii*) What strings cause the parser to succeed ?

8. Here operator-precedence relations for the grammar:

$$S \rightarrow (L)/a$$
$$S \rightarrow L,\ S/S$$

is shown as follows:

	a	(	)	,	$
a			$\gtrdot$	$\gtrdot$	$\gtrdot$
(	$\lessdot$	$\lessdot$	$\doteq$	$\doteq$	
)			$\gtrdot$	$\gtrdot$	$\gtrdot$
,	$\lessdot$	$\lessdot$	$\gtrdot$	$\gtrdot$	
$	$\lessdot$	$\lessdot$			

Parse the following sentences:
(*i*) (*a*, *a*)
(*ii*) (*a*, (*a*, *a*))

9. State the problems associated with Top-Down parsing. State and eliminate the problem associated with the following grammar for Top-Down Parsing:

$$E \rightarrow E + T/T$$
$$T \rightarrow T * F/F$$
$$F \rightarrow (E)/id$$

10. After computing First and Follow functions, construct predictive parsing table for the following grammar:

$$E \rightarrow TE'$$
$$E' \rightarrow +TE'/\in$$
$$T \rightarrow FT'$$
$$T' \rightarrow *FT'/\in$$
$$F \rightarrow (E)/id$$

11. Discuss the operator-precedence parsing algorithm. Consider the following operator grammar and precedence functions; explain the parsing of input string *id*+ *id* * *id*. Grammar is:

$$E \rightarrow E + E/E * E/(E)/id$$

precedence functions:

	+	*	id	$
f	4	2	4	0
g	3	1	5	0

❑❑❑

6

Chapter

BOTTOM-UP PARSING TECHNIQUES

INSIDE THIS CHAPTER

6.1. Introduction to Bottom-up Parsing **6.2.** Handle of a Right Sentential Form **6.3.** Shift Reduce Parsing **6.4.** Operator Precedence Parsers **6.5.** LR Parsers **6.6.** A Comparison of Predictive Parser with Shift-Reduce Parsers.

6.1. INTRODUCTION TO BOTTOM-UP PARSING

Consider the grammar

$$S \rightarrow aTUe$$
$$T \rightarrow Tbc/b$$
$$U \rightarrow d$$

and let us find out the rightmost derivation of the sentence for the string *abbcde*.

$$S \Rightarrow aT\underline{U}e$$
$$\Rightarrow a\underline{T}de$$
$$\Rightarrow a\underline{T}bcde$$
$$\Rightarrow abbcde$$

Now let us try to read *abbcde* from left-to-right and perform the rightmost derivation in the reverse order, it can be done in four steps as follows:

(*i*) Choose the first *b* and reduce it to the left-side of the $T \rightarrow b$ production to produce the sentential form:

aTbcde

(*ii*) Now reduce *Tbc* by the left-hand side of the production $T \rightarrow Tbc$ to produce the sentential form *aTde*.

(*iii*) Now reduce *d* by left-hand side of the production, $U \rightarrow d$ to produce *aTUe*.

(*iv*) Finally *aTUe* can be replaced by the left-hand side of $S \rightarrow aTUe$ production to get *S*.

If we carefully analyse the above discussion then it is very clear that in each step, we are applying the productions in reverse, replacing the right-side with the left-side, so we use the word reduce instead of produce.

Definition "So bottom-up parsing is an attempt to reduce the input string *w* to the start symbol of a grammar by tracing out the right-most derivations of *w* in reverse. This is equivalent to constructing a parse tree for the input string *w* in the reverse order."

Bottom-up parsing involves the selection of a substring that matches the right side of the production, whose reduction to the non-terminal on the left side of the production represents one step along the reverse of a right-most derivation. That is, it leads to the generation of the previous right most derivation. Clearly selecting a substring that matches the right side of production is not enough; the position of this substring in the sentential form is also important.

6.1.1. Shift-Reduce Parsing

Bottom-up parsing is also known as **shift-reduce parsing** because its two main actions are shift and reduce.

(*i*) At each shift action, the current symbol in the input string is pushed to a stack.

(*ii*) At each reduction step, the symbols at the top of the stack (this symbol sequence is the right side of a production) will replaced by the non-terminal at the left side of that production.

(*iii*) There are also two more actions: accept and error.

A shift-reduce parser tries to reduce the given input string into the starting symbol.

$$\text{A string} \xrightarrow[\text{reduced to}]{} \text{the starting symbol}$$

At each reduction step, a substring of the input matching to the right side of a production rule is replaced by the non-terminal at the left side of that production rule. If the substring is chosen correctly, the right most derivation of that string is created in the reverse order.

Rightmost Derivation: $S \overset{*}{\underset{rm}{\Rightarrow}} w$

Shift-Reduce Parser Finds: $w \underset{rm}{\Leftarrow} \dots \underset{rm}{\Leftarrow} S$

Let us see the example of shift-reduce parsing as follows:

$S \rightarrow aABb$ input string: $aa\underline{a}bb$

$A \rightarrow aA \mid a$ $a\underline{aA}bb$

$B \rightarrow bB \mid b$ $aA\underline{b}b \Downarrow$ reduction

$\underline{aABb}$

S

$$S \underset{rm}{\Rightarrow} \underline{aABb} \underset{rm}{\Rightarrow} aA\underline{b}b \underset{rm}{\Rightarrow} a\underline{aA}bb \underset{rm}{\Rightarrow} aa\underline{a}bb$$

Right Sentimental Forms

How do we know which substring to be replaced at each reduction step ?

6.2. HANDLE OF A RIGHT SENTENTIAL FORM [MDU 2008]

From the definition of bottom-up parsing it is clear that finding a substring that matches the right side of production, as well as its position in the current sentential form, are both equally important. We will define a new term "handle" to address both these problems simultaneously.

Definition "A handle is a substring that matches a right hand side of production rule in the grammar and whose reduction to the non-terminal on the left hand side of that grammar rule is a step along the reverse of a right most derivation".

A rightmost derivation in reverse can be obtained by **handle-pruning.**

For example for the grammar

$$S \rightarrow aTUe$$
$$T \rightarrow Tbc/b$$
$$U \rightarrow d$$

The string "*abbcde*" can be parsed bottom-up by the following reduction steps :

(*i*) *aTbcde*

(*ii*) *aTde*

(*iii*) *aTUe*

(*iv*) *S*

So here in this example the handles are:

1. The first (*b*) in (*abbcde*)
2. The (*Tbc*) substring in (*aTbcde*)
3. The (*d*) in (*aTde*)
4. The whole string, (*aTUe*)

TIPS

In step 1 and step 2 of the example the parser has three possible handles to choose from: if the parser chooses the wrong handle it won't be able to complete the reverse-ordered right most derivation. The main task of a bottom-up parser is to choose the correct handle at each step of the parse. There could be many choices on any step, that is the empty string can be inserted into the string of n symbols in any $n + 1$ different locations, so just a single $\in$-production in a grammar will give us many possible handles to choose from.

Example 6.1. Find the handles of the sentential forms occurring in the derivation of string *id* + *id* * *id* by the following grammar:

$$E \rightarrow E + E/E * E/id$$

Solution. Given grammar is

$$E \rightarrow E + E$$
$$E \rightarrow E * E$$
$$E \rightarrow id$$

The string can be reduced to E in following steps:

(*i*) $id + id * id$ (First *id* is handle of this sentential form)

(*ii*) $E + id * id$ (Now second *id* is handle)

(*iii*) $E + E * id$ (Now third *id* is handle for this sentential form)

(*iv*) $E + E * E$ (Now $E * E$ is handle)

(*v*) $E + E$ (Now $E + E$ is handle)

(*vi*) E (Reduced to starting symbol).

Example 6.2. Consider the following grammar and show the handle of each right sentential form for string (*a*, (*a*, *a*)).

$$S \rightarrow (L)/a$$
$$L \rightarrow L, S/S$$

Solution. The following sentential form will occur in reduction of (*a*, (*a*, *a*)) to *S*.

(*i*) (*a*, (*a*, *a*))	(handle is first *a*)
(*ii*) (*S*, (*a*, *a*))	(Now *S* is the handle)
(*iii*) (*L*, (*a*, *a*))	(Now first *a* is handle)
(*iv*) (*L*, (*S*, *a*))	(*S* is handle again)
(*v*) (*L*, (*L*, *a*))	(Now *a* is handle)
(*vi*) (*L*, (*L*, *S*))	(Now *L*, *S* is handle)
(*vii*) (*L*, (*L*))	((*L*) is handle)
(*viii*) (*L*, *S*)	(again *L*, *S* is handle)
(*ix*) (*L*)	((*L*) is handle)
(*x*) *S*	(finally string is reduced to starting non-terminal).

6.3. SHIFT REDUCE PARSERS

Most bottom-up parsers are implemented as **shift-reduce parsers**. Such a parser uses a stack to hold grammar symbols. A parser goes on shifting the input symbols

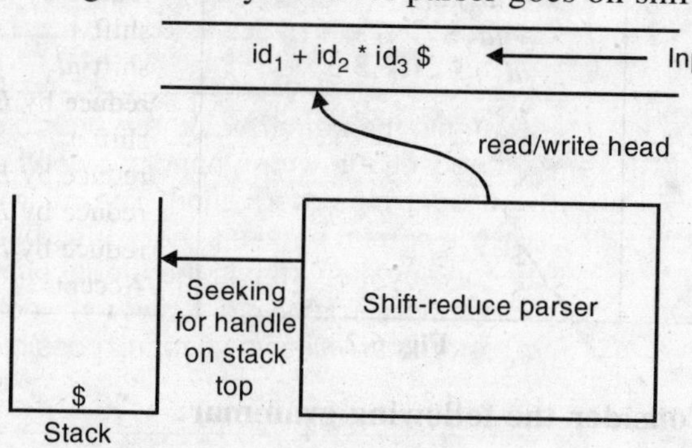

Fig. 6.1. Shift reduce parser.

onto the stack until a handle comes on the top of the stack. When a handle appears on the top of the stack, it performs reduction.

The implementation makes use of a stack to hold grammar and an input tape to hold string *w* to be parsed, which is terminated by the right end marker $, the same symbol used to mark the bottom of the stack. Parser may have following four possible actions:

- **Shift:** Move the next input symbol on to the top of the stack.
- **Reduce:** Reduce a handle on the right-most part of the stack by popping it off the stack and pushing the left-side of the production on to the right-end of the stack.
- **Accept:** Announce successful completion of parsing.
- **Error:** Signal discovery of a syntax error.

TIPS

We use $ to mark the left-end (bottom) of the stack and also the end of the input string. Initially the stack is empty. Parsing ends successfully when the input is empty and the stack contains only the start symbol.

As an example we use the following grammar:

$$E \rightarrow E + E$$
$$E \rightarrow E * E$$
$$E \rightarrow (E)$$
$$E \rightarrow id$$

Now actions of shift-reduce parser to parse the input string $id_1 * (id_2 + id_3)$, can be shown by following figure:

Stack Contents	Input String	Actions
$	$id_1 * (id_2 + id_3)$ $	shift id_1
$$id_1$	$* (id_2 + id_3)$ $	Reduce by $E \rightarrow id$
$$E$	$* (id_2 + id_3)$ $	shift *
$$E*$	$(id_2 + id_3)$ $	shift (
$$E*($	$id_2 + id_3)$ $	shift id_2
$$E * (id_2$	$+ id_3)$ $	reduce by $E \rightarrow id$
$$E * (E$	$+ id_3$ $	shift +
$$E * (E +$	id_3) $	shift id_3
$$E* (E + id_3$	) $	reduce by $E \rightarrow id$
$$E * (E + E$	) $	shift)
$$E * (E + E)$	$	reduce by $E \rightarrow E + E$
$$E * (E)$	$	reduce by $E \rightarrow (E)$
$$E * E$	$	reduce by $E \rightarrow E * E$
$$E$	$	Accept

Fig. 6.2.

Example 6.3. Consider the following grammar:

1. $E \rightarrow E + T$
2. $E \rightarrow T$
3. $E \rightarrow T * F$
4. $T \rightarrow F$
5. $F \rightarrow (E)$
6. $F \rightarrow y$

Show the shift-reduce parser action for the string $y + y + y * y$.

Solution. For the given grammar:

1. $E \rightarrow E + T$
2. $E \rightarrow T$
3. $E \rightarrow T * F$
4. $T \rightarrow F$
5. $F \rightarrow (E)$
6. $F \rightarrow y$

For the string $y + y + y * y$, following parsing table can be constructed:

Stack Contents	Input String	Actions
$	$y + y + y * y$ $	shift y
$y	$+y + y * y$ $	reduce by $F \to y$
$F	$+y + y * y$ $	reduce by $T \to F$
$T	$+y + y * y$ $	reduce by $E \to T$
$E	$+y + y * y$ $	shift +
$E+	$y + y * y$ $	shift y
$E + y	$+y * y$ $	reduce by $F \to y$
$E + F	$+y * y$ $	reduce by $T \to F$
$E + T	$+y * y$ $	reduce by $E \to T$
$E + E	$+y * y$ $	reduce by $E \to E + E$
$E	$+y * y$ $	shift +
$E +	$y * y$ $	shift y
$E + y	$* y$ $	reduce by $F \to y$
$E + F	$* y$ $	reduce by $T \to F$
$E + T	$* y$ $	reduce by $E \to T$
$E + E	$* y$ $	reduce by $E \to E + E$
$E	$* y$ $	shift *
$E*	y $	shift y
$E * y	$	reduce by $F \to y$
$E * F	$	reduce by $T \to F$
$E * T	$	reduce by $E \to T$
$E * E	$	reduce by $E \to E * E$
$E	$	Accept

Fig. 6.3.

6.3.1. Shortcomings of Shift-reduce Parsing

Shift-reduce parsing does not tell anything about the finding out the handles. Shift-reduce parsing is very efficient, when handle is identified. Handle can be easily managed with technique of shift-reduce parsing.

There are context-free grammars for which shift-reduce parsers cannot be used. Stack contents and the next input symbol may not decide action:

- **Shift/reduce conflict:** Whether make a shift operation or a reduction.
- **Reduce/reduce conflict:** The parser cannot decide which of several reductions to make.

If a shift-reduce parser cannot be used for a grammar, that grammar is called as non-LR(k) grammar.

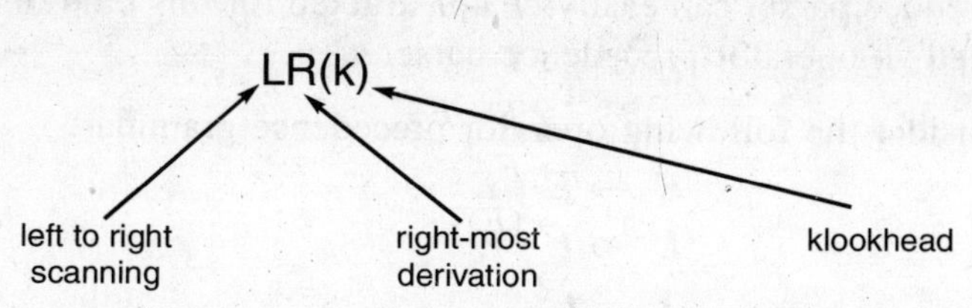

We can use several techniques to find-out the handle and on the basis of these techniques we can define different shift-reduce parsers, such as shift-reduce parsers with precendence relationship and various LR parsers.

6.3.2. Types of Shift-Reduce Parsers

There are two main categories of shift-reduce parsers as follows:

1. ***Operator-Precedence Parser***

   Simple, but only a small class of grammars.

2. ***LR-Parsers***

   Covers wide range of grammars.

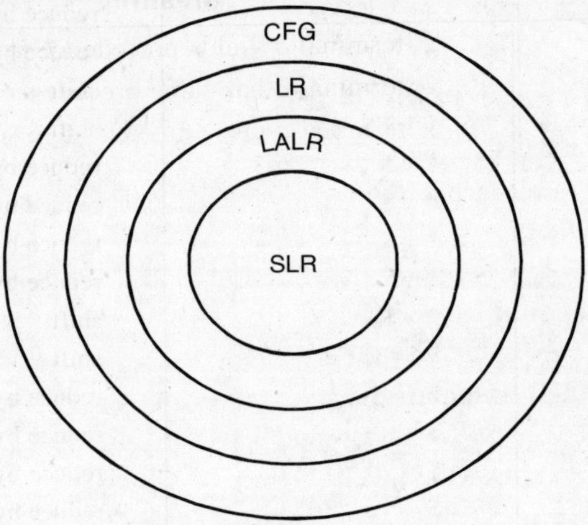

Fig. 6.4.

1. SLR – simple LR parser
2. LR – most general LR parser
3. LALR – intermediate LR parser (lookhead LR parser)

SLR, LR and LALR work same, only their parsing tables are different.

6.4. OPERATOR PRECEDENCE PARSERS

For a small class of CFG's, the principles of operator precedence can be used to build a simple bottom up parser.

Definition The CFG which have the following properties:

(*i*) No R.H.S. of any production has an $\in$

(*ii*) No two non-terminals are adjacent.

Such grammar is called operator precedence grammar.

A shift-reduce parser can easily be constructed for this kind of grammar and it is called an operator-precedence parser.

Let us consider the following operator precedence grammar:

$$E \rightarrow E + E$$
$$E \rightarrow E * E$$

$$E \to (E)$$
$$E \to -E$$
$$E \to id$$

This grammar has six terminals: + – * ( ) *id* and to eliminate the ambiguity we must establish precedence relation between certain pairs of terminals.

6.4.1. Operator Precedence Relation

The precedence relations are only established between the terminals of the grammar (and with $ markers at both ends of a string), non-terminals are ignored.

Relation	Meaning
$a \lessdot b$	terminal *a* yields precedence to *b*
$a \doteq b$	terminal *a* has same precedence as *b*
$a \gtrdot b$	terminal *a* takes precedence over terminal *b*.

- If OP_1 (operator 1) has higher precedence than OP_2, then set $OP_1 \gtrdot OP_2$ and $OP_2 \lessdot OP_1$.
- If OP_1 and OP_2 have equal precedence.

(*a*) If they have **equal associativity**

$$OP1 \doteq OP2$$

(*b*) If they have **left associativity**

$$OP_1 \gtrdot OP_2 \quad \text{and} \quad OP_2 \gtrdot OP_1$$

(*c*) If they have **right-associativity**

$$OP_1 \lessdot OP_2 \quad \text{and} \quad OP_2 \lessdot OP_1$$

For example consider the following grammar:

$$E \to E + T/T$$
$$T \to T * F/F$$
$$F \to id/(E)$$

Now to find out the precedence relation table of above grammar we will adopt the following algorithm step by step:

Step 1. Calculate the **Leads and Lasts** for every non-terminal involved in given grammar.

Here given grammar is:

$$E \to E + T/T$$
$$T \to T * F/F$$
$$F \to (E)/id$$

So we have to calculate the Leads and Lasts for *E*, *T* and *F* by applying following facts:

(*i*) For any production like $X \Rightarrow YaZ$

Lead $(X) = \{a\}$ if *Y* is $\in$ or a single non-terminal.

(*ii*) For any production like $X \Rightarrow YbZ$

Last $(X) = \{b\}$ if *Z* is $\in$ or a single non-terminal.

(*iii*) If *a* is in lead (*B*), and there is a production of the form $X \to B\alpha$, then *a* is Lead (*X*).

(*iv*) If non-terminal 1 → non-terminal 2 then every Lead and Last of non-terminal 2 will be in Leads and Lasts of non-terminal 1 also.

(*v*) If non-terminal 1 → terminal then terminal will be in Leads and Lasts of non-terminal 1.

On the basis of above facts we can calculate Leads and Lasts of *E*, *T* and *F* :

Non-terminal :	*E*	*T*	*F*
Lead :	*, +, (, *id*	*, (, *id*	(, *id*
Last :	*, +,) , *id*	*,) , *id*	) , *id*

Now let me explain the above calculation:

$$\text{Lead}(F) = \{\ (\ \ \text{by } F \to (E),\quad id\ \ \text{by } F \to id\ \}$$

$$\text{Last}(F) = \{\ (\ ,\ id\ \}$$

$$\text{Lead}(T) = \{*\} \cup \text{Lead}(F) \quad (\text{since } F \to T)$$
$$= \{*, (\ , id\ \}$$
$$\text{Last}(T) = \{*\} \cup \text{Last}(F) \quad (\text{since } F \to T)$$
$$= \{\ *,)\ , id\ \}$$
$$\text{Lead}(E) = \{+\} \cup \text{Lead}(T) \quad (\text{since } E \to T)$$
$$= \{+\} \cup \{*, (\ , id\}$$
$$= \{\ +, *, (\ , id\ \}$$
$$\text{Last}(E) = \{+\} \cup \text{Last}(T)$$
$$= \{+\} \cup \{*\ ,)\ , id\}$$
$$= \{+, *,)\ , id\ \}$$

Step 2. Now establish the precedence relation between two terminals as follows :

(*i*) $a \doteq b$: If in right side of any production $X \to \alpha a \beta b \delta$, '*a*' and '*b*' are separated by nothing or by a single non-terminal.

In $F \to (E)$,(≐) since (and) are separated by *E* only.

(*ii*) $a \lessdot b$: If *b* is in Lead (*A*) where *A* is immediately right non-terminal of *a* in any production.

$\$ \lessdot b$: If in any production $S \overset{+}{\Rightarrow} rb\delta$, *S* is starting non-terminal, and *b* is lead (*S*).

In given grammar consider

$E \to E\ \underline{+T}$ So $+ \lessdot$ Lead (*T*)

$+ \lessdot *\ , + \lessdot (\ , + \lessdot id$

$\$ \lessdot$ Lead (*E*) (since *E* is starting non-terminal)

$\$ \lessdot +\ , \$ \lessdot *\ , \$ \lessdot (\ , \$ \lessdot id$

$T \to T\ \underline{*F}$ So $* \lessdot$ Lead (*F*)

$* \lessdot ($ and $* \lessdot id$

$F \to (\underline{E})$ so $(\lessdot$ Lead (*E*)

$(\lessdot *, (\lessdot +\ , (\lessdot (\ , (\lessdot id$

So $\lessdot$ relation is defined as:

$$+ \lessdot *\ ,\ + \lessdot (\ ,\ + \lessdot id\ ,\ \$ \lessdot +\ ,\ \$ \lessdot *\ ,\ \$ \lessdot (\ ,\ \$ \lessdot id,$$

$$* \lessdot (\ ,\ * \lessdot id\ ,\ (\lessdot *\ ,\ (\lessdot +\ ,\ (\lessdot (\ ,\ (\lessdot id.$$

(*i*) $a \gtrdot b$: If 'a' is in Last (B) where B is immediately left of 'b' in any production.

$b \gtrdot \$$: if 'b' is in Last (S) where S is starting non-terminal of given grammar.

In given grammar consider:

$E \rightarrow \underline{E} + T$	So Last (E) $\gtrdot +$
	$* \gtrdot +\ ,\ + \gtrdot +\ ,\ ) \gtrdot +\ ,\ id \gtrdot +$
$T \rightarrow \underline{T} * F$	So Last (T) $\gtrdot *$
	$* \gtrdot *\ ,\ + \gtrdot *\ ,\ ) \gtrdot *\ ,\ id \gtrdot *$
$E \rightarrow \underline{(E)}$	So Last (E) $\gtrdot$)
	$+ , * ,) , id \gtrdot)$
	$+ \gtrdot)\ ,\ * \gtrdot)\ ,\ ) \gtrdot)\ ,\ id \gtrdot)$

Last (E)$\gtrdot$ \$

$$* \gtrdot \$,\ + \gtrdot \$\ ,\ ) \gtrdot \$\ ,\ id \gtrdot \$$$

So relation $\gtrdot$ is defined as follows:

$* \gtrdot +\ ,\ + \gtrdot +\ ,\ ) \gtrdot +,\ id \gtrdot +\ ,\ * \gtrdot *\ ,\ + \gtrdot *\ ,\ ) \gtrdot *\ ,\ id \gtrdot *,\ + \gtrdot)\ ,\ * \gtrdot)\ ,\ ) \gtrdot)\ ,$ $id \gtrdot)\ ,\ * \gtrdot \$\ ,\ + \gtrdot \$\ ,\ ) \gtrdot \$,\ id \gtrdot \$.$

Step 3. Now by the help of relation $\lessdot$ and $\gtrdot$ construct the precedence relation table as follows:

(*i*) First fill all the $\doteq$ relations.

(*ii*) Secondly fill all $\lessdot$ relations.

(*iii*) and lastly fill all $\gtrdot$ relations.

	+	*	(	)	*id*	\$
+	$\gtrdot$	$\lessdot$	$\lessdot$	$\gtrdot$	$\lessdot$	$\gtrdot$
*	$\gtrdot$	$\gtrdot$	$\lessdot$	$\gtrdot$	$\lessdot$	$\gtrdot$
(	$\lessdot$	$\lessdot$	$\lessdot$	$\doteq$	$\lessdot$	*X*
)	$\gtrdot$	$\gtrdot$	*X*	$\gtrdot$	*X*	$\gtrdot$
id	$\gtrdot$	$\gtrdot$	*X*	$\gtrdot$	*X*	$\gtrdot$
\$	$\lessdot$	$\lessdot$	$\lessdot$	*X*	$\lessdot$	*X*

Fig. 6.5.

6.4.2. Developing Some Other Operator-Precedence Relation

Here we describe how the operator-precedence relation are developed.

1. Unary Operators. Unary operators usually have higher precedence than binary operators:

> $-A + B$ usually means $(-A) + B$ instead of $-(A + B)$. Most unary operators are prefix operators (written before their operands), so they are right-associative: $- -A$ is $- (-A)$. Postfix unary operators are left-associative.

2. Leaves of Syntax Trees. The leaves of a syntax tree are ID tokens, numeric constants and boolean constants. Leaves must be evaluated before their values can be used by the operators so they are given higher precedence than the operators.

3. Parentheses. For the parentheses we must have (= =) so the handle of the $E \rightarrow (E)$ production can be found. A pair of parenthesis can't be removed until all operations between the pair have been performed. A pair of outer parentheses can't be removed until all pairs of inner parentheses have been removed. An operation outside a pair of parentheses can't be performed until the pair has been removed. These rules dictate the operator-precedence relations in the following table (*x* **any operator**).

	(	x	)
(	$\lessdot$	$\lessdot$	$\doteq$
x	$\lessdot$	$\gtrdot$	$\gtrdot$
)	×	$\gtrdot$	$\gtrdot$

Fig. 6.6.

Note that it's usually a syntax error (*X*) to have *a*) (combination with no intervening operator. The relation in this table can also be used for brackets and any other grouping operators.

4. Function Calls. In most languages the production for a function call is:

expression $\rightarrow$ *id* (expression_list)

where expression_list is one or more expressions separated by commas. The handle for this production includes zero or more commas between the parentheses so the relations in the following table are required:

	id	(	,	)
id	×	$\doteq$	$\gtrdot$	$\gtrdot$
(	$\lessdot$	$\lessdot$	$\doteq$	$\doteq$
,	$\lessdot$	$\lessdot$	$\doteq$	$\doteq$
)	×	×	$\gtrdot$	$\gtrdot$

Fig. 6.7.

5. Reading Array Elements. In many languages, the production for reading an element of an array is:

expression $\rightarrow$ *id* [expression_list]

The precedence relations table for function calls can be used if its parentheses are replaced by brackets.

6. End Markers. The \$-sign at the left-end of the stack is never in any handle so it should be < all following terminals. Similarly, the \$-sign at the end of the input is never shifted on to the stack so all preceding terminals should be > than it.

7. Precedence Function. Suppose the grammar for an operator-precedence parser has *n*-terminals so the table of operator-precedence relation has $n + 1$ rows and $n + 1$ columns (including the \$). Often *n* is so large that coding and decoding the table of $(n + 1)^2$ entries is a very difficult task. Fortunately, one can usually replace the table with a pair of precedence functions that are much easier to code and debug.

The idea is to try and define two functions *f* and *g*, that map the terminals into integers such that:

(*i*) $f(a) < g(b)$ whenever $a \lessdot b$;

(*ii*) $f(a) = g(b)$ whenever $a \doteq b$; and

(*iii*) $f(a) > g(b)$ whenever $a \gtrdot b$;

If *f* and *g* can be defined then the table of precedence relations is no longer needed: instead of referring to the table to determine the precedence relation between two terminals, a and b, the parser code compares $f(a)$ with $g(b)$.

Functions *f* and *g* are called **precedence functions.** Each function has only one argument with only $n + 1$ values so the two functions are much easier to code and de-bug than the large precedence table.

Note that there is some loss in error-detection capability when the precedence functions are used: they never notice any syntax errors whereas the table of precedence relations does report some of the errors. A syntax error won't be discovered until the parser tries to reduce a handle that doesn't match the right-side of any production.

As an example, the table is repeated below with values of the precedence functions in *f*-column and the *g*-row shown to the corresponding terminals:

	g	6	6	6	4	2	1	0
f		*id*	–	(	*	+	)	$
5	*id*	*X*	*X*	*X*	$\gtrdot$	$\gtrdot$	$\gtrdot$	$\gtrdot$
5	)	*X*	*X*	*X*	$\gtrdot$	$\gtrdot$	$\gtrdot$	$\gtrdot$
5	–	$\lessdot$	$\lessdot$	$\lessdot$	$\gtrdot$	$\gtrdot$	$\gtrdot$	$\gtrdot$
5	*	$\lessdot$	$\lessdot$	$\lessdot$	$\gtrdot$	$\gtrdot$	$\gtrdot$	$\gtrdot$
3	+	$\lessdot$	$\lessdot$	$\lessdot$	$\lessdot$	$\gtrdot$	$\gtrdot$	$\gtrdot$
1	(	$\lessdot$	$\lessdot$	$\lessdot$	$\lessdot$	$\lessdot$	$\doteq$	*X*
1	$	$\lessdot$	$\lessdot$	$\lessdot$	$\lessdot$	$\lessdot$	*X*	Accepted

Fig. 6.8.

TIPS

For every $\lessdot$, $\doteq$, and $\gtrdot$ entry in the table one can see that the comparison of the corresponding *f* and *g* function values matches the entry. If the parser uses the precedence functions instead of the table it must have a separate test for the accept state and it won't notice a syntax error until it tries to reduce an illegal handle.

6.4.3. Operator Precedence Parsing

In the previous section we have seen the method to find out the precedence relation.

Definition Operator precedence parsing is the bottom-up parsing using operator precedence table.

For example if grammar is:

$$E \rightarrow E + T/T$$
$$T \rightarrow T * F/F$$
$$F \rightarrow id.$$

Here we need a precedence table, which have precedence relation between <*E*, +> , <*T*, +> , <*E*, $> , <*T*, *> , <*T*, $> , <*F*, +> , <*F*, *> , <*F*, $ >, <*id*, +> , <*id*, *> , <*id*, $> , <+, *T*> , <+, *F*> , <+, *id*> , <*, *F*> , <*F*, *id*> , <$, *E*> , <$, *T*> , <$, *F*> and <$, *id*>

We can also write precedence relation using tree, symbols lower on the tree have a higher precedence. The symbol just opposite on the same branch have equal precedence.

The tree for above grammar can be as follows:

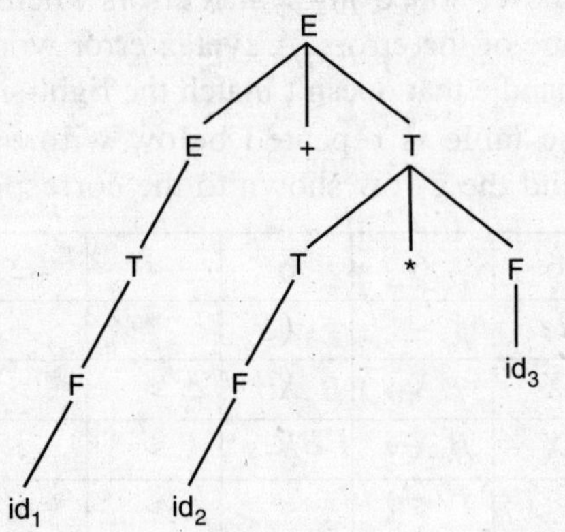

Fig. 6.9.

Now with the help of algorithm discussed in previous section and tree (given above) following precedence relation table is obtained.

	E	*T*	*F*	*id*	+	*	$
E	*X*	*X*	*X*	*X*	$\doteq$	*X*	$\gtrdot$
T	*X*	*X*	*X*	*X*	$\gtrdot$	$\doteq$	$\gtrdot$
F	*X*	*X*	*X*	*X*	$\gtrdot$	$\gtrdot$	$\gtrdot$
id	*X*	*X*	*X*	*X*	$\gtrdot$	$\gtrdot$	$\gtrdot$
+	*X*	$\lessdot$ $\doteq$	$\lessdot$	$\lessdot$	*X*	*X*	*X*
*	*X*	*X*	$\doteq$	$\lessdot$	*X*	*X*	*X*
$	$\lessdot$	$\lessdot$	$\lessdot$	$\lessdot$	*X*	*X*	*X*

Fig. 6.10.

Now let us parse: $id_1 + id_2 * id_3$ for given precedence table.

- Our very first task is to find the handle. The handle is the left most simple phrase and can be found by scanning left to right for the $\gtrdot$ relation, then back to the left until the $\lessdot$ relation is found.

- If the relation is $\lessdot$ or $\doteq$ push the following symbol on the stack. The handle will always be found on the top of the stack when $\gtrdot$ is reached. Pop the stack for the handle until $a \lessdot$ relation is found.

Now processing of the string is as follows:

$$\$ \lessdot id_1 \gtrdot + id_2 * id_3 \$$$
$$\$ \lessdot F \gtrdot + id_2 * id_3 \$$$
$$\$ \lessdot T \gtrdot + id_2 * id_3 \$$$
$$\$ \lessdot E \doteq + \lessdot id_2 \gtrdot * id_3 \$$$
$$\$ \lessdot E \doteq + \lessdot F \gtrdot * id_3 \$$$
$$\$ \lessdot E \doteq + \lessdot T \doteq * \lessdot id_3 \gtrdot \$$$
$$\$ \lessdot E \doteq + \lessdot T = * \doteq F \gtrdot \$$$
$$\$ \lessdot E \doteq + \doteq T \gtrdot \$$$
$$\$ \lessdot E \doteq + \doteq T \gtrdot \$$$
$$\$ \lessdot E \gtrdot \$$$

Accept.

Example 6.4. Consider the following grammar:

$$E \to TA$$
$$A \to +TA/\in$$
$$T \to FB$$
$$B \to *FB/\in$$
$$F \to id$$

test that grammar is operator precedence grammar or not and how string $w = id_1 + id_2 * id_3 + id_4$ can be processed in operator precedence parsing for given grammar.

Solution. The given grammar is as follows:

$$E \to TA$$
$$A \to +TA/\in$$
$$T \to FB$$
$$B \to *FB/\in$$
$$F \to id$$

$A \to \in$ and $B \to \in$ productions are in above grammar so it is not operator precedence grammar but we can convert it into operator precedence grammar by eliminating the empty production from the above grammar. So after removing $\in$-transition grammar becomes as follows:

$$E \to TA/T$$
$$A \to +TA/+T$$
$$T \to FB/F$$
$$B \to *FB/*F$$
$$F \to id$$

Now the grammar is operator precedence grammar.

Given string is $w = id_1 + id_2 * id_3 + id_4$.

Let us construct a parse tree for it as follows:

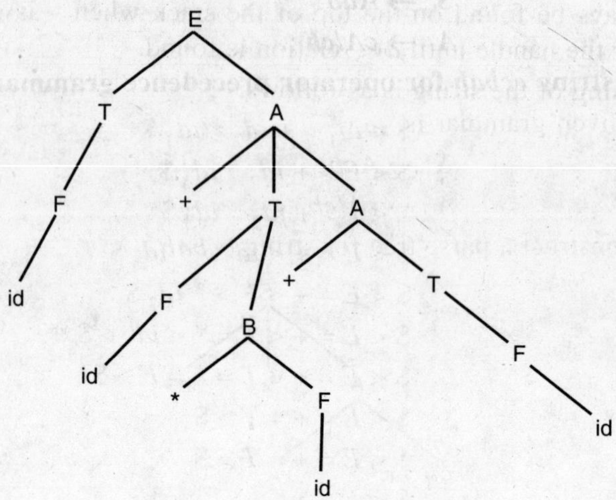

Fig. 6.11.

So yield is *id + id * id + id*

Now on the basis of above tree we can find the following operator precedence table :

	E	T	A	F	B	id	$+$	$*$	$\$$
E	×	×	×	×	×	×	×	×	$\gtrdot$
T	×	×	$\doteq$	×	×	×	$\lessdot$	$\lessdot$	$\gtrdot$
A	×	×	×	×	×	×	$\lessdot$	$\lessdot$	$\gtrdot$
F	×	×	×	×	$\doteq$	×	$\gtrdot$	$\lessdot$	$\gtrdot$
B	×	×	×	×	×	×	$\gtrdot$	$\lessdot$	$\gtrdot$
id	×	×	×	×	×	×	$\gtrdot$	$\gtrdot$	$\gtrdot$
$+$	×	$\doteq$	×	$\lessdot$	×	$\lessdot$	×	×	×
$*$	×	×	×	$\doteq$	×	$\lessdot$	×	×	×
$\$$	$\lessdot$	$\lessdot$	$\lessdot$	$\lessdot$	$\lessdot$	$\lessdot$	×	×	×

Fig. 6.12.

Now let us process the string with the help the of above precedence table:

	Handle
$id_1 + id_2 * id_3 + id_4$	
$\$ \lessdot id_1 \gtrdot + id_2 * id_3 + id_4 \$$	id_1
$\$ \lessdot \underline{F} \gtrdot + id_2 * id_3 + id_4 \$$	F
$\$ \lessdot T \lessdot + \lessdot \underline{id}_2 \gtrdot * id_3 + id_4 \$$	id_2
$\$ \lessdot T \lessdot + \lessdot F \lessdot * \lessdot id_3 \gtrdot + id_4 \$$	id_3
$\$ \lessdot T \lessdot + \lessdot F \lessdot * \doteq \underline{F} \gtrdot + id_4 \$$	F
$\$ \lessdot T \lessdot + \lessdot \underline{F \doteq B} \gtrdot + id_4 \$$	FB
$\$ \lessdot T \lessdot + \doteq T \lessdot + \lessdot id_4 \gtrdot \$$	id_4
$\$ \lessdot T \lessdot + \doteq T \lessdot + \lessdot \underline{F} \gtrdot \$$	F
$\$ \lessdot T \lessdot + \doteq T \lessdot + \doteq T \gtrdot \$$	T
$\$ \lessdot T \lessdot + \doteq \underline{T \doteq A} \gtrdot \$$	TA
$\$ \lessdot T \doteq A \gtrdot \$$	TA
$\$ \lessdot E \gtrdot \$$	Accept

Example 6.5. Consider the following grammar

$$S \rightarrow Aab$$
$$A \rightarrow cA/cb$$

and process the string *ccbab* for operator precedence grammar.

Solution. Given grammar is

$$S \rightarrow Aab$$
$$A \rightarrow cA/cb$$

Now let us construct a parse tree for string *ccbab*

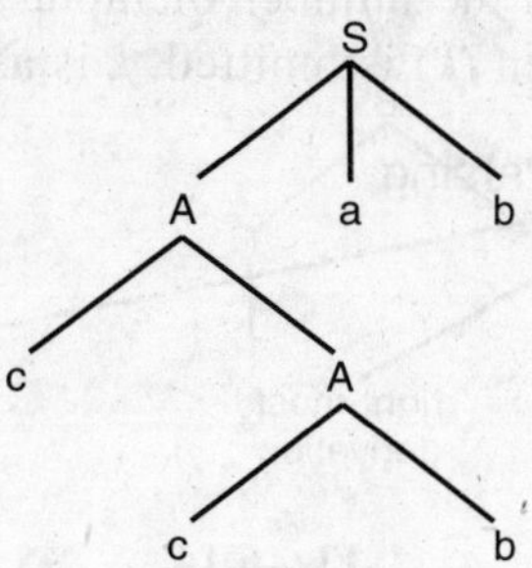

Fig. 6.13.

On the basis of above tree, we can design following precedence table:

	S	*A*	*a*	*b*	*c*	$
S						
A			≐			
a				≐		
b			⋗			⋗
c				≐	⋖	
$		⋖			⋖	

Let us process the string as follows:

	Handle
$\$\ c\ ,\ \underline{cb}\ ab\ \$$	
$\$\ c \lessdot c \doteq b \gtrdot ab\ \$$	*cb*
$\$\ c\,\underline{A}\ ab\ \$$	
$\$ \lessdot c \doteq A \gtrdot ab\ \$$	*cA*
$\$\ Aab\ \$$	
$\$ \lessdot A \doteq a \doteq b \gtrdot \$$	*Aab*
$\$\ S\ \$$	Accept

6.4.4. Advantages and Disadvantages of Operator Precedence Parsing

It have following **advantages:**

- It is very simple.
- By using the operator precedence table, these parsers can be very fast.
- We can quickly identify the next handle.

It also have some **disadvantages:**

- Unary operators are difficult to resolve.
- It is very difficult to verify that the parser and the grammar are equivalent.
- It only works on a small class of grammars.

6.5. LR PARSERS

A large class of grammars can be parsed using LR (*k*) parsers: the "*L*" stands for left-to-right scanning of the input, the "R" stands for constructing a right most derivation in reverse, and *k* is the number of input symbols of lookahead used to making parsing decision. When (*k*) is omitted, *k* is assumed to equal 1.

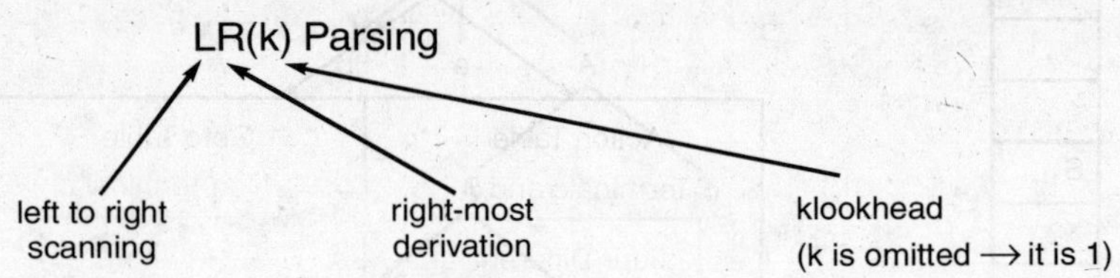

Fig. 6.14.

LR parsing has several advantages:

- It can recognize virtually all programming language constructs for which grammars can be written.
- It is the most general non-backtracking shift-reduce parsing method known.
- It can be implemented as efficiently as other shift-reduce methods.
- It can parse any grammar that a predictive parser can parse plus other grammars.
- It can detect a syntax error as early as possible on a left-to-right scan of the input.

TIPS

The main disadvantage of LR parsing is that, it is too much work to construct a parser by hand: one needs a specialized tool – an LR parser generator.

6.5.1. The LR Parsing Algorithm **[MDU 2008, 2009]**

If we see the block diagram of an LR parser in Fig. 6.15, it has an input, an output, a stack, a driver program, and a parsing table with two parts (action and goto). The driver program is the same for all LR parsers: it reads the input string one symbol at a time and maintains a stack.

The stack always maintains the following form:

$$S_0X_1\ S_1X_2\ S_2\ ...\ X_{m-1}\ S_{m-1}X_m\ S_m$$

where X_i is a grammar symbol, each S_i is the state, and S_m state is top of the stack.

The action of the driver program depends on action $[S_m, a_i]$ where a_i is the current input symbol. Following actions are possible:

1. Shift. If action $[S_m, a_i]$ = shift S, the parser shifts the input symbol, a_i, onto the stack, and then stacks state S. Now current input symbol becomes a_{i+1}.

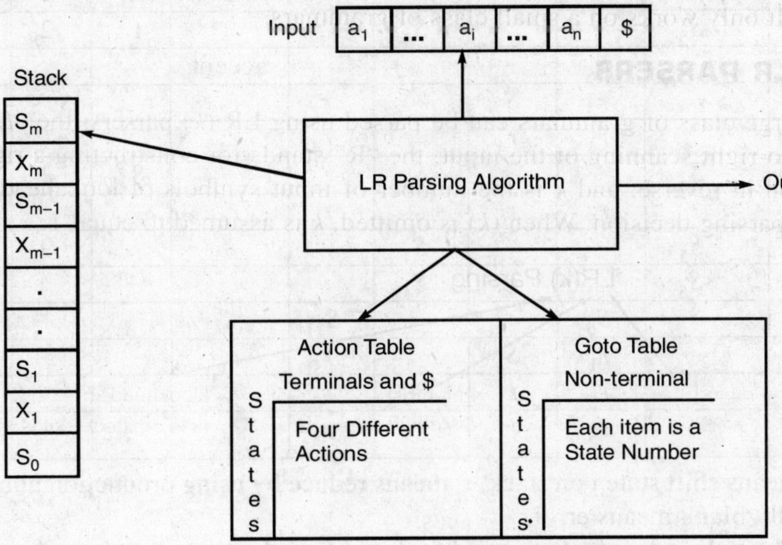

Fig. 6.15. Block diagram of LR parser.

Stack	**Input**
$S_0X_1\ S_1X_2 \ldots X_{m-r}\ S_{m-r}\ X_m\ S_m\ a_i S$	$a_{i+1}, a_{i+2} \ldots a_n$ \$

2. Reduce. If action $[S_m, a_i]$ = reduce $A \rightarrow \beta$, the parser executes a reduce move using the $A \rightarrow \beta$ production of the grammar. If β has r grammar symbols, first $2r$ symbols are poped off the stack (r state symbol and r grammar symbol). So the top of the stack now becomes S_{m-r}, then A is pushed on the stack, and then state go to $[S_{m-r}, A]$ is pushed on the stack. The current input symbol is still a_i.

Stack	**Input**
$S_0X_1\ S_1X_2 \ldots X_{m-r}\ S_{m-r}\ \underline{AS}$	$\underline{a_i}, a_{i+1} \ldots a_n$ \$

where S = Goto $[S_{m-r}, A]$

- If action $[S_m, a_i]$ = accept, parsing is completed.
- If action $[S_m, a_i]$ = error, the parser has discovered a syntax error.

For example for the given grammar:

(1) $E \rightarrow E + T$

(2) $E \rightarrow (T)$

(3) $T \rightarrow T * F$

(4) $T \rightarrow F$

(5) $F \rightarrow (E)$

(6) $F \rightarrow id$

and given parsing table:

State	Action						Go to		
	id	+	*	(	)	$	*E*	*T*	*F*
0	S_5			S_4			1	2	3
1		S_6				accept			
2		r_2	S_7		r_2	r_2			
3		r_4	r_4		r_4	r_4			
4	S_5			S_4			8	2	3
5		r_6	r_6		r_6	r_6			
6	S_5			S_4				9	3
7	S_5			S_4					10
8		S_6			S_{11}				
9		r_1	S_1		r_1	r_1			
10		r_3	r_3		r_3	r_3			
11		r_5	r_5		r_5	r_5			

where S_i means shift state *i* on stack, r_j means reduce by using production numbered *j*, and finally blank means error.

Now show the parsing of string *id* * *id* + *id* $.

Stack	Input string	Action
$	*id* * *id* + *id* $	S_5
$ *id*5	* *id* + *id* $	r_6
$ *F*3	* *id* + *id* $	r_4
$ *T*2	* *id* + *id* $	S_7
$ *T*2 * 7	*id* + *id* $	S_5
$ *T*2 * 7 *id* 5	+ *id* $	r_6
$ *T*2 * 7 *F* 10	+ *id* $	r_3
$ *T*2	+ *id* $	r_2
$ *E*1	+ *id* $	S_6
$ *E*1 + 6	id $	S_5
$ *E*1 + 6 id_5	$	r_6
$ *E*1 + 6 F3	$	r_4
$ *E*1 + 6 T9	$	r_1
$ *E*1	$	Accept

6.5.2. LR Grammars

An LR grammar is a grammar for which one can construct a parsing table. A grammar is LR if a left-to-right shift—reduce parser can recognize handles when they appear on the top of stack.

An LR parser does not need to examine the entire stack for a handle, the state symbol on the top of the stack contains all the information it needs. It can also examine the next k input symbol to help make a decision. The cases $k = 0$ and $k = 1$ are of practical interest so we only consider those cases here.

6.5.3. Augmented Grammar

Let A given grammar $G = (V_n, V_t, P, S)$

where V_n : Non-emtpy finite set of non-terminals.

V_t : Non-empty finite-set of terminals.

S : Starting non-terminal, $S \in V_N$.

P : Productions or rules of the grammar.

Now let us modify the above grammar by adding a unit production

$$S' \to S$$

to the grammar and now starting non-terminal becomes S' and grammar becomes

$$G' = (V_n \cup \{S'\}, V_t, P \cup \{S' \to S\}, S')$$

The grammar G' is called augmented grammar. We generally construct the augmented grammar to make it explicitly clear to parser when to accept the string.

6.5.4. Constructing the SLR Parsing Tables (or Tables for LR(0) Parsers)

[MDU 2007]

To construct the parsing table we must find LR(0)/LR items (or some time we say canonical collection of sets of LR(0) items).

Let us see some basics to calculate canonical collection set:

1. An item: of a grammar G is a production of G with a dot at some position of the right side, that is if $T \to XYZ$ is a production of G then items of G are:

$$T \to \cdot XYZ$$
$$T \to X \cdot YZ$$
$$T \to XY \cdot Z$$
$$T \to XYZ \cdot$$

If G has $\in$-production, $T \to \cdot$ is an item of G. Intuitively, an item indicates how much of a production has been so far in a parsing process, **that is $T \to X \cdot YZ$ indicates that a string derivable from X has been seen so far on the input and we hope to see a string derivable from YZ next on the input.**

2. The Closure operation on a set of items: If I is a set of items for a grammar then **closure** (I) is a set of items constructed from I by the following two rules:

1. Initially, every LR(0) item in I is added to closure(I).
2. If $A \to \alpha.B\beta$ is in closure(I) and $B \to \gamma$ is a production rule of G; then $B \to .\gamma$ will be in the closure(I).We will apply this rule until no more new LR(0) items can be added to closure(I).

Let us consider the following example to clear the concept.

$$E' \to E$$
$$E \to E + T$$
$$E \to T$$

$$T \rightarrow T * F$$
$$T \rightarrow F$$
$$F \rightarrow (E)$$
$$F \rightarrow id$$

If set $I = \{E' \rightarrow \cdot E\}$ then closure (I) contains the following seven items:

$$E' \rightarrow \cdot E \quad \leftarrow \text{kernel item F}$$
$$E \rightarrow \cdot E + T$$
$$E \rightarrow \cdot T$$
$$T \rightarrow \cdot T * F$$
$$T \rightarrow \cdot F$$
$$F \rightarrow \cdot (E)$$
$$F \rightarrow \cdot id$$

3. The Go to Operation. If I is the set of items and X is a grammar symbol (terminal or non-terminal), then goto (I, X) is defined to be:

go to (I, X) = closure ($\{[A \rightarrow \alpha X \cdot \beta]\}$) such that $[A \rightarrow \alpha \cdot X\beta]$ is in I.

For example I contains the following items:

$$I = \{ E' \rightarrow .E, \; E \rightarrow .E + T, \; E \rightarrow .T,$$
$$T \rightarrow .T*F, \; T \rightarrow .F,$$
$$F \rightarrow .(E), \; F \rightarrow .id \}$$

$$\text{goto}(I,E) = \{ E' \rightarrow E., E \rightarrow E. + T \}$$
$$\text{goto}(I,T) = \{ E \rightarrow T., T \rightarrow T.*F \}$$
$$\text{goto}(I,F) = \{ T \rightarrow F. \}$$
$$\text{goto}(I,() = \{ F \rightarrow (.E), E \rightarrow .E+T, E \rightarrow .T, T \rightarrow .T*F, T \rightarrow .F,$$
$$F \rightarrow .(E), F \rightarrow .id \}$$
$$\text{goto}(I,id) = \{ F \rightarrow id. \}$$

4. Constructing the Canonical Collection of sets of LR(0) items. A canonical collection of sets of items for an augmented grammar, C is constructed as follows :

(*i*) The first set in C is the closure of $\{[S' \rightarrow \cdot S]\}$, where S is starting symbol of original grammar and S' is the starting non-terminal of augmented grammar.

(*ii*) For each set I in C and each grammar symbol X where go to (I, X) is non-empty and not in C add the set go to (I, X) to C.

We can understand above steps by following algorithm:

```
begin
     C: = {Closure ({[S' → ·S]}) } ;
    repeat
        for each item set I ∈ C:
        for each grammar symbol X:
            if go to (I, X) ! = 0 and go to (I,
X) ∉ C then
            add go to (I, X) to C ;
        until no new sets of items can be added
to C ;
    return C ;
  end.
```

Let us consider an example to make the concept of canonical collection, clear.

Let us consider the following grammar:

$$E \rightarrow E + T/T$$
$$T \rightarrow T * F/F$$
$$F \rightarrow id$$

Now we will find out canonical collection as follows:

Augmented grammar is:

$$S \rightarrow E$$
$$E \rightarrow E + T/T$$
$$T \rightarrow T * F/F$$
$$F \rightarrow id$$

Initially $C = \{\text{closure of } (S \rightarrow \cdot E)\}$

$= \{$

$S \rightarrow \cdot E$

$E \rightarrow \cdot E + T$

$E \rightarrow \cdot T$

$T \rightarrow \cdot T * F$

$T \rightarrow \cdot F$

$F \rightarrow \cdot id$

$\}$. (Let us say this set I_0)

Now we will calculate go to (I_0, X), where X is grammar symbol:

go to (I_0, E) = closure $(\{S \rightarrow E\cdot$
$E \rightarrow E \cdot +T\})$
$= \{S \rightarrow E\cdot$
$E \rightarrow E \cdot +T\}$ (Let say it I_1)

I_1 ! = 0 and $I_1 \notin C$ then

$$\boxed{C = \{I_0, I_1\}}$$

go to (I, T) = closure $(\{E \rightarrow T\cdot, T \rightarrow T\cdot * F,\}$
$= \{E \rightarrow T\cdot, T \rightarrow T\cdot * F\}$ (Let say it I_2)

$I_2 \notin C$ so now

$$\boxed{C = \{I_0, I_1, I_2\}}$$

go to (I_0, F) = closure $(\{T \rightarrow F\cdot\})$
$= \{T \rightarrow F\cdot\}$ (say it I_3)

$I_3 \notin C$ then

$$\boxed{C = \{I_0, I_1, I_2, I_3\}}$$

go to (I_0, id) = closure $(\{F \rightarrow id\ \cdot\})$
$= \{F \rightarrow id\ \cdot\}$ (say it I_4)

$I_4 \notin C$ so

$$\boxed{C = \{I_0, I_1, I_2, I_3, I_4\}}$$

This is the end of first iteration and now we will calculate go to (I, X), where $I = \{I_1, I_2, I_3, I_4\}$ and X is grammar symbol.

go to $(I_1, E) = \phi$ (since there is no production in I_1 which contains $\cdot E$ in its right-side)

go to $(I_1, T) = \phi$

go to $(I_1, F) = \phi$

go to $(I_1, id) = \phi$

go to $(I_1, *) = \phi$

go to $(I_1, +)$ = Closure $(\{E \rightarrow E + \cdot T\})$

$= \{E \rightarrow E + \cdot T, T \rightarrow \cdot T * F, T \rightarrow \cdot F, F \rightarrow \cdot id\}$ (say it I_5)

$I_5 \notin C$

So $C = \{I_0, I_1, I_2, I_3, I_4, I_5\}$

go to $(I_2, E) = \phi$

go to $(I_2, T) = \phi$

go to $(I_2, F) = \phi$

go to $(I_2, id) = \phi$

go to $(I_2, +) = \phi$

go to $(I_2, *)$ = Closure $(\{T \rightarrow T * \cdot F\})$

$= \{T \rightarrow T * \cdot F, F \rightarrow \cdot id\}$ (say it I_6)

$I_6 \; ! \neq \phi$ and $I_6 \notin C$ so now C becomes

$$C = \{I_0, I_1, I_2, I_3, I_4, I_5, I_6\}.$$

go to $(I_3, E) = \phi$

go to $(I_3, T) = \phi$

go to $(I_3, F) = \phi$

go to $(I_3, id) = \phi$

go to $(I_3, +) = \phi$

go to $(I_3, *) = \phi$

go to $(I_4, E) = \phi$

go to $(I_4, T) = \phi$

go to $(I_4, F) = \phi$

go to $(I_4, id) = \phi$

go to $(I_4, +) = \phi$

go to $(I_4, *) = \phi$

Now at the end of second iteration $C = \{I_0, I_1, I_2, I_3, I_4, I_5, I_6\}$ and in third iteration we will calculate go to (I, X) where $I = \{I_5, I_6\}$.

Now go to $(I_5, E) = \phi$

go to (I_5, T) = Closure $(\{E \rightarrow E + T \cdot, T \rightarrow T \cdot * F\})$

$= \{E \rightarrow E + T \cdot$

$T \rightarrow T \cdot * F\}$ (say I_7)

$I_7 \notin C$ then $$C = \{I_0, I_1, I_2, I_3, I_4, I_5, I_6, I_7\}$$

$$\text{go to } (I_5, F) = \text{Closure } (\{T \to F \cdot \}).$$
$$= \{T \to F\cdot\} \quad \text{(it is same as } I_3)$$
$$\text{go to } (I_5, id) = \text{Closure } (\{F \to id\cdot\})$$
$$= \{F \to id\cdot\} \quad \text{(same as } I_4)$$
$$\text{go to } (I_5, *) = \phi$$
$$\text{go to } (I_5, +) = \phi$$
$$\text{go to } (I_6, E) = \phi$$
$$\text{go to } (I_6, T) = \phi$$
$$\text{go to } (I_6, F) = \text{Closure } (\{T \to T * F\cdot\})$$
$$= \{T \to T * F\cdot\} \quad \text{(say it } I_8)$$

$I_8 \notin C$ so now $C = \{I_0, I_1, I_2, I_3, I_4, I_5, I_6, I_7, I_8\}$

$$\text{go to } (I_6, id) = \text{Closure } (\{F \to id.\}) = \{F \to id.\} \text{ (same as } I_4)$$
$$\text{go to } (I_6, *) = \phi$$
$$\text{go to } (I_6, +) = \phi$$

So after end of this third iteration C is

$$\boxed{C = \{I_0, I_1, I_2, I_3, I_4, I_5, I_6, I_7, I_8\}}$$

$$\text{go to } (I_7, E) = \phi$$
$$\text{go to } (I_7, T) = \phi$$
$$\text{go to } (I_7, F) = \phi$$
$$\text{go to } (I_7, id) = \phi$$
$$\text{go to } (I_7, *) = \text{Closure } (\{T \to T * \cdot F, F \to \cdot id\}$$
$$= \{T \to T * \cdot F, F \to \cdot\, id\} \quad \text{(same as } I_6)$$
$$\text{go to } (I_7, +) = \phi$$

So still $\boxed{C = \{I_0, I_1, I_2, I_3, I_4, I_5, I_6, I_7, I_8\}}$

Now
$$\text{go to } (I_8, E) = \phi$$
$$\text{go to } (I_8, T) = \phi$$
$$\text{go to } (I_8, F) = \phi$$
$$\text{go to } (I_8, id) = \phi$$
$$\text{go to } (I_8, +) = \phi$$
$$\text{go to } (I_8, *) = \phi$$

So after fourth iteration C becomes

$$C = \{I_0, I_1, I_2, I_3, I_4, I_5, I_6, I_7, I_8\}$$

which is similar to C, which we get after third iteration, so here we must stop the process and canonical collection C is:

$$\boxed{C = \{I_0, I_1, I_2, I_3, I_4, I_5, I_6, I_7, I_8\}}$$

5. DFA for Canonical Collection. "By the definition of closure we know that closure of any grammar production $S \rightarrow \cdot E$ will contain all those production which can be derived by E including $S \rightarrow \cdot E$ itself."

TIPS

We also know that there is no significance of · operator (dot) in $S \rightarrow \cdot E$ but it gives the information that a string derivable from E can be seen and in $S \rightarrow E\cdot$, dot operator gives information that string derivable from E is already seen.

So on the basis of above two observations a NFA of items of any grammar can be constructed. Let us do it for above discussed grammar, that is closure ($S \rightarrow \cdot E$) is :

$$\begin{aligned} S &\rightarrow \cdot E \\ E &\rightarrow \cdot E + T \\ E &\rightarrow \cdot T \\ T &\rightarrow \cdot T * F \\ T &\rightarrow \cdot F \\ F &\rightarrow \cdot id \end{aligned}$$

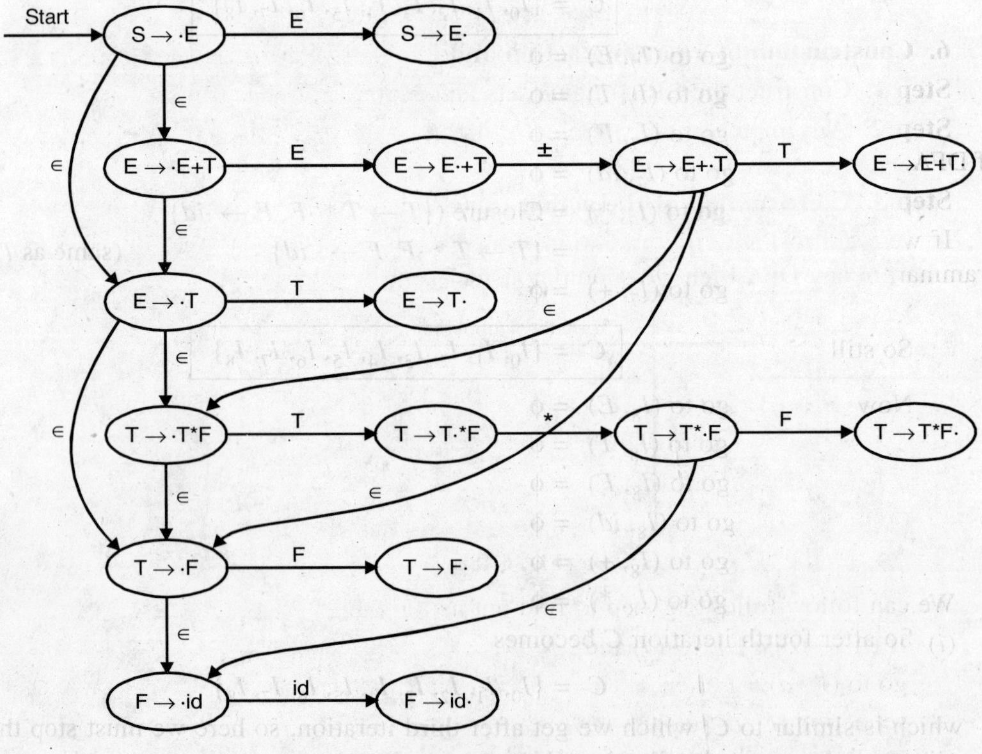

Fig. 6.16.

Now we can convert this NFA to DFA by subset construction and method of getting DFA from NFA, which we have seen in chapter 2.

The DFA is as follows:

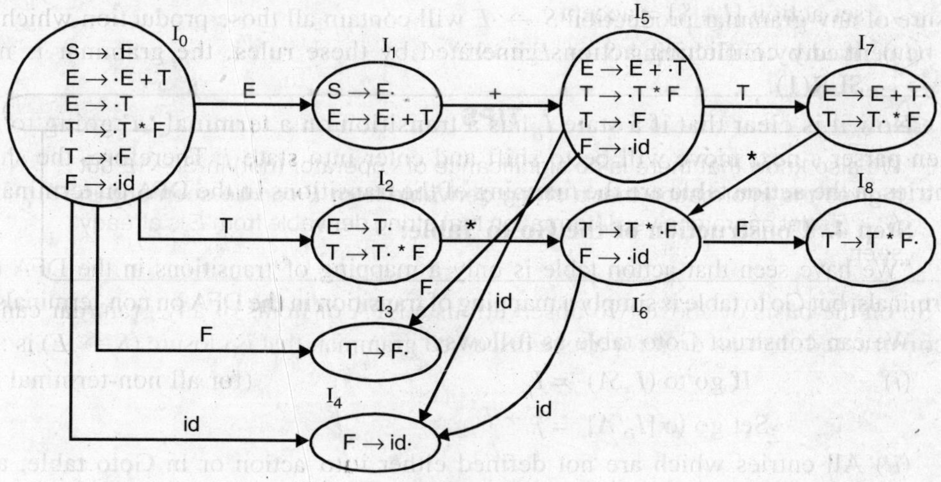

Fig. 6.17.

If we see carefully then it becomes clear that states of DFA are nothing but canonical collection set of the item set of grammar that is

$$C = \{I_0, I_1, I_2, I_3, I_4, I_5, I_6, I_7, I_8\}$$

6. Construction of a parsing action and Go To Table for an SLR parser :

Step 1. Construct the LR(0) items sets and canonical collection set.

Step 2. Any member of canonical collection set I_i corresponds to a state i of DFA.

Step 3. Construction of the action table:

If we want to know the transition from any state for any terminal symbol of grammar, in any DFA then we should consider action table:

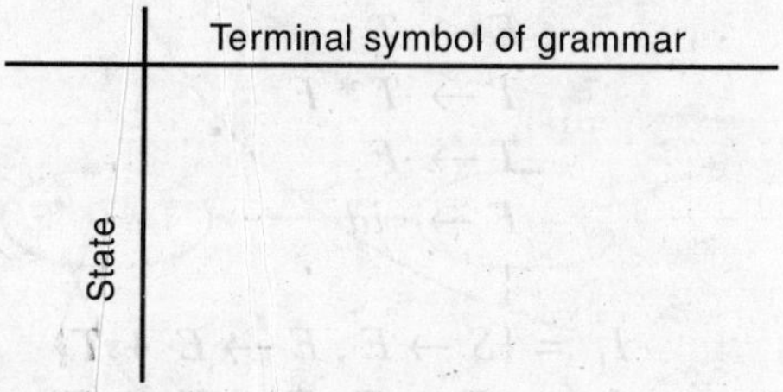

Fig. 6.18.

We can follow following step to find out action table:

(*i*) If $(A \to \alpha \cdot a\, \beta)$ is in I_i and
go to $(I_i, a) = I_j$ then
set action $[I_i, a]$ to shift j (S_j)

(*ii*) If $(A \to \alpha\cdot)$ is in I_i
then for every b in follow (A)
set action $[I_i, b]$ = reduce $A \to \alpha$ (R_K, where k is the number of production $A \to \alpha$ in grammar)

(*iii*) If $(S' \rightarrow S\cdot)$ is in I_i

set action $[I_i, \$]$ = accept.

(*iv*) If any conflicting actions generated by these rules, the grammar is not SLR(1).

Now it is clear that if a state I_i has a transition on a terminal '*a*' going to I_j, then parser's next move will be to shift and enter into state *j*. Therefore, the shift entries in the action table are the mapping of the transitions in the DFA on terminals.

Step 4. Construction of the Go to Table:

"We have seen that action table is only a mapping of transitions in the DFA on terminals, but Go to table is simply a mapping of transition in the DFA on non-terminals".

We can construct Goto table as follows:

(*i*) If go to $(I_i, A) = I_j$ (for all non-terminal *A*)

Set go to $[I_i, A] = j$

(*ii*) All entries which are not defined either into action or in Goto table, are considered as error ( ).

(*iii*) $[S' \rightarrow \cdot S]$ is the initial state.

Example 6.6. Construct a SLR parsing table for the grammar

$$S \rightarrow E$$
$$E \rightarrow E + T/T$$
$$T \rightarrow T * F/F$$
$$F \rightarrow id$$

Solution. We have already seen canonical collection set for above grammar,

$$C = \{I_0, I_1, I_2, I_3, I_4, I_5, I_6, I_7, I_8\}$$

where

$$I_0 = \{S \rightarrow \cdot E$$
$$E \rightarrow \cdot E + T$$
$$E \rightarrow \cdot T$$
$$T \rightarrow \cdot T * F$$
$$T \rightarrow \cdot F$$
$$F \rightarrow \cdot\, id$$
$$\}$$

$$I_1 = \{S \rightarrow E\cdot, E \rightarrow E\cdot + T\}$$
$$I_2 = \{E \rightarrow T\cdot, T \rightarrow T\cdot * F\}$$
$$I_3 = \{T \rightarrow F\cdot\}$$
$$I_4 = \{F \rightarrow id\cdot\}$$
$$I_5 = \{E \rightarrow E + \cdot T$$
$$T \rightarrow \cdot T * F$$
$$T \rightarrow \cdot F$$
$$F \rightarrow \cdot id$$
$$\}$$

$$I_6 = \{T \rightarrow T * \cdot F$$
$$F \rightarrow \cdot id$$
$$\}$$

$$I_7 = \{E \rightarrow E + T\cdot$$
$$T \rightarrow T\cdot * F\}$$
$$I_8 = \{T \rightarrow T * F\cdot\}.$$

We can see DFA also as follows:

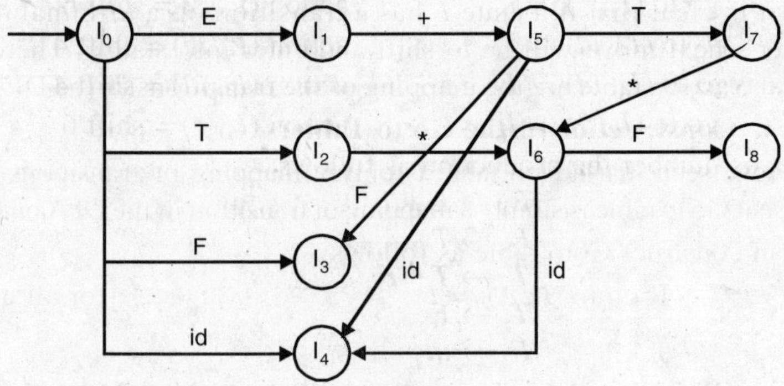

Fig. 6.19.

Transition can be seen as follows:

$$I_0 = \{S \rightarrow \cdot E$$
$$E \rightarrow \cdot E + T$$
$$E \rightarrow \cdot T * F$$
$$T \rightarrow \cdot F$$
$$F \rightarrow \cdot id\,\}$$

and $$I_1 = \{S \rightarrow E\cdot, E \rightarrow E\cdot + T\}$$

As we see position of dot operator in $I_0 = \{S \rightarrow \cdot E$ and in $S \rightarrow \cdot E + T\}$.

and $$I_1 = \{S \rightarrow E\cdot, E \rightarrow E\cdot + T\}$$

Clearly in I_0 the string derived from E are expected and in I_1 the string derived from E are parsed so clearly 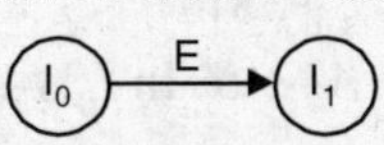

is justified. All the transition can be justified in this manner.

Now we are in situation to construct action and goto table:

	Action Table				Go to Table		
	id	+	*	$	*E*	*T*	*F*
I_0	S_4				1	2	3
I_1		S_5		Accept			
I_2		R_2	S_6	R_2			
I_3		R_4	R_4	R_4			
I_4		R_5	R_5	R_5			
I_5	S_4					7	3
I_6	S_4						8
I_7		R_1	S_6	R_1			
I_8		R_3	R_3	R_3			

Now let us analyse the entries in the table:

Here, first we will analyse the shift move of the action table:

go to (I_0, *id*) = I_4 So action (I_0, *id*) = shift 4
go to (I_1, +) = I_5 action (I_1, +) = shift 5
go to (I_2, *) = I_6 action (I_2, *) = shift 6
go to (I_5, *id*) = I_4 action (I_5, *id*) = shift 4
go to (I_6, *id*) = I_4 action (I_6, *id*) = shift 4
go to (I_7, *) = I_6 action (I_7, *) = shift 6

Now let us number the production as follows:

$E \to E + T$ ...(1)
$E \to T$ ...(2)
$T \to T * F$ ...(3)
$T \to F$ ...(4)
$F \to id$ ...(5)

Now we will analyse the reduction move of the action table:

We know that for finding reduction move we need to follow of *E*, *T* and *F* so let us calculate follow, as follows:

First (E) = First ($E + T$) $\cup$ First (T)
First (T) = First ($T * F$) $\cup$ First (F)
First (F) = $\{id\}$
So First (T) = $\{id\}$
First (E) = $\{id\}$
Follow (E) = First ($+T$) $\cup$ $\{\$\}$ = $\{+, \$\}$
Follow (T) = First ($* F$) $\cup$ Follow (E)
= $\{*, +, \$\}$
Follow (F) = $\{*, +, \$\}$

Now let us search production like $A \to \alpha$. In $I_0, I_1, ..., I_8$ and we will do reduction operation as follows:

I_0 : no $A \to \alpha$. kind production
I_1 : have $S \to E\cdot$ and follow (S) = \$ so action [I_1, \$] = Accept
I_2 : $E \to T\cdot$ is there and follow (E) = {+, \$}, So action [I_2, +] = R_2, action [I_2, \$] = R_2
I_3 : $T \to F\cdot$ is there and follow (T) = {+, *, \$}, so action [I_3, +] = R_4, action [I_3, *] = R_4, action [I_3, \$] = R_4
I_4 : $F \to id.$ is there and follow (F) = {+, *, \$} so action [I_4, +] = R_5, action [I_4, *] = R_5, action [I_5, \$] = R_5
I_5 : no $A \to \alpha \cdot$ like production
I_6 : no $A \to \alpha \cdot$ like production
I_7 : $E \to E + T\cdot$ is there and follow (E) = {+, \$} so action [I_7, +] = R_1, action [I_7, \$] = R_1
I_8 : $T \to T * F$, and follow (T) = {+, *, \$} so action [I_8, +] = R_3, action [I_8, *] = R_3, action [I_8, \$] = R_3.

Here analysis of action table finished now we will consider the Go to table:

go to $(I_0, E) = I_1$ since I_0 have $E \to \cdot E + T$ and $S \to \cdot E$
and I_1 is having $E \to E \cdot +T$ and $S \to E\cdot$
so Goto $[I_0, E] = 1$

In same manner

go to $(I_0, T) = I_2$ so Go to $[I_0, T] = 2$
go to $(I_0, F) = I_3$ so Go to $[I_0, F] = 3$
go to $(I_5, T) = I_7$ so Go to $[I_5, T] = 7$
go to $(I_5, F) = I_3$ so Go to $[I_5, F] = 3$
go to $(I_6, F) = I_8$ so Go to $[I_6, F] = 8$.

Example 6.7. For the following grammar

$$E \to E + T/T$$
$$T \to T * F/F$$
$$F \to (E)/id$$

Construct the LR(0) canonical collection and also design the SLR parsing table.

Solution. Given grammar

$$E \to E + T/T$$
$$T \to T * F/F$$
$$F \to (E)/id$$

Augmented grammar is

$$S \to E$$
$$E \to E + T/T$$
$$T \to T * F/F$$
$$F \to (E)/id$$

I_0 = {
$S \to \cdot E$
$E \to \cdot E + T$
$E \to \cdot T$
$T \to \cdot T * F$
$T \to \cdot F$
$F \to \cdot (E)$
$F \to \cdot id$
}

$S \to E\cdot$ describes the acceptance.

goto:

I_1 = goto $(I_0, E) = \{S \to E\cdot, E \to E\cdot + T\}$
I_2 = goto $(I_0, T) = \{E \to T\cdot, T \to T\cdot * F\}$
I_3 = goto $(I_0, F) = \{T \to F\cdot\}$
I_4 = goto $(I_0, () = \{F \to (\cdot E), E \to \cdot E + T, E \to \cdot T, T \to \cdot T * F, T \to \cdot F,$
$F \to \cdot(E), F \to \cdot id\}$
I_5 = goto $(I_0, id) = \{F \to id\cdot\}$
I_6 = goto $(I_1, +) = \{E \to E + \cdot T, T \to \cdot T * F, T \to \cdot F, F \to \cdot(E), F \to \cdot id\}$

$$
\begin{aligned}
I_7 &= \text{goto}(I_2, *) = \{T \to T * \cdot F,\ F \to \cdot(E),\ F \to \cdot id\} \\
I_8 &= \text{goto}(I_4, E) = \{F \to (E\cdot),\ E \to E\cdot + T\} \\
I_9 &= \text{goto}(I_7, F) = \{T \to T * F\cdot\} \\
I_{11} &= \text{goto}(I_8,)\,) = \{F \to (E)\cdot\} \\
I_{10} &= \text{goto}(I_9, F) = \{T \to T\cdot * F\}
\end{aligned}
$$

Let us see expanded item sets:

$I_0 = S \to \cdot E$	$(E, 1)$	$I_1 = \text{goto}(I_0, E)$	
$E \to \cdot E + T$	$(E, 1)$		
$E \to \cdot T$	$(T, 2)$	$I_2 = \text{goto}(I_0, T)$	
$E \to \cdot T * F$	$(T, 2)$		
$T \to \cdot F$	$(F, 3)$	$I_3 = \text{goto}(I_0, F)$	
$F \to \cdot(E)$	$((, 4)$		
$F \to \cdot id$	$(id, 5)$		
$I_1 = S \to E\cdot$			
$E \to E\cdot + T$	$(+, 6)$		
$I_2 = E \to T\cdot$			
$T \to T\cdot * F$	$(*, 7)$		
$I_3 = F \to id\cdot$			
$I_4 = F \to (\cdot E)$	$(E, 8)$		$(I_8 = \text{goto}(I_4, E)$
$E \to \cdot E + T$	$(E, 8)$		
$E \to \cdot T$	$(T, 2)$		$I_2 = \text{goto}(I_4, T)$
$T \to \cdot T * F$	$(T, 2)$		
$T \to \cdot F$	$(F, 3)$		$I_3 = \text{goto}(I_4, F)$
$F \to \cdot(E)$	$((, 4)$		$I_4 = \text{goto}(I_4, ()$
$F \to \cdot id$	$(id, 5)$		$I_5 = \text{goto}(I_4, id)$
$I_5 = F \to id\cdot$			
$I_6 = \text{goto}(I_1, +)$			
$E \to E + \cdot T$	$(T, 9)$		$I_9 = \text{goto}(I_6, T)$
$T \to \cdot T * F$	$(T, 9)$		
$T \to \cdot F$	$(F, 3)$		$I_3 = \text{goto}(I_6, F)$
$F \to \cdot(E)$	$((, 4)$		$I_4 = \text{goto}(I_6, ()$
$F \to \cdot id$	$(id, 5)$		$I_5 = \text{goto}(I_6, id)$
$I_7 = T \to T * \cdot F$	$(F, 10)$		$I_{10} = \text{goto}(I_7, *)$
$F \to \cdot(E)$	$((, 4)$		$I_4 = \text{goto}(I_7, ()$
$F \to \cdot id$	$(id, 5)$		$I_5 = \text{goto}(I_7, id)$
$I_8 = F \to (E\cdot)$	$( ), 11)$		$I_{11} = \text{goto}(I_8,)\,)$
$E \to E\cdot + T$	$(+, 6)$		$I_6 = \text{goto}(I_8, +)$
$I_9 = E \to E + T\cdot$			
$T \to T\cdot * F$	$(*, 7)$		$I_7 = \text{goto}(I_9, *)$
$I_{10} = T \to T * F\cdot$			
$I_{11} = F \to (E)\cdot$			

Now Detail FA is as follows:

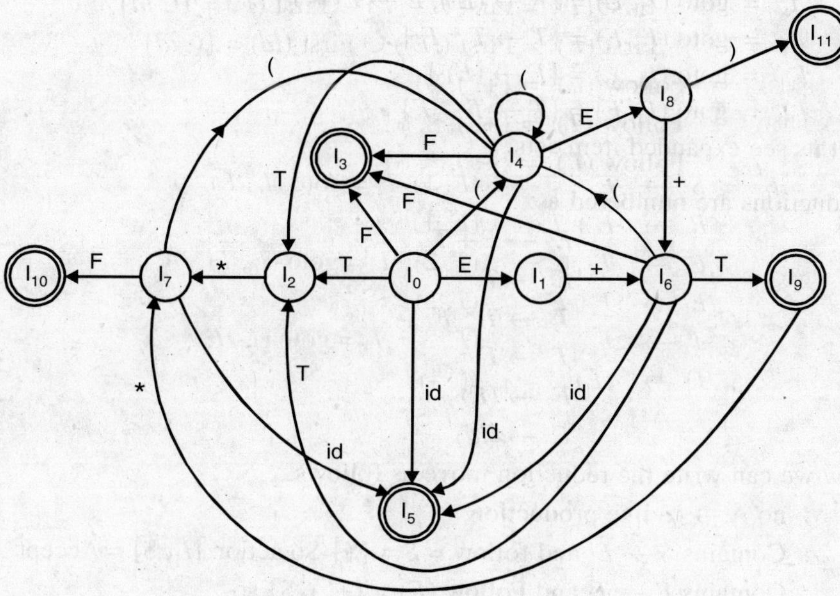

Fig. 6.20.

Construction of LR parser:

goto (I_0, ( ) = I_4, I_0: $F \to \cdot(E)$ and compare with $A \to \alpha\cdot\, a\beta$
So action [I_0, (] = shift 4

goto (I_0, id) = I_5
I_0: $F \to \cdot id$
action [I_0, id] = shift 5

goto (I_1, +) = I_6
I_1: $E \to E\cdot + T$
action [I_1, +] =shift 6

goto (I_2, *) = I_7
I_2: $T \to T\cdot * f$
action [I_2, *] = shift 7

goto (I_8,)) = I_{11}
I_8: $F \to (E\cdot)$
action [I_8,)] = shift 11

Now let us calculate Follow of E, T, F for the reduce operation:

$$E \to E + T/T$$
$$T \to T * F/F$$
$$F \to (E)/id$$

$$\text{First }(E) = \text{First }(E + T) \cup \text{First }(T) = \{(, id\}$$

$$\text{First }(T) = \text{First }(T * F) \cup \text{First }(F) = \{(, id\}$$

$$\text{First }(F) = \text{First }((E)) \cup \text{First }(id) = \{(, id\}$$

$$\text{Follow }(E) = \{+,), \$\}$$

$$\text{Follow }(T) = \{+,), *, \$\}$$

$$\text{Follow }(F) = \{+,), *, \$\}$$

Productions are numbered as:

$E \to E + T$ ...(1)

$E \to T$ ...(2)

$T \to T * F$ ...(3)

$T \to F$ ...(4)

$F \to (E)$ ...(5)

$F \to (id)$ ...(6)

Now we can write the reduction move as follows:

I_0 : no A $\to \alpha\cdot$ like production

I_1 : Contains $S \to E\cdot$ and follow $= S = \{\$\}$ So action $[I_1, \$]$ = Accept

I_2 : Contains $E \to T\cdot$ and Follow $(E) = \{+,), \$\}$ so action $[I_2, +] = r_2$, action $[I_2,)] = r_2$, action $[I_2, \$] = r_2$.

I_3 : Contains $T \to F\cdot$ and Follow $(T) = \{+,), *, \$\}$ so action $[I_3, +] = r_4$, action $[I_3,)] = r_4$, action $[I_3, *] = r_4$ and action $[I_3, \$] = r_4$

I_4 : No A $\to \alpha\cdot$ like production

I_5 : Contains $F \to id\cdot$ and Follow $(F) = \{+,), *, \$\}$ so action $[I_5, +] = r_6$, action $[I_5,)] = r_6$, action $[I_5, *] = r_6$, action $[I_5, \$] = r_6$

I_6, I_7 : does't have any $A \to \alpha \cdot$ like production.

I_8 : no $A \to \alpha \cdot$

I_9 : Contains $T \to T * F\cdot$ and Follow $(T) = \{+,), *, \$\}$ so action $[I_9, +] = r_3$, action $[I_9,)] = r_3$, action $[I_9, *] = r_3$, action $[I_9, \$] = r_3$.

I_{10} : no $A \to \alpha.$

I_{11} : Contains $F \to (E)\cdot$ and Follow $(F) = \{+,), *, \$\}$ so action $[I_{11}, +]$ = action $[I_{11},)]$ = action $[I_{11}, *]$ = action $[I_{11}, \$] = r_5$.

goto $(I_0, E) = I_1$ so Goto $[I_0, E] = 1$

goto $(I_0, T) = I_2$ so Goto $[I_0, T] = 2$

goto $(I_0, F) = I_3$ so Goto $[I_0, F] = 3$

goto $(I_4, E) = I_8$ so Goto $[I_4, E] = 8$

goto $(I_4, T) = I_2$ so Goto $[I_4, T] = 2$

goto $(I_4, F) = I_3$ so Goto $[I_4, F] = 3$

goto $(I_6, T) = I_9$ so Goto $[I_6, T] = 9$

goto $(I_6, F) = I_3$ so Goto $[I_6, F] = 3$

goto $(I_7, F) = I_{10}$ so Goto $[I_7, F] = 10$

Now by making the above discussed entries, action and Goto tables are as follows :

State	Action table						Goto table		
	id	+	*	(	)	$	*E*	*T*	*F*
I_0	S_5			S_4			1	2	3
I_1		S_6				Accept			
I_2		r_2	S_7		r_2	r_2			
I_3		r_4	r_4		r_4	r_4			
I_4	S_5			S_4			8	2	3
I_5		r_6	r_6		r_6	r_6			
I_6	S_5			S_4				9	3
I_7	S_5			S_5					10
I_8		S_6			S_{11}				
I_9		r_1	S_7		r_1	r_1			
I_{10}		r_3	r_3		r_3	r_3			
I_{11}		r_5	r_5		r_5	r_5			

6.5.5. Limitation of SLR Parser

Let us consider an example for the following grammar:

$$S \rightarrow AaAb$$
$$S \rightarrow BbBa$$
$$A \rightarrow \in$$
$$B \rightarrow \in$$

Augmented grammar is as follows:

$$S' \rightarrow S$$
$$S \rightarrow AaAb$$
$$S \rightarrow BbBa$$
$$A \rightarrow \in$$
$$B \rightarrow \in$$

Canonicals of above grammar for LR(0) items are:

$$C = \{I_0, I_1, I_2, I_3, I_4, I_5, I_6, I_7, I_8, I_9\}$$

where

$$I_0 = \{S' \rightarrow \cdot S, S \rightarrow \cdot AaAb, S \rightarrow \cdot BbBa, A \rightarrow \cdot, B \rightarrow \cdot\}$$
$$I_1 = \{S' \rightarrow S\cdot\}$$
$$I_2 = \{S \rightarrow A \cdot aAb\}$$
$$I_3 = \{S \rightarrow B \cdot bBa\}$$
$$I_4 = \{S \rightarrow Aa\cdot Ab, A \rightarrow \cdot\}$$
$$I_5 = \{S \rightarrow Bb \cdot Ba, B \rightarrow \cdot\}$$
$$I_6 = \{S \rightarrow AaA\cdot b\}$$
$$I_7 = \{S \rightarrow BbB\cdot a\}$$

$$I_8 = \{S \rightarrow AaAb\cdot\}$$
$$I_9 = \{S \rightarrow BbBa\cdot\}$$

And If we find out the parsing table for the above grammar then it will be as follows:

	Action table			Goto table		
	a	***b***	***$***	***S***	***A***	***B***
I_0	R_3 or R_4	R_3 or R_4		1	2	3
I_1			Accept			
I_2	S_4					
I_3		S_5				
I_4	R_3	R_3			6	
I_5	R_4	R_4				7
I_6		S_8				
I_7	S_9					
I_8			R_1			
I_9			R_2			

where production are numbered as follows:

$S \rightarrow AaAb$ ...(1)

$S \rightarrow BbBa$ ...(2)

$A \rightarrow \in$ ...(3)

$B \rightarrow \in$ ...(4)

It is clear from the parsing table that

action $[I_0, a] = R_3 / R_4$

and action $[I_0, b] = R_3 / R_4$

are multiple entries, so it is possible that our parser will be in indeterministic situation which is not allowed in the computer science.

So it becomes very clear that SLR is less-powerful LR parser since SLR grammars constitute a small subset of context-free grammars. **(Here power of parser is measured by the number of grammars on which it can succeed).**

Problem of Multiple entries arises when reduction may not lead to the generation of previous right-most derivations.

TIPS

Let us consider production

$S \rightarrow AaAb$

here Follow of *A* are both '*a*' and 'b' so parser can do the reduction by the $A \rightarrow \in$ or $B \rightarrow \in$ in the state I_0 if the next input symbol happens to be either '*a*' or '*b*', because both '*a*' and '*b*' are in Follow (*A*), this is reduce/reduce conflict.

But reduction will be proper only when next input symbol is '*a*', if anyhow input is *b* then parsing fails. **These kinds of problems can be solved if list of terminals (lookheads) are attached** with the items.

Now LR(0) items will be appended in following form:

$$A \rightarrow \alpha \cdot X\beta, \underline{a} \leftarrow \text{lookhead}$$

Now it is called LR(1) item, because the length of the lookhead is one.

In general we can define LR(K) item where length of lookhead is K. LR(1) item is comprised of two parts, the LR(0) item and the lookhead associated with the item.

6.5.6. Construction of LR(1) Parsing Table [MDU 2008, 2009]

Parsing table for LR(1) parsers can be constructed in the same manner as we did in the LR(0) parsers:

Step 1. LR(1) item set of a grammar G is a production of G with a dot at some position of the right side and a lookhead is also attached to it as follows:

closure(I) is: (where I is a set of LR(1) items)

(*i*) every LR(1) item in I is in closure(I)

(*ii*) Repeat for every item of the form $A \rightarrow \alpha.B\beta, a$ in closure(I) and $B \rightarrow \gamma$ is a production rule of G; then $B \rightarrow .\gamma, b$ will be in the closure(I) for each terminal b in FIRST(βa).

Let us consider the following example.

$$E \rightarrow E + T/T$$
$$T \rightarrow T * F/F$$
$$F \rightarrow id$$

Augmented grammar is

$$S \rightarrow E$$
$$E \rightarrow E + T$$
$$E \rightarrow T$$
$$T \rightarrow T * F$$
$$T \rightarrow F$$
$$F \rightarrow id$$
$$I_0 = \{\text{Closure } (\{S \rightarrow \cdot E, \$\})\}$$

Let us calculate I_0 with a explanation as follows:

$$\text{Closure } (\{S \rightarrow \cdot E, \$\})$$

Compare $S \rightarrow \cdot E, \$$ with $A \rightarrow \alpha \cdot B\beta$, a then we find that $\alpha = \in$, $B = E$, $\beta = \in$ and $a = \$$ so we will add $E \rightarrow \cdot E + T$, First (\$) that is $E \rightarrow \cdot E + T, \$$ and $E \rightarrow \cdot T, \$$

Now at this moment I_0 becomes

$$\{S \rightarrow \cdot E, \$$
$$\text{Closure } (E \rightarrow \cdot E + T, \$$
$$E \rightarrow \cdot T, \$)\}$$

So let us consider $E \rightarrow \cdot E + T, \$$, here $\alpha = \in$, $B = E$, $\beta = +T$ and $a = \$$ so we will add $E \rightarrow \cdot E + T$, First ($+T\$$) and $E \rightarrow \cdot T$, First ($+T\$$). In $E \rightarrow \cdot T, \$$, $\alpha = \in$, $B = T$, $\beta = \in$ and $a = \$$ so we will add $T \rightarrow \cdot T * F, \$$ and $T \rightarrow \cdot F\$$ for $E \rightarrow \cdot T, \$$

So now I_0 becomes:

{

$$S \rightarrow \cdot E, \$$

$$E \rightarrow \cdot E + T, \$$$
$$E \rightarrow \cdot T, \$$$
$$\text{closure } (E \rightarrow \cdot E + T, +$$
$$E \rightarrow \cdot T, +$$
$$T \rightarrow \cdot T * F, \$$$
$$T \rightarrow \cdot F, \$$$
$$) \}.$$

Now for:

- $E \rightarrow \cdot E + T, +$ we will add $E \rightarrow \cdot E + T, +$ and $E \rightarrow \cdot T, +$
- $E \rightarrow \cdot T, +$ we will add $T \rightarrow \cdot T * F, +$ and $T \rightarrow \cdot F, +$
- $T \rightarrow \cdot T * F, \$$ we will add $T \rightarrow \cdot T * F, *$ and $T \rightarrow \cdot F, *$
- $T \rightarrow \cdot F, \$$ we will add $F \rightarrow \cdot id, \$$

So now I_0 becomes

$$\{$$
$$S \rightarrow \cdot E, \$$$
$$E \rightarrow \cdot E + T, \$$$
$$E \rightarrow \cdot T, \$$$
$$T \rightarrow \cdot T * F, \$$$
$$T \rightarrow \cdot F, \$$$
$$\text{Closure } (E \rightarrow \cdot E + T, +$$
$$E \rightarrow \cdot T, +$$
$$T \rightarrow \cdot T * F, +$$
$$T \rightarrow \cdot F, +$$
$$T \rightarrow \cdot T * F, *$$
$$T \rightarrow \cdot F, *$$
$$T \rightarrow \cdot id, \$)$$

Now for:

- $E \rightarrow \cdot E + T, +$ and $E \rightarrow \cdot T, +$ we will get same production as we got in previous state.
- $T \rightarrow \cdot T * F, +$ we will add $T \rightarrow \cdot T * F, *$ and $T \rightarrow \cdot F, *$ (which are already there).
- $T \rightarrow \cdot F, +$ we will add $F \rightarrow \cdot id +$
- $T \rightarrow \cdot T * F, *$ we will add $T \rightarrow \cdot T * F, *$ and $T \rightarrow \cdot F *$
- $T \rightarrow \cdot F, *$ we will add $F \rightarrow \cdot id, *$

So now I_0 is:

$$\{$$
$$S \rightarrow \cdot E, \$$$
$$E \rightarrow \cdot E + T, \$$$
$$E \rightarrow \cdot T, \$$$
$$T \rightarrow \cdot T * F, \$$$

$$T \rightarrow \cdot F, \$$$
$$E \rightarrow \cdot E + T, +$$
$$E \rightarrow \cdot T, +$$
$$T \rightarrow \cdot T * F, +$$
$$T \rightarrow \cdot F, +$$
$$T \rightarrow \cdot T * F, *$$
$$T \rightarrow \cdot F, *$$
$$T \rightarrow \cdot id, \$$$

Closure ($T \rightarrow \cdot T * F, *$

$$T \rightarrow \cdot F, *$$
$$F \rightarrow \cdot id+$$
$$T \rightarrow \cdot T * F, *$$
$$T \rightarrow \cdot F, *$$
$$F \rightarrow \cdot id, * \quad) \quad \}$$

Now finally we can say that I_0 is as follows, since every production of closure ( ) is either repeated or not eligible for calculation.

$$\begin{aligned} I_0 = \{ \; & S \rightarrow \cdot E, \$ \\ & E \rightarrow \cdot E + T, \$ \\ & E \rightarrow \cdot T, \$ \\ & T \rightarrow \cdot T * F, \$ \\ & T \rightarrow \cdot F, \$ \\ & E \rightarrow \cdot E + T, + \\ & E \rightarrow \cdot T, + \\ & T \rightarrow \cdot T * F, + \\ & T \rightarrow \cdot F, + \\ & T \rightarrow \cdot T * F, * \\ & T \rightarrow \cdot F, * \\ & F \rightarrow \cdot id, \$ \\ & F \rightarrow id, + \\ & F \rightarrow id, * \quad \} \end{aligned}$$

We can also write as follows:

$$\begin{aligned} I_0 = \{ \; & S \rightarrow \cdot E, \$ \\ & E \rightarrow \cdot E + T, \$/+ \\ & E \rightarrow \cdot T, \$/+ \\ & T \rightarrow \cdot T * F, \$/+/* \\ & T \rightarrow \cdot F, \$/+/* \\ & F \rightarrow \cdot id, \$/+/* \\ & \quad \}. \end{aligned}$$

Step 2. Constructing the Canonical Collection of Sets of LR(1) items: A canonical collection of sets of items for an augmented grammar, C is constructed as follows:

(*i*) The first set in C is the closure of $\{[S' \rightarrow \cdot S, \$]\}$, where S is the starting non-terminal of normal grammar.

(*ii*) For each set I in C and each grammar symbol X where goto (I, X) is non-empty and not in C, add the set goto (I, X) in C.

We can construct C by following algorithm:

```
begin
    C:= {closure ({[S' → S·], $});
    repeat
        for each item set I ∈ C:
            for each grammar symbol X:
                if goto (I, X) ! = 0 and goto (I, X) ∉ C then
            add goto (I, X) to C;
        until no new sets of items can be added to C;
        return C;
    end.
```

Now let us consider example, which we consider the limitations of SLR parser:

Augmented grammar is:

$$S' \rightarrow S$$
$$S \rightarrow AaAb$$
$$S \rightarrow BbBa$$
$$A \rightarrow \in$$
$$B \rightarrow \in$$

I_0 = Closure $(\{S' \rightarrow S, \$\})$

= $\{ S' \rightarrow \cdot S, \$$ (Compare with $A \rightarrow \alpha \cdot B\beta$, $\alpha = \in$, $B = S$, $\beta = \in$ and $a = \$$)

= $S \rightarrow \cdot AaAb\ \$$ so add

$S \rightarrow \cdot BbBa\ \$$ closure $(S \rightarrow \cdot AaAb, \$, S \rightarrow \cdot BbBa, \$)$

$A \rightarrow \cdot, a$ (By comparing $S \rightarrow \cdot AaAb$, \$ and $S \rightarrow \cdot BbBa$, \$ with $A \rightarrow \alpha \cdot B\beta$, a we get $A \rightarrow \cdot, a$ and $B \rightarrow \cdot, b$)

$B \rightarrow \cdot, b \}$

So I_0 = $S' \rightarrow .S, \$$

$S \rightarrow .AaAb, \$$

$S \rightarrow .BbBa, \$$

$A \rightarrow ., a$

$B \rightarrow ., b$

I_1 = goto $(\{I_0, S\})$ = closure $(\{S' \rightarrow S\cdot, \$\}) = \{S' \rightarrow S\cdot, \$\}$

I_2 = goto $(\{I_0, A\})$ = closure $(\{S \rightarrow A\cdot aAb, \$\}$

= $\{S \rightarrow A\cdot aAb, \$\}$

$I_3 \;=\; \text{goto}(\{I_0, B\}) = \text{closure}(\{S \to B{\cdot}bBa, \$\})$

$\;\;\;\;= \{S \to B{\cdot}bBa, \$\}$

$I_4 \;=\; \text{goto}(\{I_2, a_3)\} = \text{closure}(\{S \to Aa{\cdot}Ab, \$\})$

$\;\;\;\;= \{S \to Aa{\cdot}Ab, \$$

$\;\;\;\;\;\;A \to {\cdot}, b$

$\;\;\;\;\;\;\}$

$I_5 \;=\; \text{goto}(\{I_3, b\}) = \;\; \text{closure}(\{S \to Bb{\cdot}Ba, \$\})$

$\;\;\;\;= \{S \to Bb{\cdot}Ba, \$$

$\;\;\;\;\;\;B \to {\cdot}, a$

$\;\;\;\;\;\;\}.$

(I_5 is calculating for comparing $S \to Bb{\cdot}Ba$, \$ with $A \to \alpha{\cdot}B\beta$, a so $\alpha = Bb$, $B = B$, $\beta = a$ and $a = \$$.

So we will add $\;\; B \to {\cdot}$, First $(a\,\$)\cdot$)

$I_6 = \text{goto}(\{I_4, A\}) = \text{closure}(\{S \to AaA \cdot b, \$\})$

$= \{S \to AaA{\cdot}b, \$\}$

$I_7 = \text{goto}(\{I_5, B\}) = \text{closure}(\{S \to BbB \cdot a, \$\})$

$= \{S \to BbB{\cdot}a, \$\}$

$I_8 = \text{goto}(\{I_6, b\}) = \text{closure}(\{AaAb{\cdot}, \$\})$

$= \{S \to AaAb{\cdot}, \$\}$

$I_9 = \text{goto}(\{I_7, a\}) = \text{closure}(\{S \to BbBa{\cdot}, \$\})$

$= \{S \to BbBa{\cdot}, \$\}$

Step 3. Constructing the FA from the

$$C = \{I_0, I_1, I_2, I_3, I_4, I_5, I_6, I_7, I_8, I_9\}$$

We have already learnt to design FA from C in the algorithm of finding SLR parsing table.

Let us discuss it again briefly as follows:

$I_0 \;=\; \{\; S' \to {\cdot}S, \$$ (string derived from S is expected ....)

$\;\;\;\;S \to {\cdot}AaAb, \$$

$\;\;\;\;S \to {\cdot}BbBa, \$$

$\;\;\;\;A \to {\cdot}, a$

$\;\;\;\;B \to {\cdot}, b$

$\;\;\;\;\}.$

Now if we want to decide the transition from I_0 for S, A and B then we get I_1, I_2 and I_3 respectively. Reason is very simple:

$I_1 \;=\; \{S' \to S{\cdot}, \$\}$ (string derived from S has been parsed)

$I_2 \;=\; \{S \to A{\cdot}aAb, \$\}$ (string derived from first A has been parsed)

$I_3 \;=\; \{S \to B{\cdot}bBa, \$\}$ (string derived from first B has been parsed)

- I_0 will transit to I_1 on S

- I_0 will transit to I_2 on A
- I_0 will transit to I_3 on B

Other transitions are also decided into same manner.

So FA is as follows:

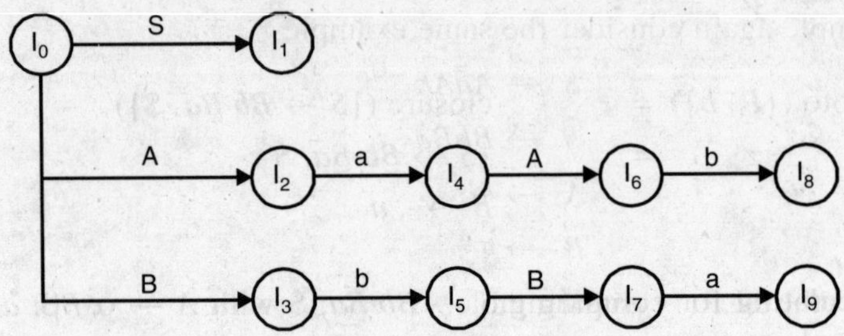

Fig. 6.21.

Step 4. Construction of Action table for SLR(1) parser: By following No. of steps action table can be constructed:

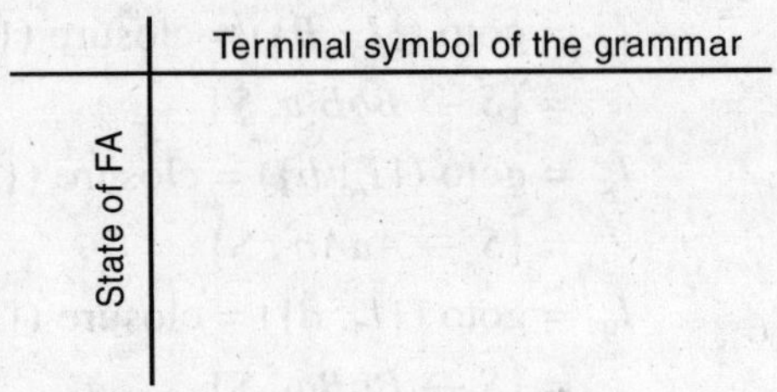

Fig. 6.22.

(*i*) If $(A \rightarrow \alpha \cdot a\beta, b)$ is in I_i and goto $(I_i, a) = I_j$
set action $[I_i, a]$ to shift to state j (S_j)

(*ii*) If $(A \rightarrow \alpha\cdot, b)$ is in I_i
and goto $[I_i, b]$ = reduce $A \rightarrow \alpha$ (R_K, where K is number of production $A \rightarrow \alpha$ in non-augmented grammar, $A \neq S'$)

(*iii*) If $[S' \rightarrow S\cdot, \$]$ is in I_i then
set action $[I_i, \$]$ = **"Accept"**.

(*iv*) If any conflicting actions generated by these rules, the grammar is not LR(1).

Step 5. Construction of the Goto table for the SLR(1) parser: By the following number of steps, Goto table can be constructed

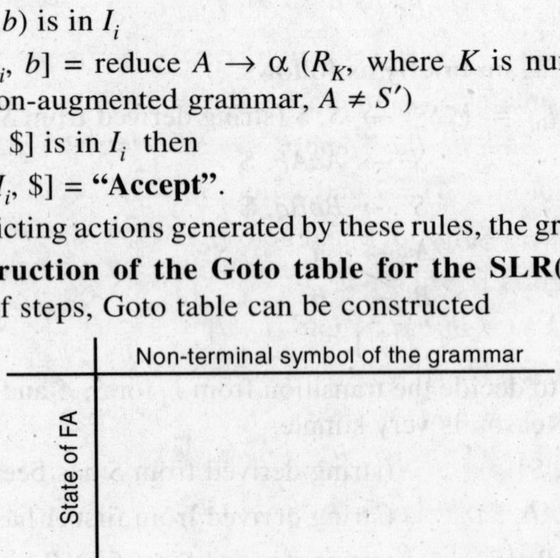

Fig. 6.23.

For every I_i in C do

```
For every non-terminal A do
if goto (Ii, A) = Ij then
set Goto [Ii, A] = j
```

Clearly Goto table is a mapping of transitions in the DFA on non-terminals.

For example again consider the same example

$$S \to AaAb \quad ...(1)$$

$$S \to BbBa \quad ...(2)$$

$$A \to \in \quad ...(3)$$

$$B \to \in \quad ...(4)$$

and construct the SLR(1) parsing table.

We have already calculated C for this grammar and also got the DFA for C.

So now it is easy to construct the parsing table as follows:

I_0 : Contains $A \to \cdot$, a and $B \to \cdot$, b

So action [I_0, a] = R_3 and action [I_0, b] = R_4

I_0 also contains $S' \to \cdot S$, \$, $S \to \cdot AaAb$, \$ and $S \to \cdot BbBa$, \$ from DFA we know

goto [I_0, S] = I_1 so Goto [I_0, S] = 1

goto [I_0, A] = I_2 so Goto [I_0, A] = 2

goto [I_0, B] = I_3 so Goto [I_0, B] = 3

I_1 : Contains $S' \to S\cdot$, \$ so
set action [I_1, S] = accept

I_2 : Contains $S \to A\cdot aAb$, \$ and goto [I_2, a] = I_4
so action [I_2, a] = S_4.

I_3 : Contains $S \to B\cdot bBa$, \$ and goto [I_3, b] = I_5 so set action [I_3, b] = S_5.

I_4 : Contains $A \to \cdot$, b so set action [I_4, b] = R_3 (that is reduce by $A \to \cdot$)
I_4 also contains $S \to Aa\cdot ab$, \$ and goto [I_4, A] = I_6 so set Goto [I_4, A] = 6

I_5 : Contains $S \to Bb\cdot Ba$, \$ and $B \to \cdot$, a
so set Goto [I_5, B] = 7 and action [B, a] = R_4

I_6 : Contains $S \to AaA\cdot b$, \$ and goto [I_6, b] = I_8
so set action [I_6, b] = S_8.

I_7 : Contains $S \to BbB\cdot a$, \$ goto [I_7, a] = I_9
so set action [I_7, a] = I_9

I_8 : Contains $S \to AaAb\cdot$, \$ so
set action [I_8, \$] = R_1

I_9 : Contains $S \to BbBa\cdot$, \$ so
set action [I_9, \$] = R_2.

Now let us put it in matrix format as follows:

	Action table			Goto table		
	a	***b***	**$**	***S***	***A***	***B***
I_0	R_3	R_4		1	2	3
I_1			Accept			
I_2	S_4					
I_3		S_5				
I_4		R_3			6	
I_5	R_4					7
I_6		S_8				
I_7	S_9					
I_8			R_1			
I_9			R_2			

There is no shift/reudce or no reduce/reduce conflict, so, it is a LR(1) grammar.

6.5.7. LALR Parsing **[MDU 2009]**

It becomes very clear that SLR(1) parsers require very less space than **LR** parsers.

LALR stands for **LookAhead LR.** LALR parsers are often used in practice because LALR parsing tables are smaller than LR(1) parsing tables. The number of states in SLR and LALR parsing tables for a grammar G are equal. But LALR parsers recognize more grammars than SLR parsers. ***yacc*** creates a LALR parser for the given grammar. A state of LALR parser will be again a set of LR(1) items.

$$\textbf{Canonical LR(1) Parser} \xrightarrow[\text{Shrink of states}]{} \textbf{LALR Parser}$$

This shrink process may introduce a **reduce/reduce** conflict in the resulting LALR parser (so the grammar is NOT LALR). But, this shrink process does not produce a **shift/reduce** conflict.

Shift/Reduce Conflict

We say that we cannot introduce a shift/reduce conflict during the shrink process for the creation of the states of a LALR parser.Assume that we can introduce a shift/reduce conflict. In this case, a state of LALR parser must have:

$$A \rightarrow \alpha.,a \text{ and } \quad B \rightarrow \beta.a\gamma,b$$

This means that a state of the canonical LR(1) parser must have:

$$A \rightarrow \alpha.,a \text{ and } \quad B \rightarrow \beta.a\gamma,c$$

But, this state has also a shift/reduce conflict. i.e. The original canonical LR(1) parser has a conflict. (Reason for this, the shift operation does not depend on lookaheads).

Reduce/Reduce Conflict

But, we may introduce a reduce/reduce conflict during the shrink process for the creation of the states of a LALR parser.

I_1: $A \rightarrow \alpha.,a$ I_2: $A \rightarrow \alpha.,b$

$B \rightarrow \beta.,b$ $B \rightarrow \beta.,c$

$\Downarrow$

I_{12}: $A \rightarrow \alpha.,a/b$ $\rightarrow$ reduce/reduce conflict

$B \rightarrow \beta.,b/c$

6.5.7.1. *Constructing the LALR Parsing Table* [MDU 2009]

LALR parsing table can be obtained by following steps:

(*i*) Very first we will obtain LR(1) items.

(*ii*) After this we will calculate canonical collection as we did in SLR(1) parsing table.

(*iii*) Now we will combin those canonicals which have same LR(0) (since LR(1) is made off LR(0) and lookaheads) but having different lookaheads.

(*iv*) **Now we will construct Action Table on reduced collection.**

We will follow following steps:

(*a*) If $(A \rightarrow \alpha \cdot a\beta, b)$ is in I_i and goto $(I_i, a) = I_j$ set action $[I_i, a]$ to shift j (S_j)

(*b*) If $(A \rightarrow \alpha\cdot, b)$ is in I_i and goto $[I_i, \text{b}]$ = reduce $A \rightarrow \alpha$ (R_K, where K is number of production $A \rightarrow \alpha$ in non-augmented grammar).

(c) If $[S' \rightarrow S\cdot, \$]$ is in I_i then

set action $[I_i, \$]$ = "Accept"

1, Construction of Goto Table: We can do it by following algorithm:

For every I_i in C d0

For every non-terminal A do

if goto $(I_i, A) = I_j$ then

set Goto $[I_i, A] = j$

Let us discuss an example to make the LALR parser clear:

Example 6.8. Consider the following grammar:

$$S \rightarrow AA$$
$$A \rightarrow aA$$
$$A \rightarrow b$$

and construct the LALR parsing table.

Solution. Given grammar is

$$S \rightarrow AA$$
$$A \rightarrow aA$$
$$A \rightarrow b$$

The augmented grammar is

$$S' \rightarrow S$$
$$S \rightarrow AA$$
$$A \rightarrow aA$$
$$A \rightarrow b$$

Now we will find the canonical collection of sets of LR(1) items:

$$I_0 = \text{closure}(S' \to \cdot S, \$) = \{ S' \to \cdot S, \$$$
$$S \to \cdot AA, \$$$
$$A \to \cdot aA, a/b$$
$$A \to \cdot b, a/b$$
$$\}$$

$$I_1 = \text{goto}(I_0, S) = \text{closure}(S' \to S\cdot, \$) = \{S' \to S\cdot, \$\}$$
$$I_2 = \text{goto}(I_0, A) = \text{closure}(S \to A \cdot A, \$)$$
$$= \{S \to A \cdot A, \$$$
$$A \to \cdot aA, \$$$
$$A \to \cdot b, \$$$
$$\}$$
$$I_3 = \text{goto}(I_0, a) = \text{closure}(A \to a\cdot \text{A}, a/b)$$
$$\{A \to a \cdot A, a/b$$
$$A \to \cdot aA, a/b$$
$$A \to \cdot b, a/b$$
$$\}$$
$$I_4 = \text{goto}(I_0, b) = \text{closure}(A \to b\cdot, a/b) = \{A \to b\cdot, a/b\}$$
$$I_5 = \text{goto}(I_2, \text{A}) = \text{closure}(S \to AA\cdot, \$) = \{S \to AA\cdot, \$\}$$
$$I_6 = \text{goto}(I_2, a) = \text{closure}(A \to a\cdot A, \$) = \{A \to a\cdot A, \$$$
$$A \to \cdot aA, \$$$
$$A \to \cdot b, \$$$
$$\}$$
$$I_7 = \text{goto}(I_2, b) = \text{closure}(A \to b\cdot, \$) = \{A \to b\cdot, \$\}$$
$$I_8 = \text{goto}(I_3, \text{A}) = \text{closure}(A \to aA\cdot, a/b) = \{A \to aA\cdot, a/b\}$$
$$I_9 = \text{goto}(I_6, \text{A}) = \text{closure}(A \to aA\cdot, \$) = \{A \to aA\cdot, \$\}$$

rest of goto (I_i, X) (where X is grammar symbol) are same as $I_0, I_1, I_2, I_3, ..., I_9$ so

$$C = \{I_0, I_1, I_2, I_3, I_4, I_5, I_6, I_7, I_8, I_9\}$$

If we carefully analyse then LR(0) items of I_3 and I_6 are same but they differ only in their lookaheads.

$$I_3 = \{A \to a\cdot A, a/b$$
$$A \to \cdot a\text{A}, a/b$$
$$A \to \cdot b, a/b\}$$
$$I_6 = \{A \to a\cdot A, \$$$
$$A \to \cdot aA, \$$$
$$A \to \cdot b, \$$$
$$\}$$

Clearly I_3 and I_6 are same in their LR(0) items but differ in their lookaheads, so we can combine them and called as I_{36}.

$$I_{36} = \{A \to a\cdot A, a/b/\$$$
$$A \to \cdot a\text{A}, a/b/\$$$
$$A \to \cdot b, a/b/\$$$
$$\}$$

So now C becomes as follows:

$$C = \{I_0, I_1, I_2, I_{36}, I_{47}, I_5, I_{89}\}$$

where $I_0 = \{ S' \to \cdot S, \$$
$S \to \cdot AA, \$$
$A \to \cdot aA, a/b$
$A \to \cdot b, a/b$
$\}$

$I_1 = \{ S' \to S\cdot, \$\}$

$I_2 = \{ S \to A\cdot A, \$$
$A \to \cdot aA, \$$
$A \to \cdot b, \$\}$

$I_{36} = \{ A \to a\cdot A, a/b/\$$
$A \to \cdot aA, a/b/\$$
$A \to \cdot b, a/b/\$$
$\}$

$I_{47} = \{ A \to b\cdot, a/b/\$\}$

$I_5 = \{ S \to AA\cdot, \$\}$

$I_{89} = \{ A \to aA\cdot, a/b/\$\}$

Let us design the FA as follows:

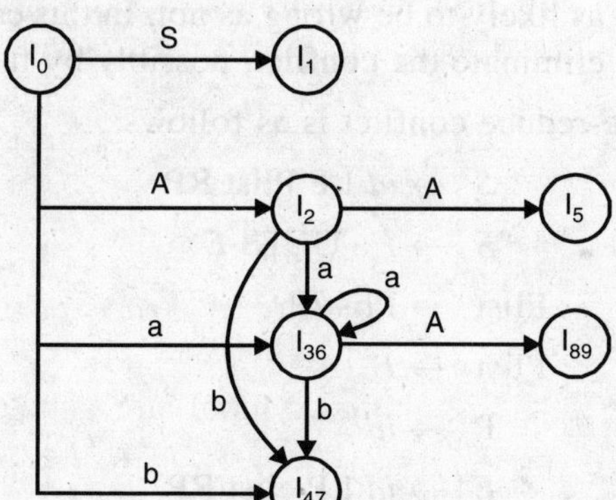

Fig. 6.24.

LALR parsing table is as follows:

	Action table			Goto table	
	a	b	$\$$	S	A
I_0	S_{36}	S_{47}		1	2
I_1			Accept		
I_2	S_{36}	S_{47}			5
I_{36}	S_{36}	S_{47}			89
I_{47}	R_3	R_3	R_3		
I_5			R_1		
I_{89}	R_2	R_2	R_2.		

6.5.8. Conflicts in Shift-Reduce Parsing

"Conflicts" occur when an ambiguity in the grammar creates a situation where the parser does not know which step to perform at a given point during parsing. There are two kinds of conflicts that occur.

1. Shift-Reduce Conflicts: A shift reduce conflict occurs when the grammar indicates that different successful parses might occur with either a shift or a reduce at a given point during parsing.

The vast majority of situations where this conflict occurs can be correctly resolved by shifting.

For example consider the following grammar:

$$S \rightarrow \text{if } E \text{ then } S$$

$$S \rightarrow \text{if } E \text{ then } S \text{ else } S$$

In many languages two nested "if" statement produce a situation where an "else" clause could legally belong to either "if". The usual rule (to shift) attaches the else to the nearest (that is inner) if statement.

2. Reduce–Reduce Conflicts: A reduce-reduce conflict occurs when the parser has two or more handles at the same time on the top of the stack. Whatever choice the parser makes is just as likely to be wrong as not. In this case it is usually best to rewrite the grammar to eliminate the conflict, possibly by factoring.

Example for reduce-reduce conflict is as follows:

$$S \rightarrow id \text{ LP Plist RP} \quad ...(1)$$

$$S \rightarrow F \cdot \text{GETS } E \quad ...(2)$$

$$\text{Plist} \rightarrow \text{Plist/P} \quad ...(3)$$

$$\text{Plist} \rightarrow P \quad ...(4)$$

$$\text{P} \rightarrow id \quad ...(5)$$

$$E \rightarrow id \text{ LP elist RP} \quad ...(6)$$

$$E \rightarrow id \quad ...(7)$$

$$\text{elist} \rightarrow \text{elist}/E \quad ...(8)$$

$$\text{elist} \rightarrow E \quad ...(9)$$

By the point the stack holds .....*id* LP *id* the parser will know which rule to use to reduce the *id*: (5) or (7).

6.6. A COMPARISON OF PREDICTIVE PARSER WITH SHIFT-REDUCE PARSERS

Both Predictive parser and shift-reduce parser read the input from left-to-right and maintain a stack of grammar symbol but their parsing operations are decidedly different as shown in following table:

Predictive Parser	Shift-Reduce Parser
Top-down (LL) Parser	Bottom-up (LR) parser
Stack predicts what is to come.	Stack shows what has been seen so far.
The stack initially contains the start symbol of the grammar.	The stack is initially empty
The stack is empty when the accept state is reached.	The stack contains the start symbol of the grammar when the accept state is reached.
Input tokens are popped off the stack.	Input tokens are pushed on the stack.
Left sides of productions are popped off the stack.	Right sides of productions are popped off the stack
Right sides of productions are pushed on the stack.	Left sides of productions are pushed on the stack.

But there are several reasons, which make LR parsers more attractive over predictive parsers:

(*i*) An LR parser can be constructed to recognize virtually all programming language constructs for which a CFG can be written.

(*ii*) The LR parsing method is the most general, non backtracking shift-reduce method known. Yet it can be implemented as efficiently as any other method.

(*iii*) The class of grammars that can be parsed by using the LR method is a proper superset of the class of grammars that can be parsed with a predictive parser.

(*iv*) The LR can quickly detect a syntactic error while left-to-right scanning of input.

EXERCISE

1. Obtain the canonical collection of sets of LR(1) items for the following grammar :

$$S \rightarrow SB / Ca$$
$$B \rightarrow Bb / \in$$
$$C \rightarrow aB / c$$

2. Find the LR(1) parsing table for the following grammar:

$$T \rightarrow \text{int}$$
$$L \rightarrow L,\ id / id$$

3. Construct an LALR(1) parsing table for the following grammar:

$$D \rightarrow L:\ T$$
$$L \rightarrow L,\ id\ /\ id$$
$$T \rightarrow \text{integer}$$

4. Construct an SLR(1) parsing table for the following grammar:

$$S \rightarrow A\ )$$
$$A \rightarrow A,\ P / (\ P,\ P$$
$$P \rightarrow \{\text{num, num}\}$$

5. Consider the following grammar:

$$S \rightarrow AS\ /\ b$$
$$A \rightarrow SA\ /\ a$$

(*i*) List all the LR(0) items for the above grammar.
(*ii*) Construct an NFA whose states are the LR(0) items from (*i*).
(*iii*) Is the grammar SLR ? If so construct the SLR parsing table.

6. Consider the following grammar:

$$E \rightarrow E + T / T$$
$$T \rightarrow TF / F$$
$$F \rightarrow F * / a / b$$

(*i*) Construct the SLR parsing table for this grammar.
(*ii*) Construct the LALR parsing table.

7. Show that following grammar:

$$S \rightarrow AaAb/BbBa$$
$$A \rightarrow \in$$
$$B \rightarrow \in$$

is LL(1) but not SLR(1).

8. Show that following grammar

$$S \rightarrow Aa/bAc/Bc/bBa$$
$$A \rightarrow d$$
$$B \rightarrow d$$

is LR(1) also design the LR(1) parsing table.

9. Construct an SLR parsing table for the grammar:

$$E \rightarrow E \ \underline{\textbf{sub}} \ R \ / \ E \text{ sub } E \ / \ \{E\}/c$$
$$R \rightarrow E \ \underline{\textbf{sub}} \ E \ / \ E$$

resolve the parsing action conflict.

10. Construct the LALR parsing table for the following grammar:

$$S \rightarrow AA$$
$$A \rightarrow aA / b$$

11. Discuss algorithm for computation of the sets of LR(1) items. Also show that the following grammar is LR(1) but not LALR(1)

$$S \rightarrow Aa / bAc / Bc / bBa$$
$$A \rightarrow d$$
$$B \rightarrow d$$

❑❑❑

SECTION-C

- Syntax-Directed Translation
- Symbol Table Management
- Error-Detection and Recovery

SYLLABUS

Syntax Directed Translations : Syntax Directed Definition, Construction of Syntax Trees, Syntax Directed Translation Scheme, Implementation of Syntax Directed Translation, Three Address Code, Quadruples and Triples.

Symbol Table and Error Detection and Recovery : Symbol Tables, Its Contents and Data Structure for Symbol Tables; Trees, Arrays, Linked Lists, Hash Tables. Errors, Lexical Phase Error, Syntactic Phase Error, Semantic Error.

7

Chapter

SYNTAX-DIRECTED TRANSLATION

INSIDE THIS CHAPTER

7.1. Introduction of Syntax-Directed Translation **7.2.** Basic Definition of Attributes and Semantic Rules **7.3.** Two-Pass Vs One Pass Compilation and Syntax Directed Translation **7.4.** Syntax-Directed Definition **7.5.** Syntax Tree **7.6.** S-Attributed Definitions **7.7.** Implementation Issues of Syntax-Directed Translation **7.8.** Bottom-Up Evaluation of S-Attributed Definitions **7.9.** Implementation of Syntax-Directed Translation.

7.1. INTRODUCTION OF SYNTAX-DIRECTED TRANSLATION

We know that every programming language contains the constructs, so it is very important to know the rules of the translations of programming constructs. Whenever a construct encounters in programming language, it is evaluated/ Translated according the sementic rules define in that particular programming language. Translation may be generation one of the intermediate code, object code, or adding the information in symbol table about constructs type.

TIPS

Syntactic structure of programming language specify the properties of construct. We know that CFG is used to specify the syntactical structure of programming language. We adds attributes with grammar symbol and semantic rules with productions to make the translation of constructs easy.

Modern compilers use syntax-directed translation to interleaves the action of the compiler phases.

The syntax analyzer directs the whole process during the parsing of the source code.

- Calls the lexical analyzer whenever syntax analyzer wants another token.
- Perform the action of semantic analyzer.
- Perform the action of the intermediate code generator.

Clearly syntax directed translation, make, user's life easy by hiding many implementation detail and free the user from having to specify explicity the order in which translation takes place. Translation schemes basically indicate the order in which semantic rules are to be evaluated. The translation of token streams takes place by evaluatic the semantic rules.

It is not always necessary to follow the structure of syntax-directed translation as shown in Fig. 7.1.

Syntax-directed definitions can be implemented in a single pass by evaluating semantic rules during parsing itself.

TIPS

The actions of semantic analyzer and the intermediate code generator require the passage of information up and /or down the parse tree. We think of this information as attributes attached to the nodes of the parse tree and the parser moving this information between parent nodes and children nodes as it performs the productions of the grammar.

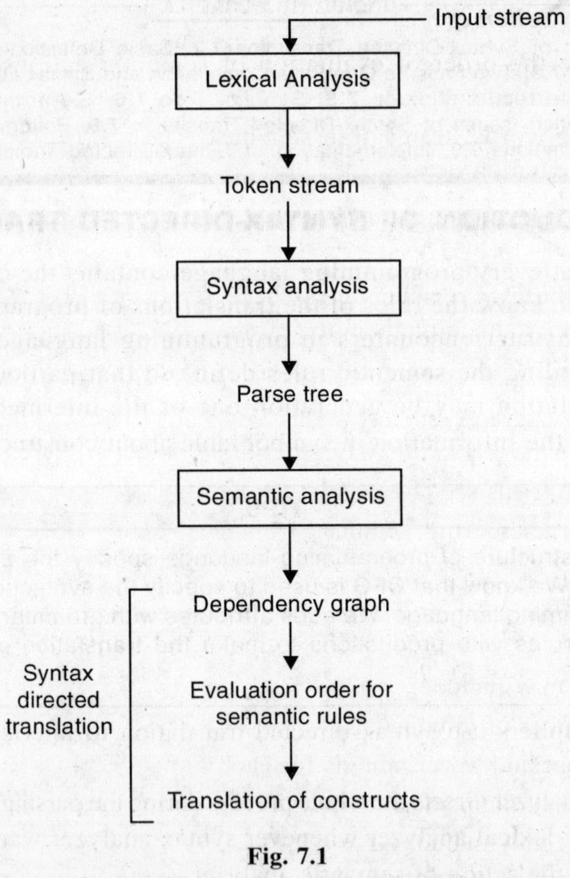

Fig. 7.1

Values of these attributes (added with grammar symbol) are evaluated by the semantic rules associated with the production rules. Evaluation of these semantic rules:

(*i*) may generate intermediate codes

(*ii*) may put information into the symbol table

(*iii*) may perform type checking

(*iv*) may issue error messages

(*v*) may perform some other activities

(*vi*) in fact, they may perform almost any activities.

An attribute may hold almost anything like a string, a number, a memory location, and a complex record.

When we associate semantic rules with productions, we use two notations:

1. Syntax-Directed Definitions

(*i*) Give high-level specifications for translations

(*ii*) Hide many implementation details such as order of evaluation of semantic actions.

(*iii*) We associate a production rule with a set of semantic actions, and we do not say when they will be evaluated.

2. Translation Schemes

(*i*) Indicate the order of evaluation of semantic actions associated with a production rule.

(*ii*) In other words, translation schemes give a little bit information about implementation details.

7.2. BASIC DEFINITION OF ATTRIBUTES AND SEMANTIC RULES

Definition

Attributes. Attributes are associated information with language construct by attaching them to grammar symbols representing that construct.

An attribute can represent any thing (reasonable) that we choose, that is a string, number, type, memory location, code fragment etc.

For example let us see an attribute for an identifier, it may include:

Name: Character string, obtained from scanner.

Scope:

Type: may be:

- Integer
- Array ; may include:
  - — number of dimensions
  - — upper and lower bounds for each dimension
  - — type of elements
- Record ; may include:
  - — name and type of each field.
- Function ; may include:
  - — number of parameters
  - — types of parameters (in order)
  - — type of return value
  - — entry point in memory
  - — size of stack frame.

Definition

Semantic Rules. Values for attributes are computed using semantic rules associated with grammar productions. The semantic rules also impose an evaluation order on the attributes.

For example:

Production	**Semantic Rule**
$E \longrightarrow E_1 + T$	$E \cdot \text{Val} := E_1 \text{Val} + T \cdot \text{Val}$
$E \longrightarrow T$	$E \cdot \text{Val} := T \cdot \text{Val}$
$T \longrightarrow T_1 * F$	$T \cdot \text{Val} := T_1 \cdot \text{Val} \times F \cdot \text{Val}$
$T \longrightarrow F$	$T \cdot \text{Val} := F \cdot \text{Val}$
$F \longrightarrow (F)$	$F \cdot \text{Val} := E \cdot \text{Val}$
$F \longrightarrow \text{intcon}$	$F \cdot \text{Val} := \text{intcon} \cdot \text{Val}$

7.3. TWO-PASS VS ONE PASS COMPILATION AND SYNTAX DIRECTED TRANSLATION

In case of **Two-pass** compilation following steps should be followed to generate the code

Step 1. Parse input and use semantic rules to:

(*a*) Process declaration into symbol table.

(*b*) Construct syntax tree.

Step 2. Traverse syntax tree:

(*a*) Check types

(*b*) Make storage allocation decision

(*c*) Generate code.

One-pass compilation includes following: Parse input and use semantic rules to

(*a*) process declarations into symbol table

(*b*) check types

(*c*) make storage allocation decisions

(*d*) generate code.

Example 7.1. Compute the type of an expression that includes both arithmatic and boolean operators. (The type is INT, BOOL, or ERROR) given CFG is

exp ⟶ exp + exp
exp ⟶ exp and exp
exp ⟶ exp = = exp
exp ⟶ true
exp ⟶ false
exp ⟶ int
exp ⟶ (exp)

Solution.

Production	**Translation Rules**
1. exp ⟶ exp + exp	if (($\text{exp}_2 \cdot \text{trans}$ = = INT) and ($\text{exp}_3 \cdot \text{trans}$ = = INT) then $\text{exp}_1 \cdot \text{trans}$ = INT else $\text{exp}_1 \cdot \text{trans}$ = ERROR

2. exp ⟶ exp and exp	if ((exp_2 trans = = BOOL) and ($exp_3 \cdot$ trans = = BOOL) then $exp_1 \cdot$ trans = = BOOL else $exp_1 \cdot$ trans = ERROR.
3. exp ⟶ exp = = exp	if (($exp_2 \cdot$ trans = = $exp_3 \cdot$ trans) and $exp_2 \cdot$ trans ! = ERROR)) then $exp_1 \cdot$ trans = BOOL else $exp_1 \cdot$ trans = ERROR
4. exp ⟶ true	exp · trans = BOOL
5. exp ⟶ false	exp · trans = ERROR
6. exp ⟶ int	exp · trans = INT
7. exp ⟶ (exp)	$exp_1 \cdot$ trans = $exp_2 \cdot$ trans

Note. Here first exp is written as exp_1, second as exp_2 and third as exp_3.

7.4. SYNTAX-DIRECTED DEFINITION

A syntax-directed definition generalizes a content-free grammar by associating a set of attributes with each node in a parse tree. Each attribute gives some information about the node. For example, attributes associated with an expression node may give its value, its type or its location in memory, etc.

As an example, suppose the grammar contains the production:

$X \longrightarrow YZ$ so node X in a parse tree has nodes Y and Z as children and further suppose the nodes X, Y and Z have associated attributes $X \cdot a$, $Y \cdot a$, and $Z \cdot a$ respectively.

If the semantic rule $\{X \cdot a := Y \cdot a + Z \cdot a\}$ is associated with the $X \longrightarrow YZ$ production then the parser should add the 'a' attributed of node Y and Z together and set the 'a' attribute of node X to their sum.

TIPS

The attributes may be of two types: synthesized attributes and inherited attributes.

Each production rule is associated with a set of semantic rules. Semantic rules set up dependencies between attributes which can be represented by a dependency graph. This dependency graph determines the evaluation order of these semantic rules. Evaluation of a semantic rule defines the value of an attribute. But a semantic rule may also have some side effects such as printing a value.

7.4.1. Synthesized Attributes

TIPS

An attribute at a node is synthesized if its value is computed from the attributed values of the children of that node in the parse tree.

For example:

$E \longrightarrow E_1 + E_2$ has semantic rule as $E \cdot \text{Val} = E_1 \cdot \text{Val} + E_2 \cdot \text{Val}$ and can be written as:

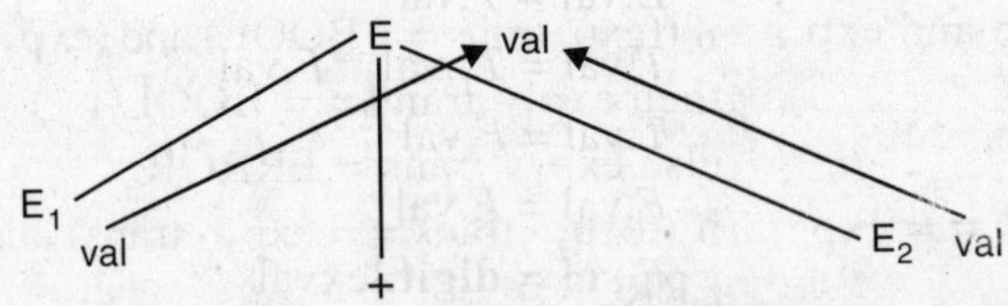

Fig. 7.2.

Here $E \cdot$ Val is synthesized attribute. Let us see another figure to make the concept clear.

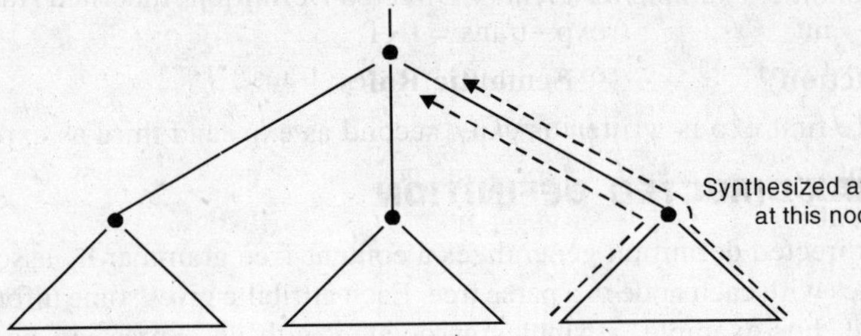

Fig. 7.3.

TIPS

Synthesized attribute can be evalulated by a single bottom-up traversal of the parse tree.

7.4.2. Inherited Attributes

Definition An attribute at a node is inherited if its value is computed from **attribute values at the siblings and/or parent** of that node in the parse tree.

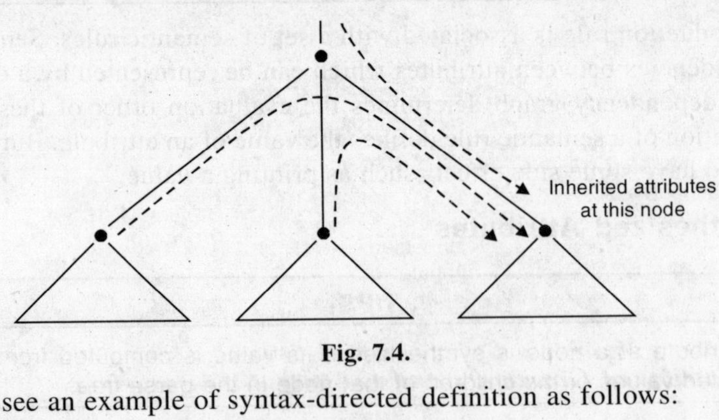

Fig. 7.4.

Let us see an example of syntax-directed definition as follows:

Production	**Semantic Rules**
$L \rightarrow E$ return	print(E.val)
$E \rightarrow E_1 + T$	E.val = E_1.val + T.val

$E \rightarrow T$	E.val = T.val
$T \rightarrow T_1 * F$	T.val = T_1.val * F.val
$T \rightarrow F$	T.val = F.val
$F \rightarrow (E)$	F.val = E.val
$F \rightarrow$ **digit**	F.val = **digit**.lexval

Symbols E, T and F are associated with a synthesized attribute *val*. The token **digit** has a synthesized attribute *lexval* (it is assumed that it is evaluated by the lexical analyzer).

Let us see another example for Syntax-Directed Definition–Inherited Attributes as follows:

Production	**Semantic Rules**
$D \rightarrow T\ L$	L.in = T.type
$T \rightarrow$ **int**	T.type = integer
$T \rightarrow$ **real**	T.type = real
$L \rightarrow L_1$ **id**	L_1.in = L.in, addtype(**id**.entry,L.in)
$L \rightarrow$ **id**	addtype(**id**.entry,L.in)

Symbol *T* is associated with a synthesized attribute *type*. Symbol *L* is associated with an inherited attribute *in*.

7.4.3. Attribute Grammars

Every grammar symbol is associated with a set of attributes. Semantic rules specify how each attribute is to be computed.

The attributes of a grammar symbol are partitioned into two sets: inherited and synthesized. That is for any particular grammar symbol, **a given attribute can not be inherited in same places and synthesized in others.**

For example $A \longrightarrow XYZ$

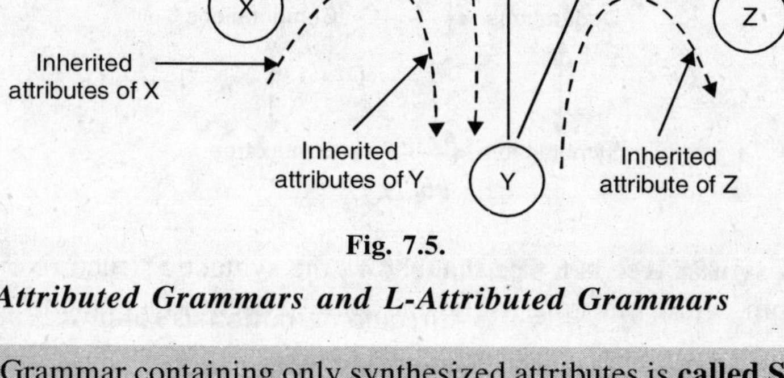

Fig. 7.5.

7.4.3.1. S-Attributed Grammars and L-Attributed Grammars

Definition "Grammar containing only synthesized attributes is **called S-attributed.**"

Synthesized attributes can be conveniently handled during bottom up parsing as it builds the parse tree bottom up.

> **Definition** "On the other hand grammar for which the attributes can always be evaluated by a depth-first *L*-to-*R* traversal of the parse tree, **is called *L*-attributed grammar".**

In case of *L*-attributed grammar, all the attributes can be conveniently handled during LL(1) parsing because the parse tree is built depth-first *L*-to-*R*.

Every S-attributed definition is *L*-attributed. For example, consider the following program segment:

```
int a, b, c ;
a + * c ;
```

Symbol Table

Name	Type	Addr
a	int	1
b	int	2
c	int	3

Syntax Tree

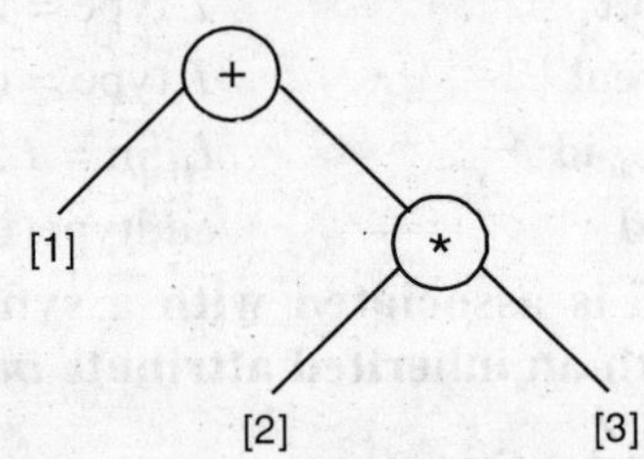

Fig. 7.6.

We will develop semantic rules for constructing symbol table from the declarations and constructing syntax tree for the expression.

- Inherited attributed needed to propagate the type to each declared variable.
- Synthesized attributed needed to construct syntax tree for an expression from syntax trees of subexpression.

7.5. SYNTAX TREE

The complete program can be viewed as follows:

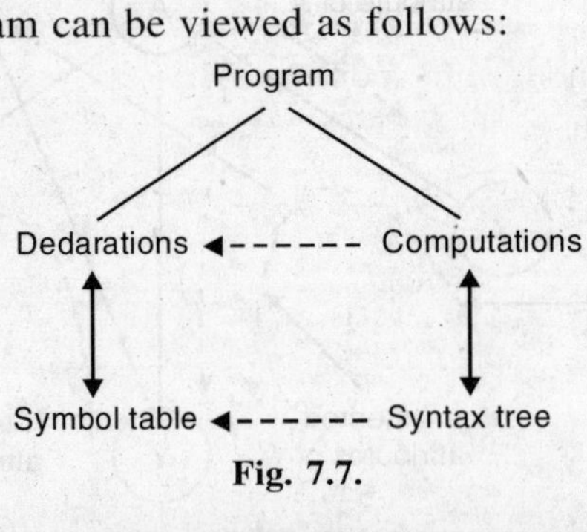

Fig. 7.7.

> **Definition** "A **syntax tree** is a tree that shows the syntactic structure of a program, while omitting irrelevant detail present in a parse tree".

Each node of the syntax tree represents "what to do" at that point, that is computation. The children of the node correspond to the objects to which that computation is applied.

Example 7.2. For the given grammar:

$$E \longrightarrow E + T/T$$
$$T \longrightarrow T * F/F$$
$$F \longrightarrow (E)/id$$

design parse tree and syntax tree for string

$$W = id + id * id.$$

Solution. Given grammar is:

$$E \longrightarrow E + T/T$$
$$T \longrightarrow T * F/F$$
$$F \longrightarrow (E)/id$$

Parse tree:

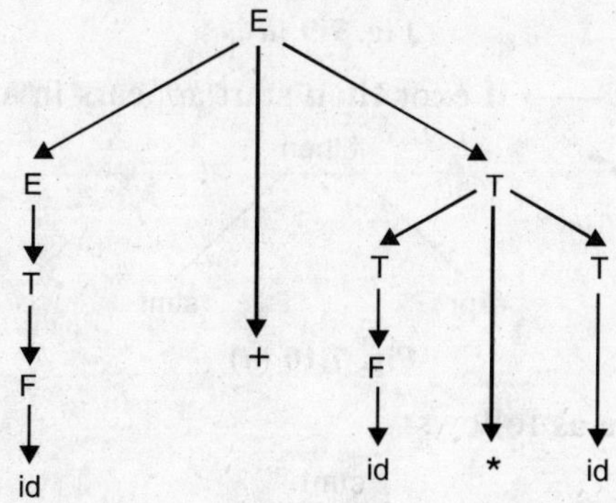

Fig. 7.8.

7.5.1. Structure of Syntax Trees

Expression. In case of expression leaves, internal nodes and childrens can be:

- Leaves: identifiers or constants.
- Internal nodes: labelled with operations.
- Children: of a node are its operands.

Statement. In case of statements structure will be as follows:

- A node's label indicates what kind of statement it is
- The children of a node correspond to the components of the statement.

Let us see some examples as follows:

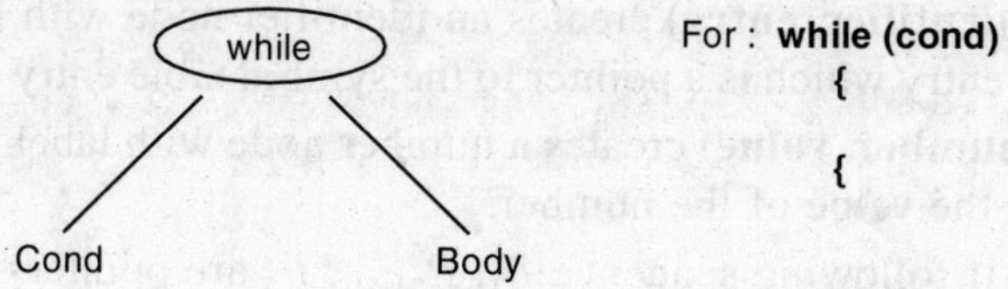

Fig. 7.9 (*a*).

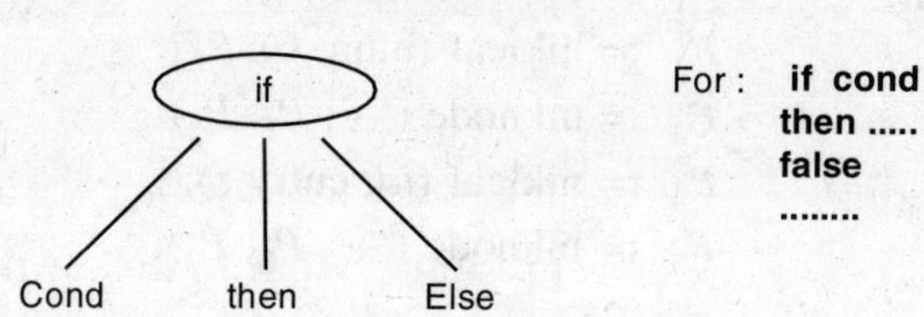

Fig. 7.9 (*b*).

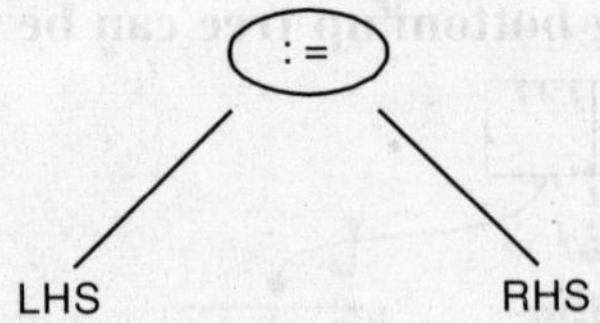

For :
LHS := RHS kind statement

Fig. 7.9 (*c*).

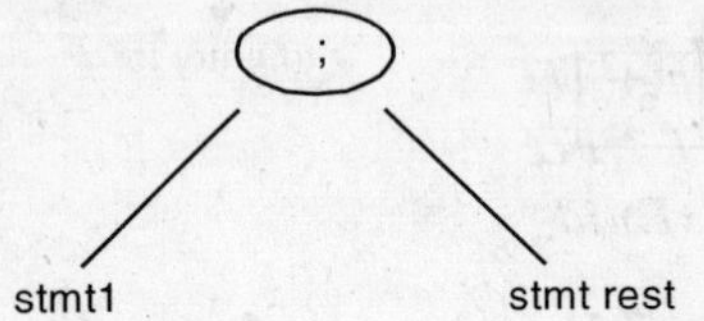

For :
stmt1; stmt rest kind statements

Fig. 7.9 (*d*).

The production stmt ⟶ **if** expr **then** start appears in a syntax tree like:

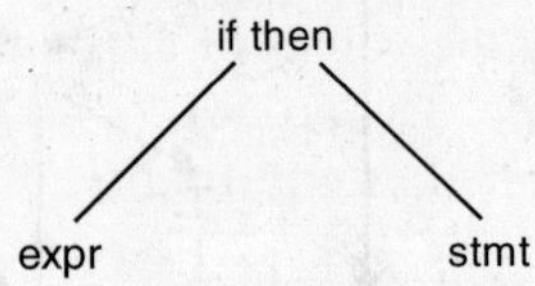

Fig. 7.10 (*a*)

But in parse tree like as follows:

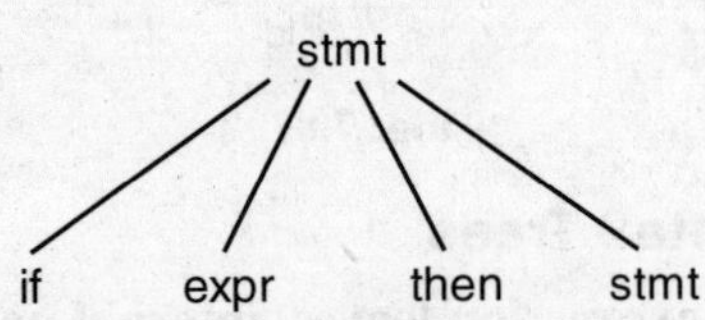

Fig. 7.10 (*b*)

7.5.2. Construction of Syntax Tree

Each node in a syntax tree can be implemented as a record with several fields. The following functions are used to creates the node of syntax tree:

1. MKnode (Operator, left, right) creates an operator with label OP and with two fields containing pointers to left and right.

2. MKleaf (identifier, entry) creates an identifier node with label *id* and with a field containing entry which is a pointer to the symbol table entry for the identifier.

3. MKleaf (number, value) creates a number node with label number and with a field containing the value of the number.

For example, if following sequence $P_1, P_2, ..., P_5$ are pointers to nodes:

P_1 := mkleaf (*id*, entry *x*)

P_2 := mkleaf (num, 6);

P_3 := mknode ('–', P_1, P_2)

P_4 := mkleaf (*id*, entry *z*);

P_5 := mknode ('+', P_3, P_4);

Then for expression $x - 6 + z$ following **bottom up tree can be constructed :**

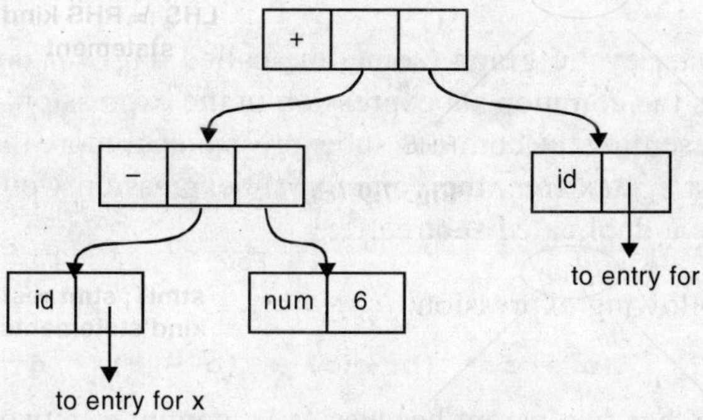

Fig. 7.11.

7.5.3. Annotated Parse Tree

Definition: A parse tree showing the values of attributes at each node is called an annotated parse tree. The process of computing the attribute values at the nodes is called **annotating or decorating** the parse tree.

For example, consider the following semantic rules for given productions:

Production	Semantic Rules
$L \longrightarrow En$	print $(E \cdot \text{val})$
$E \longrightarrow E_1 + T$	$E \cdot \text{val} := E_1 \cdot \text{val} + T \cdot \text{val}$
$E \longrightarrow T$	$E \cdot \text{val} := T \cdot \text{val}$
$T \longrightarrow T_1 * F$	$E \cdot \text{val} := T_1 \cdot \text{val} * F \cdot \text{val}$
$T \longrightarrow F$	$T \cdot \text{val} := F \cdot \text{val}$
$F \longrightarrow (E)$	$F \cdot \text{val} := E \cdot \text{val}$
$F \longrightarrow \text{digit}$	$F \cdot \text{val} := \text{digit} \cdot \text{lexval}$

Now let us draw the parse tree with the values of attributes for expression $3 * 5 + 4n$.

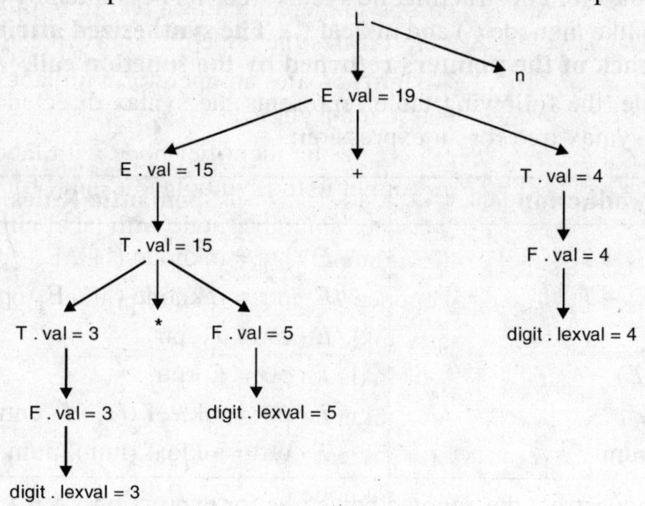

Fig. 7.12.

Figure 7.12 shows the annated parse tree for expression $3 * 5 + 4n$.

7.5.4. Directed Acyclic Graph for Expression

Definition A directed acyclic graph (sometime called dag) for an expression identifies the common subexpression in the expression. A node in a dag representing the common subexpression has more than one "parent", in a syntax tree, the common subexpression would be represented as a duplicated subtree.

Let consider following expression:

a + a * (b – c) + (b – c) * d

Here leaf for 'a' has two parent because 'a' is common in two sub expressions a and $a * (b-c) \cdot (b-c)$ is also common in $a * (b-c)$ and $(b-c) * d$ subexpressions. Now let us design the Dag for given expression as follows:

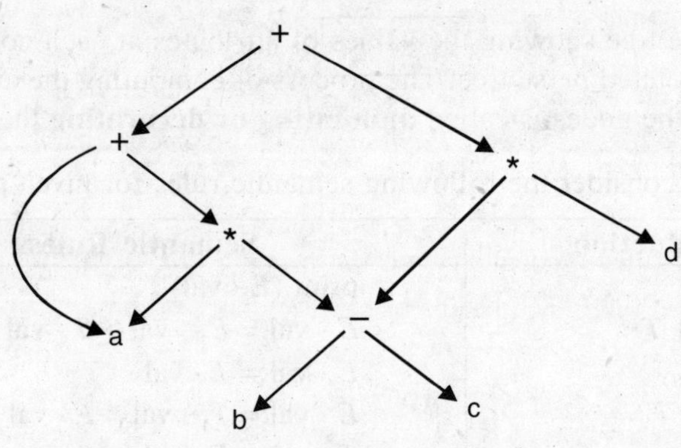

Fig. 7.13.

7.5.5. A Syntax Directed Definition For Constructing Syntax Trees

Semantic rules for constructing the syntax tree can be written by the help of the basic functions like mknode ( ) and mkleaf ( ). **The synthesized attribute ptr for E and T keeps track of the pointers returned by the function call.**

For example, the following table represents the syntax-directed definition for constructing a syntax tree for an expression:

Production	Semantic Rules
$E \longrightarrow E_1 + T$	$E \cdot$ ptr:= mknode ('+', $E_1 \cdot$ ptr, $T \cdot$ ptr)
$E \longrightarrow E_1 - T$	$E \cdot$ ptr:= mknode ('–', $E_1 \cdot$ ptr, $T \cdot$ ptr)
$E \longrightarrow T$	$E \cdot$ ptr:= $T \cdot$ ptr
$T \longrightarrow (E)$	$T \cdot$ ptr:= $E \cdot$ ptr
$T \longrightarrow id$	$T \cdot$ ptr:= mkleaf (id, $id \cdot$ entry)
$T \longrightarrow$ num	$T \cdot$ ptr:= mkleaf (num, num, val)

Now let us construct the annoted parse tree for expression $x + 6 + z$, as shown in following figure:

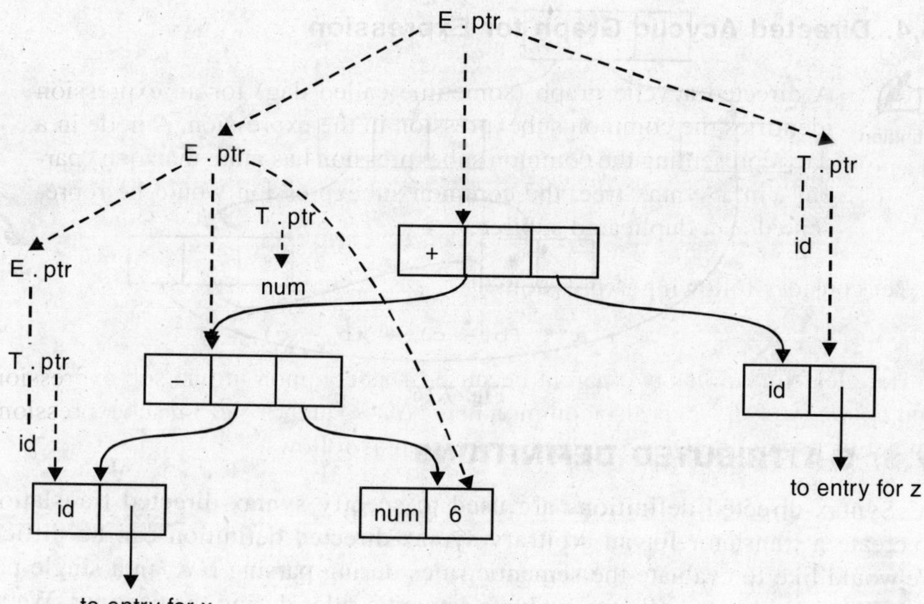

Fig. 7.14.

Here semantic rules associated with the productions $T \longrightarrow id$ and $T \longrightarrow$ num define attribute *T*.nptr to be a pointer to a new leaf or an identifier and a number, respectively.

Example 7.3. For the semantic rules discussed in section 7.5.5, find the Dag for expression *a* + *b* * (*b* + *a*).

Solution. Expression is $a + b * (b + a)$

Let us construct the parse tree first as follows:

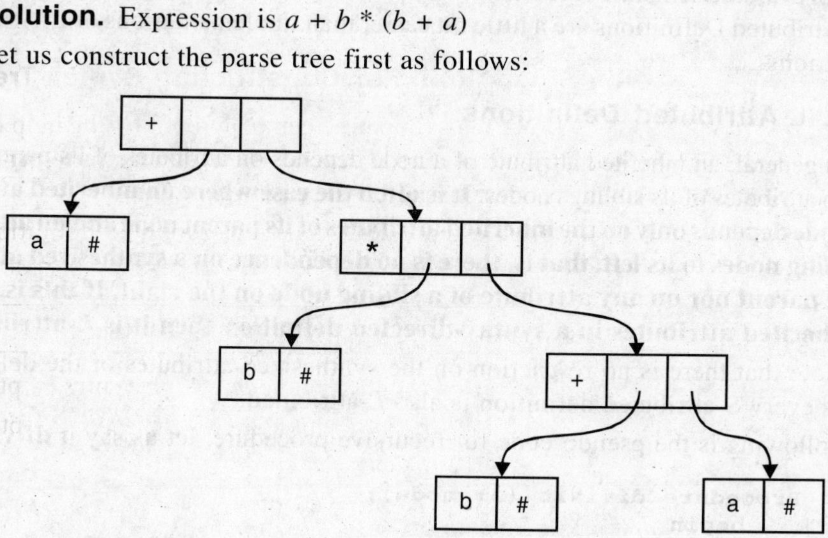

Fig. 7.15.

In expression *b* is common in $b * (b + a)$ and $(b + a)$, so as a is also common in $a + b * (b + a)$ and $(b + a)$, Dag is as follows:

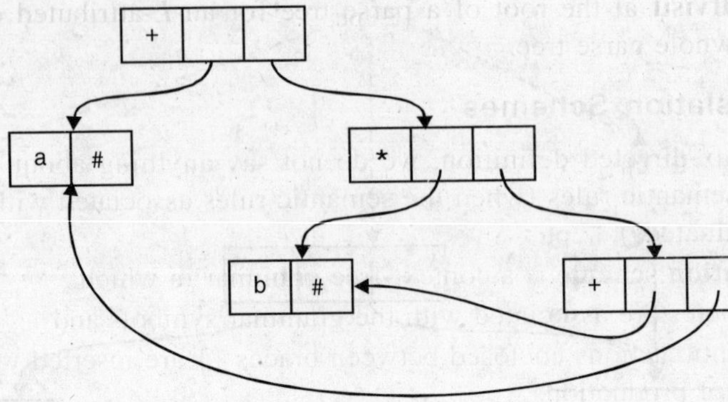

Fig. 7.16.

7.6. S-ATTRIBUTED DEFINITIONS

Syntax-directed definitions are used to specify syntax-directed translations. To create a translator for an arbitrary syntax-directed definition can be difficult. We would like to evaluate the semantic rules during parsing (*i.e.*, in a single pass, we will parse and we will also evaluate semantic rules during the parsing). We will look at two sub-classes of the syntax-directed definitions:

- **S-Attributed Definitions:** Only synthesized attributes used in the syntax-directed definitions.
- **L-Attributed Definitions:** In addition to synthesized attributes, we may also use inherited attributes in a restricted fashion.

To implement S-Attributed Definitions and L-Attributed Definitions are easy (we can evaluate semantic rules in a single pass during the parsing). Implementations of S-attributed Definitions are a little bit easier than implementations of L-Attributed Definitions.

7.6.1. L-Attributed Definitions

In general, an inherited attribute of a node depends on attributes of its parent node and on attributes of its siblings nodes. **It is often the case where an inherited attribute of a node depends only on the inherited attributes of its parent node and on attributes of sibling nodes to its left, that is, there is no dependence on a synthesized attribute of the parent nor on any attribute of a sibling node on the right. If this is true of all inherited attributes in a syntax-directed definition then it is *L*-attributed.**

Note that there is no restriction on the synthesized attributes of the definition, that is every S-attributed definition is also *L*-attributed.

Following is the pseudo-code for recursive procedure, let us say it **df visit.**

```
Procedure dfvisit (n: node);
  begin
     for each child m of n in left-to-right order do begin
     evaluate inherited attributes of m;
     dfvisit (m);
     end; /* end of for loop */
     evaluate synthesized attributes of n;
  end
```

Calling **dfvisit** at the root of a parse tree for an L-attributed definition will annotate the whole parse tree.

7.6.2. Translation Schemes

In a syntax-directed definition, we do not say anything about the evaluation times of the semantic rules (when the semantic rules associated with a production should be evaluated?).

A **translation scheme** is a context-free grammar in which:

(*i*) Attributes are associated with the grammar symbols and

(*ii*) Semantic actions enclosed between braces {} are inserted within the right sides of productions.

For example: $A \rightarrow \{ \dots \}\ X\ \{ \dots \}\ Y\ \{ \dots \}$

Semantic Actions

A translation scheme is a convenient way of descrbing an L-attributed definition. As an example, assume the grammar has a production: A $\longrightarrow XY$ and further assume that A, X and Y, have inherited attributes $A \cdot i$, $X \cdot i$, and $Y \cdot i$, and synthesized attributes $A \cdot s$, $X \cdot s$, and $Y \cdot s$, respectively. Because we have an L-attributed definition:

- $X \cdot i$ can only be a function of $A \cdot i$ that is $X \cdot i = f(A \cdot i)$.
- $Y \cdot i$ can only be a function of $A \cdot i$, $X \cdot i$ and $X \cdot s$ that is $Y \cdot i = g(A \cdot i, X \cdot i, X \cdot s)$ and
- $A \cdot s$ is a function of $A \cdot i, X \cdot i, X \cdot s, X \cdot i$ and $X \cdot s$; that is A $\cdot\, s := h(A \cdot i, X \cdot i, X \cdot s, Y \cdot i, Y \cdot s)$.

A translation scheme would embed the following semantic action in the production $A \longrightarrow XY$ as follows:

$$A \longrightarrow \{X \cdot i := f(A \cdot i);\}\ X$$
$$\{Y \cdot i = g(A \cdot i, X \cdot i, X \cdot s);\ \}\ Y$$

TIPS

Note the careful placement of the semantic action in the production: if any semantic action is moved later in the production then an inherited attribute of a child won't be evaluated in time and if any action is moved earlier in the production it will try to use an argument that has not been evaluated.

There is no special problem with $\in$-production in the grammar. For example, assume $A \longrightarrow \in$ is a production in the grammar and assume that A has an inherited attribute, $A \cdot i$, and a synthesized attribute $A \cdot s$, that is a function, f, of $A \cdot i$. Then the translation scheme contains

$$A \longrightarrow \{A \cdot s := f(A \cdot i);\}$$

7.6.2.1. *Translation Scheme with Synthesized Attributes*

We know that grammar will have either terminal or symbols so clearly synthesized attribute will be attached with, either terminal or nonterminal.

- Synthesized attribute of a terminal are contained in the terminal symbol itself.
- Synthesized attribute associated with a nonterminal symbol is computed after seeing everything it derives.

For example:

Productions	**Translation scheme**
$E \longrightarrow E_1 + T$	$\{E \cdot \text{tree} = \text{mktree (PLUS, } E_1 \cdot \text{tree, } T \cdot \text{tree)}\}$
$E \longrightarrow T$	$\{E \cdot \text{tree} = T \cdot \text{tree}\}$
$T \longrightarrow T * F$	$\{T \cdot \text{tree} = \text{mktree (TIMES, } T \cdot \text{tree, } F \cdot \text{tree)}\}$
$T \longrightarrow F$	$\{T \cdot \text{tree} = F \cdot \text{tree}\}$
$F \longrightarrow id$	$\{F \cdot \text{tree} = \text{mknode (idnode, symbol\_lookup (}id \cdot \text{name))}\}$
$F \longrightarrow$ int const	$\{F \cdot \text{tree} = \text{mknode (intconstnode, intconst} \cdot \text{value)}\}$

7.6.2.2. *Translation Scheme with Inherited Attributes*

Inherited attributed associated with a non-terminal is computed before encountering the nonterminal.

For example:

$$
\begin{aligned}
E &\longrightarrow T\,\{R \cdot \text{itree} = T \cdot \text{stree}\} \\
&\quad\;\; R\{E \cdot \text{stree} = R \cdot \text{stree}\} \\
R &\longrightarrow +T\{R_1 \cdot \text{itree} = \text{mktree ("+", } R \cdot \text{itree, } T \cdot \text{stree)}\} \\
&\quad\;\; R_1\,\{R \cdot \text{stree} = R_1 \cdot \text{stree}\} \\
R &\longrightarrow -T\,\{R_1 \cdot \text{itree} = \text{mktree ("–", } R \cdot \text{itree, } T \cdot \text{stree)}\} \\
&\quad\;\; R_1\,\{R \cdot \text{stree} = R_1 \cdot \text{stree}\} \\
R &\longrightarrow \in \{R \cdot \text{stree} = R \cdot \text{itree}\} \\
T &\longrightarrow id\,\{T \cdot \text{stree} = \text{mknode (idnode, symbtab\_lookup (}id \cdot \text{name))}\} \\
T &\longrightarrow \text{intconst}\,\{T \cdot \text{stree} = \text{mknode (intconstnode, int const. value)}\}
\end{aligned}
$$

When we design a translation scheme, we must ensure that an attribute value is available when the action refers to it. This requires following:

- An inherited attribute of a symbol on the right side of a production must be computed in an action immediately preceeding (to the left of) that symbol, because it may be referred to by an action computing the inherited attribute of that symbol to the right of (following) it.
- An action that computes the synthesized attribute of a nonterminal on the left side of the production should be replaced at the end of the right side of the production, because it might refer to the attributes of any of the right-side grammar symbols. Therefore, unless they are computed, the left synthesized attribute of a nonterminal on the left cannot be computed.

Following example satisfies all the conditions, discussed above:

$$D \longrightarrow L\,\{L \cdot \text{type:=}\; T \cdot \text{type}\}\; L;$$

$$T \longrightarrow \text{int}\ \{T \cdot \text{type:= int}\}$$
$$T \longrightarrow \text{real}\ \{T \cdot \text{type:= real}\}$$
$$L \longrightarrow \{L_1 \cdot \text{type} = L \cdot \text{type}\}\ L_1,\ id\ \{\text{enter}\ (id \cdot \text{ptr}, L \cdot \text{type});\}$$
$$L \longrightarrow id\{\text{enter}\ (id \cdot \text{ptr}, L \cdot \text{type});\}$$

7.7. IMPLEMENTATION ISSUES OF SYNTAX-DIRECTED TRANSLATION

There are two main issues:

1. Triggering execution of semantic actions; how can parsing actions be made to trigger execution of semantic rules ?
2. Managing and accessing attributes values: where should the attribute values be held and how should they be accessed ?

Solution vary according to the type of parser: bottom-up vs. top-down.

7.7.1. Top-down Translation

This section describes how *L*-attributed definitions can be implemented with predictive parsers. Translation schemes are used instead of syntax-directed definitions so the order in which semantic actions and attribute evaluations should occur is shown explicitly.

When we eliminate the left recursion from the grammar (to get a suitable grammar for the top-down parsing) we also have to change semantic actions.

1. Eliminating left Recursion from a Translation Scheme. Most arithmatic operators are left associative so it is natural to use left-recursive grammars for expressions; also there are other language constructs best described with left-recursive grammars. But left recursion must be eliminated before a predictive parser can parse a grammar. We already know to remove the left recursion but what do we do when the grammar of a translation scheme is left-recursive ? Can every semantic action and attribute evaluation of a translation scheme be put in its proper place when we eliminate left recursion from its grammar ? Let us consider an example: A left-recursive grammar for a list of digits separated by plus and minus sign is shown below.

$$E \longrightarrow E + T$$
$$E \longrightarrow E - T$$
$$E \longrightarrow T$$
$$T \longrightarrow 0$$
$$T \longrightarrow 1$$
$$T \longrightarrow 2$$
$$T \longrightarrow 3$$
$$T \longrightarrow 4$$
$$T \longrightarrow 5$$
$$T \longrightarrow 6$$
$$T \longrightarrow 7$$
$$T \longrightarrow 8$$
$$T \longrightarrow 9$$

The parse tree for 9 – 5 + 2 is also shown:

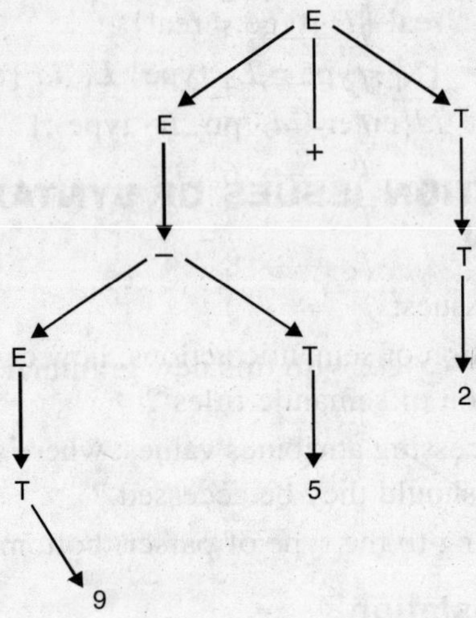

Fig. 7.17.

Note the chain of *E*-nodes going down toward the left from the root of the parse tree. Addition and substraction are left-associative so to evaluate 9 – 5 + 2 properly we should go through the chain of *E*-nodes from he bottom up to the root. A translation scheme needs only a synthesized attribute (val) to properly a list of digits separated by plus and minus sign:

$$E \longrightarrow E_1 + T \{E \cdot \text{val} := E_1 \cdot \text{val} + T \cdot \text{val}\}$$
$$E \longrightarrow E_1 - T \{E \cdot \text{val} := E_1 \cdot \text{val} - T \cdot \text{val}\}$$
$$E \longrightarrow T \{E \cdot \text{val} := T \cdot \text{val}\}$$
$$T \longrightarrow 0 \{T \cdot \text{val} := 0\}$$
$$T \longrightarrow 1 [T \cdot \text{val} := 1\}$$
$$T \longrightarrow 2 \{T \cdot \text{val} := 2\}$$
$$T \longrightarrow 3 \{T \cdot \text{val} := 3\}$$
$$T \longrightarrow 4 \{T \cdot \text{val} := 4\}$$
$$T \longrightarrow 5 \{T \cdot \text{val} := 5\}$$
$$T \longrightarrow 6 \{T \cdot \text{val} := 6\}$$
$$T \longrightarrow 7 \{T \cdot \text{val} := 7\}$$
$$T \longrightarrow 8 \{T \cdot \text{val} := 8\}$$
$$T \longrightarrow 9 \{T \cdot \text{val} := 9\}$$

Eliminating left recursion from the grammar shown above produces the grammar shown below:

$$E \longrightarrow TR$$
$$R \longrightarrow +TR$$
$$R \longrightarrow -TR$$
$$R \longrightarrow \in$$
$$T \longrightarrow 0$$
$$T \longrightarrow 1$$

$$T \longrightarrow 2$$
$$T \longrightarrow 3$$
$$T \longrightarrow 4$$
$$T \longrightarrow 5$$
$$T \longrightarrow 6$$
$$T \longrightarrow 7$$
$$T \longrightarrow 8$$
$$T \longrightarrow 9$$

The parse tree for 9 – 5 + 2 with this new grammar is also shown in Fig. 7.18.

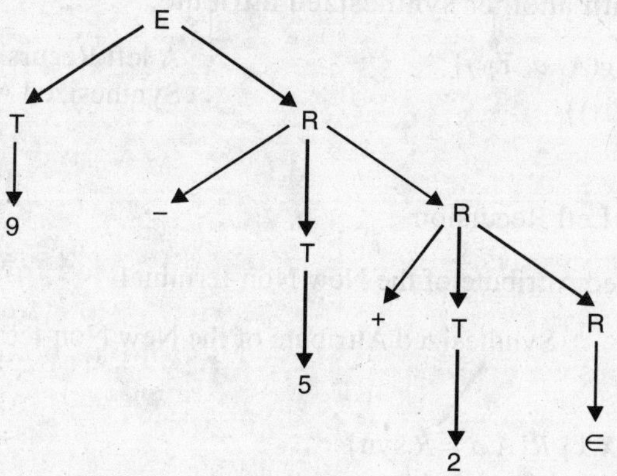

Fig. 7.18.

Note that new parse tree has a chain of *R*-nodes going down toward the right from its root whereas the first parse tree has a chain of *E*-nodes going down toward the left from its root. Addition and subtraction are still left-associative so to properly evaluate 9 – 5 + 2.

We must now go down through the chain of *R*-nodes from the root toward the $R \longrightarrow \in$ node at the bottom.

A translation scheme with this new grammar needs an inherited attribute (in) to properly evaluate a list of digits separated by plus and minus signs and the scheme sends the final result into $R \longrightarrow \in$ node at the bottom of the chain. The final result should really be sent to the root of the parse tree so the translation scheme also needs a synthesized attribute (syn) to move the final result from the $R \longrightarrow \in$ node back up the chain of *R* nodes:

Inherited attribute   Synthesized attribute

$$E \longrightarrow T\,\{R \cdot \text{in:= } T \cdot \text{val}\}\ R\ \{E \cdot \text{val:= } R \cdot \text{syn}\}$$
$$R \longrightarrow +T\,\{R_1 \cdot \text{in:= } R \cdot \text{in} + T \cdot \text{val}\ \}\ R_1\ \{R \cdot \text{syn:= } R_1 \cdot \text{syn}\}$$
$$R \longrightarrow -T\,\{R_1 \cdot \text{in:= } R \cdot \text{in} - T \cdot \text{val}\}\ R_1\ \{R \cdot \text{syn:= } R_1 \cdot \text{syn}\}$$
$$R \longrightarrow \{R \cdot \text{syn:= } R \cdot \text{in}\}$$
$$T \longrightarrow 0\ \{T \cdot \text{val:= } 0\}$$

....

$$T \longrightarrow 9\{T \cdot \text{val:= } 9\}$$

2. General Case. In general, there may be both right-associative operators and left-associative operators in a translation scheme. Right-associative operators pose no problem because they don't introduce left recursion. Left-associative operators make the scheme left-recursive but the left recursion can be easily eliminated from the grammar. Eliminating the left recursion changes parse trees by replacing each chain of nodes going down toward the left recursion changes parse trees by replacing each chain of nodes going down toward the left with a chain of nodes going down to the right. Each synthesized attribute that was originally evaluated going up the original chain is replaced by an inherited attribute that is evaluated going down the new chain. The result of the evaluation can be sent back up the new chain with another synthesized attribute.

$A \rightarrow A_1\ Y\ \{A.a = g(A_1.a,\ Y.y)\}$

$A \rightarrow X\ \{A.a = f(X.x)\}$

A left Recursive Grammar with Synthesized Attributes (a, y, x).

$\Downarrow$ Eliminate Left Recursion

Inherited Attribute of the New Non-terminal

Synthesized Attribute of the New Non-terminal

$A \rightarrow X\ \{R.\text{in} = f(X.x\}\ R\{A.a = R.\text{syn}\}$

$R \rightarrow Y\ \{R_1.\text{in} = g(R.\text{in}.Y.y)\}\ R_1\{R.\text{syn} = R_1.\text{syn}\}$

$R \rightarrow e\ \{R.\text{syn} = R.\text{in}\}$

We can evaluate attributes as follows:

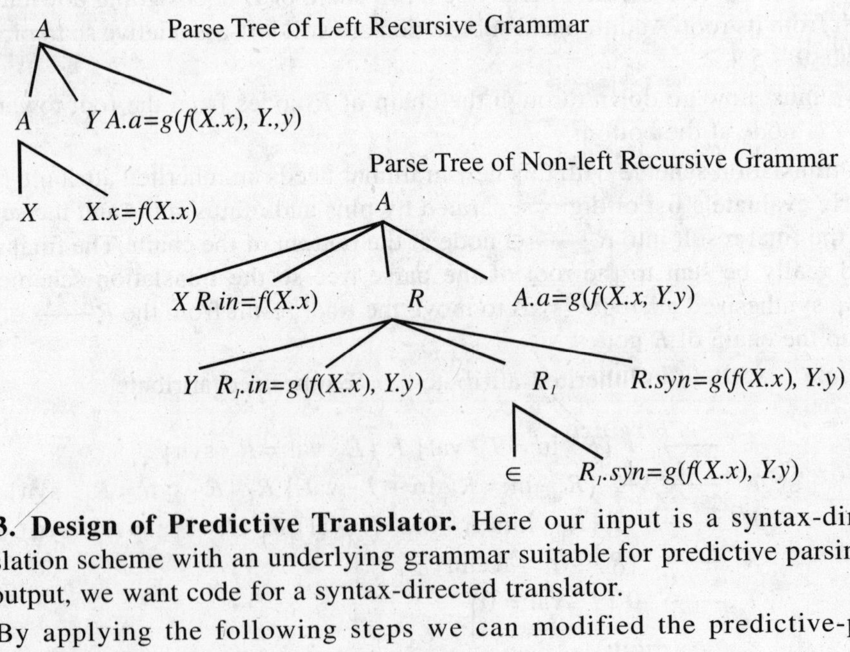

3. Design of Predictive Translator. Here our input is a syntax-directed translation scheme with an underlying grammar suitable for predictive parsing and the output, we want code for a syntax-directed translator.

By applying the following steps we can modified the predictive-parser construction.

Step 1. For each nonterminal X, construct a function that has a formal parameter for each inherited attribute of X and that returns the values of the synthesized attribute of X.

Step 2. Code for each nonterminal decides what production to use based on the current input symbol. (We have already seen it into construction of recursive decent parsing).

Step 3. By the code (associated with each production) we consider the tokens, nonterminals, and actions on he right side of production from left to right.

Step 4. For each token Y with synthesized attributed y, we will save the value y in the variable declared for $Y \cdot y$. Then we generate a call to match token Y and advance the input.

Step 5. For nonterminal X, generate an assignment $C := X(x_1, x_2, ..., x_k)$ with a function call on the right side, where $x_1, x_2, ..., x_k$ are the variables for the inherited attributes of X and C is the variable for the synthesized attribute of X.

Step 6. Finally we will copy the code into the parser, replacing each reference to an attribute by the variable for that attribute.

7.8. BOTTOM-UP EVALUATION OF S-ATTRIBUTED DEFINITIONS

We put the values of the synthesized attributes of the grammar symbols into a parallel stack. When an entry of the parser stack holds a grammar symbol X (terminal or non-terminal), the corresponding entry in the parallel stack will hold the synthesized attribute(s) of the symbol X. We evaluate the values of the attributes during reductions.

$A \rightarrow XYZ \quad A.a = f(X.x, Y.y, Z.z)$ where all attributes are synthesized.

The same is not true for inherited attributes as semantic rules for their evaluation is embedded inside the right hand side of a production.

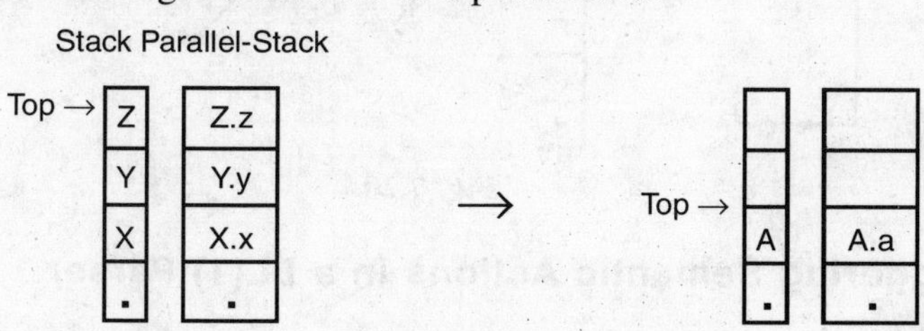

Fig. 7.19.

Production	Semantic Rules
$L \rightarrow E$ **return**	print(val[top-1])
$E \rightarrow E_1 + T$	val[ntop] = val[top-2] + val[top]
$E \rightarrow T$	
$T \rightarrow T_1 * F$	val[ntop] = val[top-2] * val[top]
$T \rightarrow F$	
$F \rightarrow (E)$	val[ntop] = val[top-1]
$F \rightarrow$ **digit**	

At each shift of **digit**, we also push **digit.lexval** into *val-stack*. At all other shifts, we do not put anything into *val-stack* because other terminals do not have attributes (but we increment the stack pointer for *val-stack*). where are we going to hold inherited attributes? Following may be the probable solution for this problem:

(*a*) We will convert our grammar to an equivalent grammar to guarantee to the followings.

(*b*) All embedding semantic actions in our translation scheme will be moved into the end of the production rules.

(*c*) All inherited attributes will be copied into the synthesized attributes (most of the time synthesized attributes of new non-terminals).

(*d*) Thus we will be evaluate all semantic actions during reductions, and we find a place to store an inherited attribute.

The Canonical LR(0) Collection for the grammar can be calculated as follows:

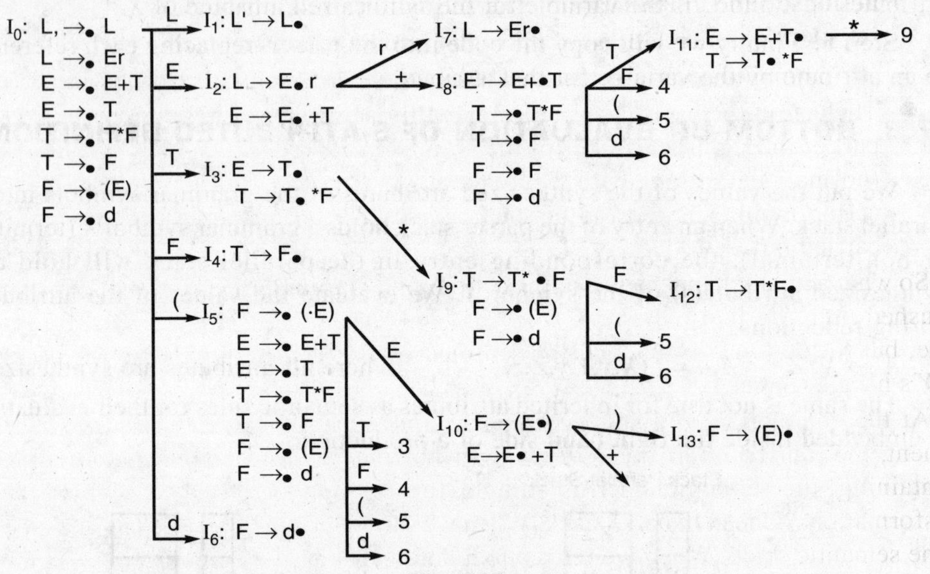

Fig. 7.20.

7.8.1. Triggering Semantic Actions in a LL(1) Parser

- Unlike the bottom-up parser, there are no distinct parsing events which can be used to trigger the execution of semantic actions.
- Augment the grammar with marker non-terminals whose only purpose is to trigger execution of semantic actions.

When a production rule is applied, these markers are pushed along with the rest of the symbols on to the syntax stack in reverse order.

When a marker is popped from the syntax stack, the corresponding semantic action is executed.

7.8.1.1. Managing Attributes in a LL(1) Parser

The syntax stack does not parallel the semantic stack - syntax stack contains what we expect to see in the future while the semantic stack contains attributes of constructs that have already been seen.

For each production applied, reserve positions in the semantic stack to hold attributes for the left hand side non-terminal and right hand side symbols.

Save these positions in the syntax stack to allow access to attributes.

7.9. IMPLEMENTATION OF SYNTAX-DIRECTED TRANSLATION

After studying the basics of syntax-directed translation, our next task is to convert it into a program that implement the input-output mapping desired. Ideally we would like to have a generator to produce this program automatically.

In Syntax-directed translation, we maintain a **semantic stack** alongwith the parse stack. The basic rules governing the semantic stack are as follows:

- When a terminal is parsed, its attribute is pushed on the stack.
- When a nonterminal is parsed, its translation is pushed on the stack.
- Values are pushed onto the semantic stack (and popped off) by adding **actions** to the grammar rule. The action for a rule must:
  - POP the translations of all right-hand-side nonterminals
  - Compute and push the translation of the left-hand-side nonterminals.
- The action themselves are represented by **action numbers,** which become part of the right-hand sides of the grammar rules. They are pushed onto the (normal) stack along with the terminal and non-terminal symbols. When an action number is the top-of-stack symbol, it is popped and the action is carried out.

So what actually happens is that the action for a grammar rule $X \longrightarrow Y_1 Y_2 \ldots Y_n$ is pushed onto the (normal) stack when the derivation step $X \longrightarrow Y_1 Y_2 \ldots Y_n$ is made, but the action is not actually performed until complete derivation for all of the Y's have been carried out.

At the completion of a successful parse, the semantic stack should have one element, the translation of the start symbol. When doing bottom-up (LR) parsing, maintaining, the semantic stack is straight forward; we essentially "mirror" the transformations done to th parse stack. When we shift a terminal, we push its attribute on the semantic stack. When we reduce to a non-terminal, we POP the items on the semantic stack corresponding to the items popped from the parse stack, and then PUSH the computed translation for the non-terminal on the semantic stack.

Unfortunately, when doing top-down parsing (LL(1)) or recursive descent), performing syntax directed translation is not so easy, since the translation is performed bottom-up. We solve this problem by adding action to our grammar, specifying how translation is to be performed for each production. In LL(1) parsing, these actions (more specifically, numbers representing the actions) are pushed on the parse stack alongwith the corresponding production, and then executed when popped. These actions are responsible for popping all the items on the semantic stack corresponding to terminals and non-terminals on the right-hand side of the production, and then pushing the computed translation.

Example 7.4. For the given CFG (for the counting-parentheses) with translation rules, write the translation actions.

CFG	Translation rules
1. exp $\longrightarrow \in$	exp · trans = 0

2. exp ⟶ (exp) exp1 · trans = exp2 · transt 1
3. exp ⟶ [exp] exp1 · trans = exp2 · trans

Solution. Here, we are given the syntax directed translation for the language of balanced parentheses and square brackets. The translation of a string in the language is the number of parenthesis pairs in the string.

CFG	Translation Rules
1. exp ⟶ ∈	exp · trans = 0
2. exp ⟶ (exp)	exp1 · trans = exp2 · trans +1
3. exp ⟶ [exp]	exp1 · trans = exp2 · trans

The first step is to replace the translation rules with **translation actions.** Each action must:

- Pop all the right hand side nonterminals translation from the semantic stack.
- Compute and push the left-hand side nonterminals translation.

Here are the translation actions:

CFG	Translation Actions
1. exp ⟶ ∈	Push (0);
2. exp ⟶ (exp)	exp 2 trans = POP ( ) ; push (exp 2 – trans + 1);
3. exp ⟶ [exp]	exp 2 trans = POP ( ); push (exp 2 trans);

Next, each action is represented by a unique action number, and those action numbers become part of the program rules:

CFG with Actions

1. exp ⟶ ∈ #1
2. exp ⟶ (exp) # 2
3. exp ⟶ [exp] # 3

Here #1: push (0);
#2: exp 2 trans = pop ( ); push (exp 2 trans +1);
#3: exp 2 trans = pop ( ); push (exp 2 trans);

7.9.1. Semantic Actions

There are three type of semantic actions necessary to give meaning to the syntactic structure. These actions are:

1. Setting attributes in the symbol table that is number of bits in a **BIT** variable, the dimensions of an **ARRAY** variable and more commonly setting type bits, that is real, integer, external, label, actual, etc.
2. Writing intermediate code, such as 4-tuple to indicate an arithmatic operation, a label, a jump, a conditional test, etc.
3. Passing of semantic information through the semantic stack.

TIPS

The syntax and semantics of a language are not orthogonal. It is not always possible to create the syntax and then later somehow define the semantics.

For example, consider the following set of productions:
Suppose in a language, the statement for declaring variable is

ABC, AAA, DFG INTEGER

and the following syntax was used to "parse" this string.

1. decl ⟶ var
2. decl ⟶ decl, var
3. declstat ⟶ decl integer

Then the type of the declared variables is determined that is production 3 is recognized, all of the preceding variables have been lost and their attributes in the symbol table cannot be set.

Example 7.5. Consider the following production for declarations,

decstat ⟶ VECTOR var ELEMENTS const

Solution. This statement in the language would translate into something like :

VECTOR AQZ ELEMENTS 50

Then suppose we were using a bottom-up parser, the syntax and semantics stack would look something like the following:

Before Reduction		After Reduction	
Syntax	**Semantics**	**Syntax**	**Semantics**
Const	157		
ELEMENTS	0		
var	113		
VECTOR	0	decstat	4 157
-	-	-	-
-	-	-	-
-	-	-	-
-	-	-	-

Let us explain the process:

In the BEFORE stack the 113 is the symbol table location of AQZ, and the 157 is the symbol table location of 50. When reducing this production, the symbol table entry of AQZ needs to have its attributes set to vector, which in this example is represented by the numeric code 4, and the first dimension set to the contents of the symbol table entry at location 157 which contain 50. At this instant in time the symbol table entry at location 113 has the symbol name of AQZ; the type is vector; the number of elements is 50.

Example 7.6. Now consider the following production for executable statements,

factor ⟶ factor * primary

Solution. When this reduction is made, the stack will look something like,

Before Reduction		After Reduction	
Syntax	**Semantics**	**Syntax**	**Semantics**
Primary	I$015		
*	0		
Factor	R$036	factor	R$040

where R$036 is the 36th real intermediate result name and I$015 is the 15th integer intermediate result name. The partial string

--------*Y*/*X* * (I + 2)-------

could have been the string that represents the stack where *Y*/*X* is syntactically the **factor** and (I + 2) is the **primary.**

The semantic for the above production might be 1.

Example 7.7. Compute the type of an expression that includes both arithmetic and boolean operators. (The type is INT, BOOL, or ERROR)

Solution.

CFG	**Translation rules:**
1. exp ⟶ exp + exp	if ((exp$_2$ · trans = INT) and – (exp$_3$ · trans = INT then exp$_1$ · trans = INT else exp$_1$ · trans = ERROR
2. exp ⟶ exp and exp	if((exp$_2$ · trans = = BOOL) and (exp$_3$ · trans = = BOOL) then exp$_1$ · trans = BOOL else exp$_1$ · trans = ERROR
3. exp ⟶ exp = = exp	if (exp$_2$ · trans = = exp$_3$ · trans) and (exp$_2$ · trans ! = ERROR)) then exp$_1$. trans = BOOL else exp$_1$ · trans = ERROR
4. exp ⟶ true	exp · trans = BOOL
5. exp ⟶ false	exp · trans = BOOL
6. exp ⟶ int	exp · trans = INT
7. exp ⟶ (exp)	exp$_1$ · trans = exp$_2$ · trans

Input: (2 + 2) = 4

Annotated Parse tree:

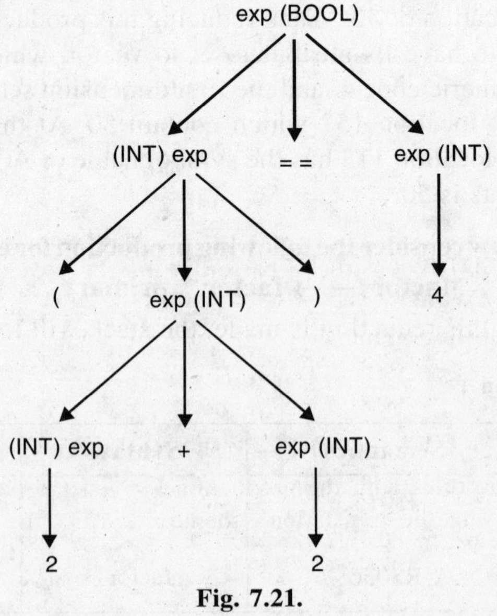

Fig. 7.21.

Example 7.8. Write a procedure for Depth-First Transversals.

Solution. A depth-first transversal of a parse tree is one way of evaluating attributes. Note that a syntax directed definition does not impose any particular order as long as order compute attribute of parent after all its children's attributes.

```
PROCEDURE visit (n: node)
BEGIN
FOR each child m of n, from left to right
Evaluate semantic rules at node n
END
```

EXERCISE

1. Write the syntax-directed translation to go alongwith the LR parser for the following :

$$S \rightarrow AE$$
$$A \rightarrow DS \text{ while}$$
$$D \rightarrow \text{do}$$

2. Construct an annotated parse tree for input expression (4 * 7 + 1) * 2, according to given syntax-directed definition:

Production	Semantic Rule
$L \rightarrow En$	print (E . val)
$E \rightarrow E_1 + T$	E . val:= E_1 . val + T . val
$E \rightarrow T$	E . val:= T . val
$T \rightarrow T_1 * F$	T . val:= T_1 . val × F . val
$T \rightarrow F$	T . val:= F . val
$F \rightarrow (E)$	F . val:= E . val
$F \rightarrow$ digit	F . val:= digit . lexval

3. Consider the following SDTS (syntax-directed translation schemes):

$E \rightarrow E_1 * T$ $\{E . \text{val} = E_1 . \text{val} * T . \text{val}\}$
$E \rightarrow T$ $\{E . \text{val} = T . \text{val}\}$
$T \rightarrow T_1 - F$ $\{T . \text{val} = T_1 . \text{val} - F . \text{val}\}$
$T \rightarrow F$ $\{T . \text{val} = F . \text{val}\}$
$F \rightarrow 2$ $\{F . \text{val} = 2\}$
$F \rightarrow 4$ $\{F . \text{val} = 4\}$

Using the above SDTS, construct a parse tree for the epxression: 4 – 2 – 4 * 2, also compute its E . val.

4. Construct the following grammar:

$$E \rightarrow E + T / T$$
$$T \rightarrow T * F / F$$
$$F \rightarrow (E) / id$$

Associate semantic rules with the production for construction of syntax tree for an expression. Using the translation scheme construct the syntax tree for the expression $a + b * c$.

❑❑❑

Chapter 8

SYMBOL TABLE MANAGEMENT

INSIDE THIS CHAPTER

8.1. Introduction to Symbol Table **8.2.** Information Caching **8.3.** What Should be Stored ?
8.4. Symbol Table Organization **8.5.** Scoping.

8.1. INTRODUCTION TO SYMBOL TABLE [MDU 2008]

Most computer programs which interact with humans have at least some places where symbolic input is allowed. For example, job control languages or command shells allow the use of commands such as "Is", "cat" or "exit" along with symbolic program and file names, and programming languages allow symbolic identifiers and keywords. **For a program to process such input, it must have some way to associate such symbolic names with their abstract values. Mechanisms which maintain such associations are called *symbol tables*.**

The problems of constructing and managing a symbol table are relatively independent of the nature of the values which it associates with symbolic names, and these problems remain essentially the same whether names are associated with integer memory addresses, as they are in a simple assembly language, or with data types, storage locations, and other information needed by a high level language compiler.

Any data storage component of a program which may be specified without reference to how it is implemented is called an abstract data type. In most common programming languages, programmers view the predefined data types such as integer and real as abstract data types, using them without reference to the number of bits or the representation system used for values of these types. When a user defines a new type, the user must give a concrete representation for the type, but modern programming methodologies encourage the user to isolate that representation from its uses, allowing the user to use the new type abstractly.

Modern programming languages such as Java, C++, Ada and Modula 2 provide facilities which allow the explicit declaration and use of abstract data types, but even when explicit declaration is not possible, it is always possible to hide the data structures representing an abstraction behind a wall of procedures and functions. In any case, a clear understanding of the abstract data types involved is one of the most powerful tools for modularizing large programs.

On the basis of above discussion, we can discuss the formal definition of symbol table as follows:

Definition "A symbol table is a data structure used by a compiler to keep track of scope, life and binding information about names. These names are used in the source program to identify the various program elements, like variables, constants, procedures, and the labels of statements. The symbol table is searched every time a name is encountered in the source text. When a new name or new information about an existing name is discovered, the content of the symbol table changes.

Therefore, a symbol table is must have an efficient mechanism for accessing the information held in the table as well as for adding new entries to the symbol table."

In any case, the symbol table is a useful abstraction to aid the compiler to ascertain and verify the semantics, or meaning, of a piece of code. It will keep track of the names, types, locations and properties of the symbols encountered in the program. The type system and the code generation system rely on the symbol table to provide them with information about symbols encountered elsewhere in the code. It makes the compiler more efficient, since the file doesn't need to be re-parsed to discover previously processed information.

This chapter will focus on two of the most important functions of the symbol table in a compiler. First, they cache useful information about various symbols from the source code program for later use during code generation. Second, they provide the type checking mechanism that determines the semantic correctness of a program.

8.2. INFORMATION CACHING

Here we introduce the issues of which information to cache and where to store it. One difficulty we have in answering this question is, of course, that the information we need to store depends on the language, and on the architecture to which the language is being compiled. In this book we discuss statement driven declarative languages like ALGOL, C++, or Pascal.

The first step in designing the symbol tables is to decide what information needs to be recorded. Generally we only record information about identifiers. Most tokens in the program, such as keywords or symbols, have a fixed meaning, so we need not record information about them. Identifiers, on the other hand, do not have a fixed meaning. They can represent different constructs or have different semantics in different files and even in the same file. In order to determine what their meaning is in a given context, we must record information about these identifiers so we can determine its meaning elsewhere in the compilation process.

TIPS

Having determined that symbol tables are useful, the question invariably arises, "How does one build the symbol table ?" There are two hidden questions in this seemingly innocent inquiry. "What information needs to be stored ?" is the first and "How will the information be organized ?" is the second.

8.3. WHAT SHOULD BE STORED ? [MDU 2007]

Exactly what information is stored in the symbol table depends on many things. The programming language will determine much of the information that is stored, but the target architecture will also influence what data is stored. In fact, some assumptions about how to produce code can affect what values are stored in the table. Rather than attempt to list all the possible structures and values that might appear in a symbol table, this section will provide a general overview of the classes of information typically stored.

When one considers the range of information that can be stored about identifiers, it seems logical to classify identifiers in terms of the constructs they represent. In other words, different information will need to be stored for constants, variables, procedures, enumerations, type definitions and so on. What follows is a description of various common declarative language constructs and typical classes of information symbol tables would record for those constructs.

(*i*) Constants: Constants are identifiers that represent a fixed value - one that can never be changed. Since programmers will wish to access these values by name, the name must be stored. Naturally, the corresponding value must also be stored. Finally, since the values must be used properly in the type system, type information is also included. Unlike variables or procedures, no run-time location needs to be stored for constants. These are typically placed right into the code stream by the compiler at compilation time.

(*ii*) Variables: Variables are identifiers whose value may change between executions and during a single execution of a program. They represent the contents of some memory location. The symbol table needs to record both the variable's name as well as its allocated storage space at runtime. Typically this location is stored as an offset relative to some position (which can be turned into an absolute location at link or load time). Just as with constants, their type information should be recorded for later use by the type system.

(*iii*) Types (user defined): A user defined type is typically a conglomeration of 1 or more existing types. (Some languages allow you to define recursive structures). Types are accessed by name and reference a type definition structure. Each structure will record important information about itself, like its size, the names of its members (if it's a record) or its upper and lower bounds (if its an array). What information is stored will depend on what type is being defined.

(*iv*) Sub-programs: Procedures, functions, and methods are named segments of code. Naturally, the symbol table should record a procedure's name. The type they return (if any) should be noted. When sub-programs are accessed at run time it is typically by their location in the code stream (or some handle to that location), thus the location of the code generated for a given procedure should also be recorded.

The (formal) parameters and local variables of a function are separate identifiers (variables) in their own right, and should be stored in separate records. Thus, they are treated much like the fields of a user defined record. They are stored as a list of variable records separated, but accessible from, the main procedure record.

***(v)* Classes :** Classes are abstract data types which restrict access to its members and (encapsulation) provides convenient language-level polymorphism. They are really a special case of user defined types, and are structurally no different. But it may be convenient to store information about classes above and beyond that required for other user defined types. This includes the location of the default constructor and destructor, and the address of the virtual function table.

***(vi)* Inheritance:** There are many different ways to perform inheritance, and a symbol table record is needed to keep track of which classes are being inherited and exactly how inheritance is performed. A compiler might consider whether *shared* or *non-shared* inheritance is specified for a given class. In C++ the keywords *public, private* and *protected* modify the visibility of inherited items, and may be recorded with the inheritance information. A reference to the participating classes (parent and child) could also be recorded in an inheritance structure.

Table 8.1 summarizes the data field requirements for symbol table records that store information about various identifier declarations.

TABLE 8.1. Common Symbol Table Information Related to Identifiers.

	Name	Scope	Type	Size	Offset	Value	Other
Variable	4	4	4	4	4	-	
Constant	4	4	4	-	-	4	
Type	4	4	4	4	-	-	
Function	4	4	4	4	4	-	return type
Base Class	4	4	-	4	-	-	inheritance method

Languages that allow a programmer to specify arbitrarily complex types will need some dynamically extensible system for defining types. It is also convenient to encapsulate the "variable-sized" information about program constructs like subprograms. The most intuitive method involves making "trees" of user-defined-type records.

***(i)* Arrays:** Arrays represent a collection of uniformly typed elements that may be randomly accessed by index. For each dimension of an array, the compiler will need to know about such things as the lower boundary of the array (the lowest valid index), the upper bound (the largest valid index), the index size, index type, the total size, and the type of the elements contained. When many different types can be used to index an array, the index type and size will also be recorded. Finally, the total amount of space to be allocated for each dimension of an array should be stored.

***(ii)* Records:** Records represent a collection of possibly heterogeneous members which can be accessed by name. The symbol table probably needs to record each of the record's members. The compiler also needs to know the size of the record (how much space to allocate for all of the members). Each of the fields of a record will probably be a reference to another symbol table record—like a variable or a type—which may in turn reference another record or array. In this manner, very complex user-defined structures can be created.

***(iii)* Sub-Programs:** In many languages procedures are not a user definable type. However, their parameters and local variables are often stored the same way fields of a record are. It may be convenient to treat this aspect of a procedure the same way records are. A structure for storing the "complex" information about procedures would record the parameters, the local variables, and the amount of stack space occupied by a procedure at run time.

***(iv)* Class:** Just like a record, the fields of a class can be conveniently stored in a separate record. Classes will also store their methods, constructors, destructors and virtual function table in this complex-information structure.

***(v)* Module:** Stores the module size, its name, parent, its members, and a time stamp. The time stamp is used to guarantee that load time that the models have been compiled in the correct order or are all up to date.

Table 8.2 summarizes the data member requirements for symbol table records that store data about complex types and complex language constructs.

TABLE 8.2. Data fields for Complex Type Information.

	Name	Member Names	Size	Index Type	Element Type	Other
Array	-	-	4	4	4	array index bounds
Record	-	4	4	-	-	
Subprogram	-	4	4	-	-	
Module	4	4	4	-	-	
Class	-	4	4	-	-	

8.4. SYMBOL TABLE ORGANIZATION [MDU 2008, 2009]

There are two major activities to relate the symbol table, namely: inserting values and accessing stored information. Value insertion occurs primarily when the parser processes declarations, while the value retrieval occurs elsewhere, generally in the expressions and statements of the language.

Definition

Value insertion operations include: creating a record for a new symbol, assigning values of fields of the records and making adjustments to the internal scoping.

Definition

Value retrieval operations consist of searching backward in the scope to find an identifier by name, examining the formal parameters of a procedure to match them with the actual arguments and examining values in a found record to compare with actual values found in the code.

It will be useful in discussing the underlying implementation(s) for a symbol table to think of it as an abstract data type. We will present the "Symbol Table" abstract data type in terms of the data it stores and the manipulations (or functions) that may be applied to the data. The data it stores is a collection of the various records we discussed in the previous section. The typical operations and the ones that will affect the underlying data structures are:

(1) Find a record by name (that is access a variable as in "x = 5")
(2) Find a record by name in a specific scope (that is access a field of a record as in "a.x = 5")
(3) Find a record by its relationship with another record (that is get the type of a variable or the parameters of a function)
(4) Insert a new record
(5) Update an existing record

We can split the symbol table implementation into two layers where each layer represents a certain data abstraction. The bottom layer is an associative array, and is concerned with storing records and retrieving records by name. This corresponds to operations 1, 2 and 4 in the previous list. The top layer organizes the records into groups and determines how the scoping of the language is maintained. It is responsible for operations 3 and 5 in the previous list, as well as providing convenient access to the operations of the lower level.

The bottom layer of the symbol table is solely concerned with the storage and retrieval of symbol table records *by name.* It has two basic operations: add (or insert) and retrieve (or find). There may be many varieties of find: find by name alone, find by name and construct-type, find by name and type, find by name in a limited scope, etc. These operations (find-by-name, insert, etc.) are also the principle operations of the "associative array" abstract data type. A conventional array addresses records by number; an associative array addresses records by label.

Unlike conventional arrays, associative arrays have no obvious implementation on modern computer architectures, so symbol table designers will need to develop algorithms to implement this abstract data type. There are many different data structures that provide functions to add, find (and remove) records using labels as an index. These include linked lists, binary-trees and hash tables. Some are easy to implement, others provide efficient methods. Which implementation a particular compiler uses will depend on the needs of the individual compiler and compiler writer.

(1) The first option we discuss is also the easiest to implement: **AN UNORDERED LINKED-LIST.** Each record in the linked list would store its name, the other relevant record data, and a pointer to the next and previous record in the list. The insert operation merely places new symbol table records at the end of the list in the order they are encountered. The retrieve operation is equally simple. This list is traversed, and the first (or last) record matching the retrieval criteria is returned. The list is usually traversed from back to front under the assumption that the more recently declared symbols are more likely to be used in a program. On the plus side, linked lists are easy to implement and insertions can be performed in constant time. On the negative side, searching the list occurs in linear time. In Fig. 8.1(*a*) and (*b*) show the library and resulting symbol table structure for lists.

(a)

```
library A;
proc B (X, Y: int)
   var  I, J: int
begin
end proc;
end library.
```

(b)

(c)

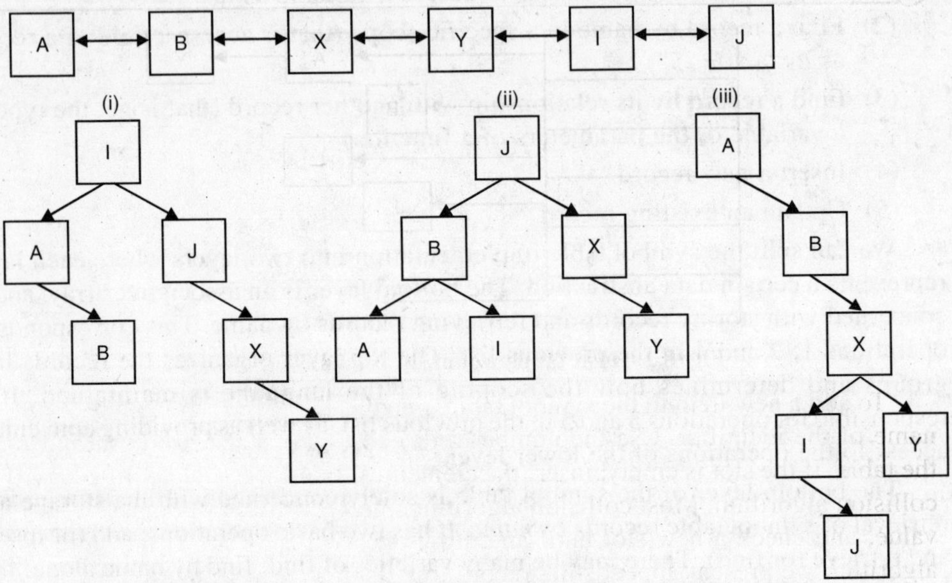

Fig. 8.1. A library (*a*) and two methods of storing the symbol table (*b*) as a linked list and (c) as a binary tree.

Since searching for information by name is one of the most common operations for a symbol table (much more common than record insertion), a better approach is often desired.

(2) The next logical step is to move to a **BINARY TREE**. Search times for binary trees are proportional to log(n). Algorithms for building and searching binary trees are explained in any data structures book. By way of quick review: all nodes to the left of a given node occur lexicographically before that node, and all nodes to the right occur after. To find a node, one simply compares the target name to the node name and takes the left or right child accordingly. Unfortunately, in order to add a node to the correct place in the tree requires traversing the tree to find the spot where the new record should go (to maintain the ordering of the nodes). This means that the insertion time is the same as the lookup time, roughly O (log (n)) [see the Fig. 8.1(i)]

Even worse, the O (log (n)) time is based on a *balanced* binary tree like the one in Fig. 8.1(ii). If the nodes are inserted in near alphabetic order, the basic binary tree algorithm will produce a tree that looks like a linked list. Figure 8.1(iii) shows the tree produced by the basic algorithm on our example program. As an additional solution, many algorithms exist which keep binary trees approximately balanced, at the cost of additional algorithm complexity.

TIPS

When speed really becomes an issue, symbol table implementers turn to **HASH TABLES**; most large scale or professional compilers take this route. In the best case, hash tables give a constant lookup time ($O(1)$) and a constant search time. Worst case time depends on the particular hash table implementation, but can be as bad a linear insertion and linear lookup if all the names hash to the same location.

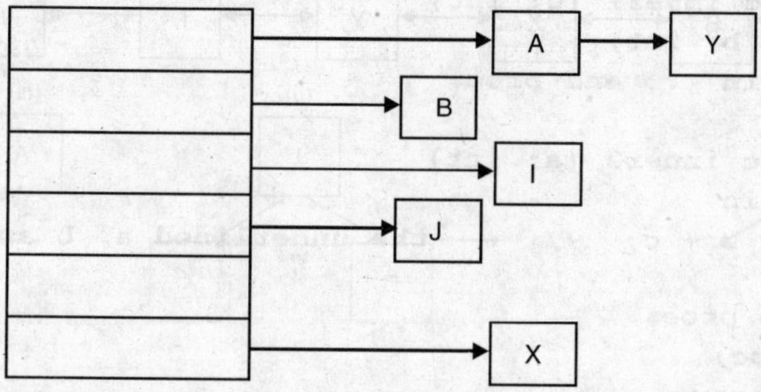

Fig. 8.2. Hash table organization of a symbol table.

To add a new item to the symbol table, one must apply the hash algorithm to the name of the identifier. Use the number returned by the algorithm as an index into the table. If the slot is empty, insert the element. If the slot is already filled apply the collision algorithm. Most collision algorithms will repetitively compute a new hash value, and check a new slot until an empty one is found. The collision resolution algorithm we present appends the record to a list of records stored at that location.

To lookup a name in the symbol table, the compiler uses the same approach as insertion, only instead of checking if the slot is empty, it checks if the slot contains a record with the correct name. (that is hash the name, check the slot, if it is empty, no record exists, if it contains a record with the correct name, return it, otherwise apply the collision algorithm, and repeat the process).

8.5. SCOPING

There are at least two popular approaches to deal with scope in structured languages. In this book we'll call them "scope-by-number" and "scope-by-location". As their names imply, one uses values (numbers) stored in each record to determine the scope of the identifier, while the other uses the structure of the symbol table to determine scoping. (It's kind of an explicit vs. implicit representation issue).

Let's look briefly at the purpose of scoping in the symbol table. In general, programs may have several constructs (variables, procedures, constants, etc.), some of which have the same name as other constructs. The language allows these synonyms to be present because they can be distinguished based on context information. For instance, in C++ a variable '*x*' in a procedure will shadow a global variable '*x*' for code in the procedure, but in other procedures the global variable will be visible. This means that when a compiler looks up an identifier by name it has to retrieve the one with the correct name *that was declared most recently within the nested blocks currently visible within the language.* The following code segment given below:

```
library scoping;
var a, b, c: int
type x is record
        a, b, c: int;
    end record
proc outer ( )
    var a, b: int;

    proc inner1 (c: int)
    var b: int;
    begin ... end proc;

    proc inner2 (a: int)
    begin
    a:= b + c;   //  ←—the underlined a, b and c are
used
    end proc;
 end proc;
 end library.
```

When the compiler processes the expression pointed to by the arrow, it should select the underlined variables. Namely, the a that is a parameter to procedure inner2, the 'b' that is a local variable of the procedure outer, and the global variable 'c'. The 'b' and 'c' from innerl should be skipped because they are not in the set of nested blocks visible at the arrow, and the 'a' in outer is shadowed by inner2's parameter. The 'a', 'b' and 'c' from type x are outside the nested scoping too, and the 'a' and 'b' in scoping are shadowed by the 'a' and 'b' in outer.

The scoping layer does not add operations to the table, rather, it determines how the find method is implemented. That is, it adds another layer to this lookup function which can find the "correct" identifier from among all the identifiers with the same name.

8.5.1. Scope-by-Number

The first method, "scope-by-number" (or single table), stores all the identifiers in the same table. When multiple identifiers have the same name, they are stored in a common "bucket". Each nesting block is assigned a number by the compiler. All identifiers declared within that block store that block's number as their scope. Thus, the scope of identifiers stored in the same bucket can be distinguished by the scope number. When searching for the correct identifier, a stack of scope numbers (representing the nested blocks), along with the name is used to determine which identifier should be retrieved. Figure 8.3 shows how the code, given above might appear in a "scope-by-number" structured symbol table.

In Fig. 8.3, bold items indicate the definition of a scope number, normal font items represent the recording of a scope number. In this case, when the compiler searches for c in the expression $a := b + c$, it will find 'c' in the hash table, skip the first and second entries in the bucket because their scope number does not appear in the search stack, and accept the last entry because the scope number 1 is also in the search stack.

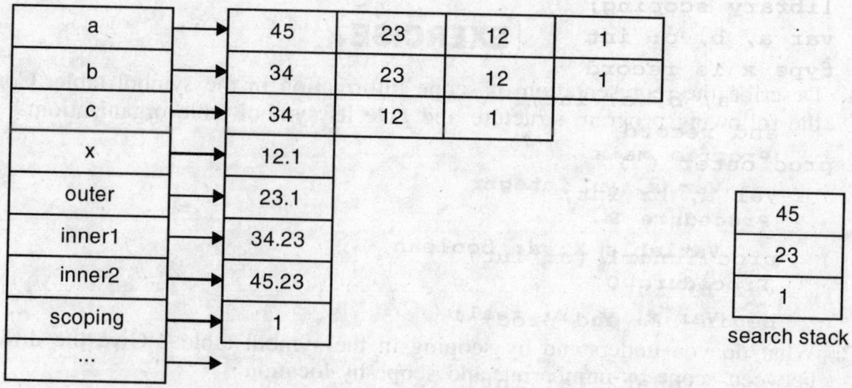

Fig. 8.3. Symbol-table scoping in a simple hash table.

8.5.2. Scope-by-Location

The second method, "scope-by-location" (or nested tables), creates separate tables for each scope block in the language. This block structured scoping approach will be the approach taken by the implementation section of this book.

New records are inserted into the data structure (bottom layer) for the current level of scope. When a new level of scope is introduced (we enter a new block — be it record, class or method), a new bottom layer associative array is created that is referenced by a record in the previous (outer) list. No buckets are needed. Instead a more versatile "find" routine is required.

The find routine checks for records in the associative array for the current level of scope first. If it finds a matching identifier, that one is returned. Otherwise it recursively searches the next (outer) layer of scope until either an appropriate record is found or the outermost layer is exhausted. The first matching identifier is returned. If no match is found the return value should reflect this.

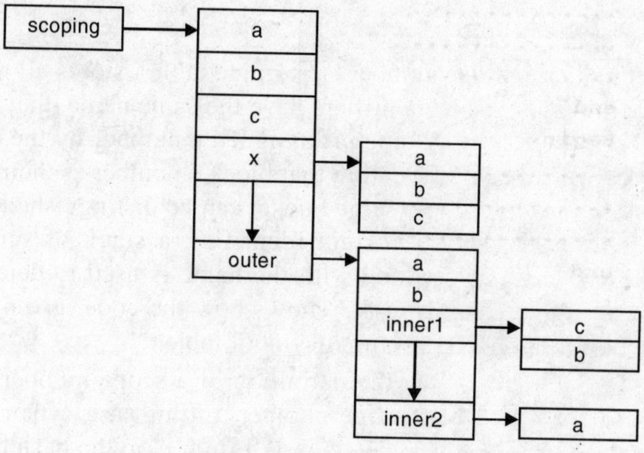

Fig. 8.4. Example of a table using block structured scoping.

If no records are deleted, the scoping layer creates a tree of associative arrays that very closely resembles the structure of the original program.

EXERCISE

1. Describe the representation of scope information in the symbol table. Consider the following program structure and give its symbol table organization:

```
Program main
    Var x, y: integer
Procedure P
    Variable X, a: boolean
Procedure Q
    Var x, y, z: real;
```

2. What do you understand by scoping in the symbol table ? Give the difference between scope-by-numbering and scope-by-location.
3. What should be stored in symbol table ?
4. Discuss the symbol table organization, also give the difference between binary tree and hashing organization of symbol table.
5. What is the significance of symbol table at runtime and compile time.
6. Discuss the symbol table design issues in reference of java language.
7. Give the scope information in the symbol table for following programming fragment :

```
Program main
var x, y: integer:
procedure P:
var x, a: boolean;
procedure q
var x, y, z: real;
Begin
..............
..............
..............
end
Begin
..............
..............
..............
end
Begin
..............
..............
..............
end
```

❑❑❑

9 Chapter

ERROR DETECTION AND RECOVERY

INSIDE THIS CHAPTER

9.1. Introduction to Error Handling **9.2.** Categories and Goals of Error Handling **9.3.** Error Recovery Strategies **9.4.** Recovery from Lexical Phase Errors **9.5.** Recovery from Syntax Error **9.6.** Error Recovery in Predictive Parsing **9.7.** Error Recovery in LR Parsing **9.8.** Recovery from Semantic Errors.

9.1. INTRODUCTION TO ERROR HANDLING [MDU 2007]

"A good compiler should be able not only to detect errors but also to recover from them."

Error recovery is an important feature of any compiler, through which compiler can read and execute the complete program even it have some errors. A good compiler should be able modify the input, when it find error in lexical analysis phase. An even more sophisticated compiler may attempt to correct the erroneous input by making a guess as to what the user intended.

Reporting the error in correct and more centric message is another important feature of any good compiler. **A compile error message should contain certain number of properties:**

- Message should report the errors in original source program rather than in terms of some internal representation of source program.
- Error message should not be complicated.
- Error message should be very specific and should fix the errors at correct positions.
- There should be no duplicacy of error messages, that is same error should not be reported again and again.

9.2. CATEGORIES AND GOALS OF ERROR HANDLING [MDU 2007]

Errors can be categorises as follows:

1. Lexical : Ill-formed numeric literals and identifiers. Some time a character is encountered, which is not in the language.

2. Syntactic: Some time programmar commits the mistake during coding process. For example unbalanced parenthesis and punctuation.

3. Semantic: Some time errors of declaration and scope are takes place. For example undeclared or multiple-declared identifiers. Type mismatch is another example of semantic errors.

4. Logical: It is next to impossible for the compilers to handle the logical errors. Infinite loops is the example of logical error. In these type of errors code is correct syntactically but it does not operate as indeed.

A good compiler have following **goals for error handling:**

- Detect the presence of errors and produce, "meaningful" diagnostics.
- To recover quickly enough to be able to detect subsequent errors.
- Error handling components should not significantly slow down the compilation of syntactically correct programs.

9.3. ERROR RECOVERY STRATEGIES

Error recovery strategies can be divided as follows:

1. Phrase Level Recovery. Replace, delete, or insert a token as a prefix to the input which will enable the parser to continue. Inserting tokens can lead to infinite loops.

2. Error Productions. Add rules to the grammar that describe the erroneous syntax. This strategy can resolve many, but not all potential errors.

- If we have a good idea of the common errors that might be encountered, we can augment the grammar with productions that generate erroneous constructs.
- When an error production is used by the parser, we can generate appropriate error diagnostics.
- Since it is almost impossible to know all the errors that can be made by the programmers, this method is not practical

3. Global Correction. Replace incorrect input with input that is correct and requires the fewest changes to create. This requires technique that are costly in terms of time and space.

- Ideally, we would like a compiler to make as few changes as possible in processing incorrect inputs.
- We have to globally analyze the input to find the error.
- This is an expensive method, and it is not in practice.

4. Panic Mode. Discard tokens until a, "synchronization" token is found, such token are often delimiters, such as, ';', '}' , ')', 'end', whose role in the source program are clear. This method can fail to detect multiple errors in close proximity.

In case of panic mode error recovery strategy, an error is detected when:

(*a*) *For a table based implementation:* given non-terminal 'A' on the stack top, and 'a' as the next input token, parsing table lookup (A, a) returns an error.

(*b*) *For a table based implementation:* The terminal on top of the stack does not match the next input token.

(*c*) *For a recursive descent implementation:* A specific token is expected to be next in input in the input, but is not.

9.4. RECOVERY FROM LEXICAL PHASE ERRORS

Definition A **lexical error** is a sequence of characters that does not match the pattern of any token.

The most common causes of lexical error are:

- The addition of an extraneous character.
- The removal of a character that should be present.
- The replacement of a character with an incorrect character.
- The transposition of two characters.

The theoretically best way of handling a lexical error is to find the closet character sequence that does match a pattern but this will take too much time and is never used in a practical compiler. The easiest way to handle a lexical error is to store the bad character in the symbol table and return an ERROR token to the syntax analyzer with a pointer to the symbol table entry. To simplify the coding projects none of their test files will have any lexical errors.

Another way the parser can help the lexical analyzer can improve its ability to recover from errors is to make its list of legitimate token available to the error recovery routine. The error-recovery routine can then decide whether a remaining input's prefix matches one of these tokens closely enough to be treated as the token.

9.5. RECOVERY FROM SYNTAX ERROR

For some examples of common syntax errors consider the following pascal program :

```
(1)  Program prmax (input, output);
(2)  Var
(3)  x, y: integer;
(4)  Function max (i: integer ; j: integer): integer;
(5)  {return maximum of integers i and j}
(6)  begin
(7)  if i > j then max: = i
(8)  else max: = j
(9)  end;
(10) begin
(11) read ln (x, y);
(12) write ln (max (x, y))
(13) end.
```

Errors in punctuation are common. For example: using a comma instead of a semicolon in the argument list of a function declaration (line 4); leaving out a mandatory semicolon at the end of a line (line 4); or using an extraneous semicolon before an else (line 7).

Operators errors often occur: for example, using = instead of: = (line 7 or 8),

Keywords may be misspelled: **writelin** instead of **writeln** (line 12). A **begin** or **end** may be missing (line 9).

Usually difficult to repair.

A common technique is to print the offending line with a pointer to the position of the error. The parser might add a diagnostic message like "semicolon missing at position" if it knows what the likely error is.

9.6. ERROR RECOVERY IN PREDICTIVE PARSING

An error may occur in the predictive parsing (LL(1) parsing in following cases:

- If the terminal symbol on the top of stack does not match with the current input symbol.
- If the top of stack is a non-terminal *A*, the current input symbol is a, and the parsing table entry *M*[*A*,*a*] is empty.

The parser do the following in case of error:

- The parser should be able to give an error message (as much as possible meaningful error message).
- It should be recover from that error case, and it should be able to continue the parsing with the rest of the input.

9.6.1. Panic-Mode Error Recovery in LL(1) Parsing

In panic-mode error recovery, we skip all the input symbols until a synchronizing token is found. All the terminal-symbols in the follow set of a non-terminal can be used as a synchronizing token set for that non-terminal.

So, a simple panic-mode error recovery for the LL(1) parsing:

- All the empty entries are marked as *synch* to indicate that the parser will skip all the input symbols until a symbol in the follow set of the non-terminal A which on the top of the stack. Then the parser will pop that non-terminal A from the stack. The parsing continues from that state.
- To handle unmatched terminal symbols, the parser pops that unmatched terminal symbol from the stack and it issues an error message saying that unmatched terminal is inserted.

Let us see the following example to understand the panic-mode recovery:

$S \rightarrow AbS \mid e \mid \varepsilon$
$A \rightarrow a \mid cAd$

Follow(S)={\$}
Follow(A)={b,d}

	a	b	c	d	e	\$
S	$S \rightarrow AbS$	sync	$S \rightarrow AbS$	sync	$S \rightarrow e$	$S \rightarrow \varepsilon$
A	$A \rightarrow a$	sync	$A \rightarrow cAd$	sync	sync	sync

stack	input	output
\$$S$	aab\$	$S \rightarrow AbS$
\$$SbA$	aab\$	$A \rightarrow a$
\$$Sba$	aab\$	
\$$Sb$	ab\$	Error: missing b, inserted
\$$S$	ab\$	$S \rightarrow AbS$
\$$SbA$	ab\$	$A \rightarrow a$
\$$Sba$	ab\$	
\$$Sb$	b\$	
\$$S$	\$	$S \rightarrow \varepsilon$
\$	\$	accept

stack	input	output
\$S	$ceadb$\$	$S \rightarrow AbS$
\$$SbA$	$ceadb$\$	$A \rightarrow cAd$
\$$SbdAc$	$ceadb$\$	
\$$SbA$	$eadb$\$	Error: unexpected e (illegal A)
(Remove all input tokens until first b or d, pop A)		
\$$Sbd$	db\$	
\$$Sb$	b\$	
\$$S$	\$	$S \rightarrow \varepsilon$
\$	\$	Accept

9.7. ERROR RECOVERY IN LR PARSING

We have already seen that LR parsing table contains some the entries as error entry.

An LR parser will detect an error when it consults the parsing action table and finds an error entry. All empty entries in the action table are error entries. Errors are never detected by consulting the goto table. An LR parser will announce error as soon as there is no valid continuation for the scanned portion of the input. A canonical LR parser (LR(1) parser) will never make even a single reduction before announcing an error.

The SLR and LALR parsers may make several reductions before announcing an error. But, all LR parsers (LR(1), LALR and SLR parsers) will never shift an erroneous input symbol onto the stack.

To get recover from error in LR parsing we can adopt following two techniques:

1. Panic Mode Error Recovery in LR Parsing

- Scan down the stack until a state **s** with a goto on a particular nonterminal **A** is found. (Get rid of everything from the stack before this state s).
- Discard zero or more input symbols until a symbol **a** is found that can legitimately follow A. The symbol a is simply in FOLLOW(A), but this may not work for all situations.
- The parser stacks the nonterminal **A** and the state **goto[s,A]**, and it resumes the normal parsing.
- This nonterminal A is normally is a basic programming block (there can be more than one choice for A). stmt, expr, block, ...

2. Phrase-Level Error Recovery in LR Parsing

- Each empty entry in the action table is marked with a specific error routine.
- An error routine reflects the error that the user most likely will make in that case.
- An error routine inserts the symbols into the stack or the input (or it deletes the symbols from the stack and the input, or it can do both insertion and deletion).
  - (*a*) missing operand
  - (*b*) unbalanced right parenthesis

For example to recover from an construct error that starts with an operator, the error recovery routine will push an imaginary id onto the stack and cover it with the appropriate state. While doing this, the error entries in a particular state that call for a particular reduction on some input symbols are replaced by that reduction.

For example consider the following grammar and given SLR parsing table:

$$E \rightarrow E + E$$

$$E \rightarrow E * E$$

$$E \rightarrow \text{id}$$

	Action Table				Go to Table
	id	**+**	*****	**$**	**E**
I_0	S_2				1
I_1		S_3	S_4	Accept	
I_2		R_3	R_3	R_3	
I_3	S_2				5
I_4	S_2				6
I_5		S_3/R_1	S_4/R_1	R_1	
I_6		S_3/R_2	S_4/R_2	R_2	

The shift-reduce conflict can be resolved by giving higher precedence to * and using left associativity. Now parsing table becomes as follows:

	Action Table				Go to Table
	id	**+**	*****	**$**	**E**
I_0	S_2				1
I_1		S_3	S_4	Accept	
I_2		R_3	R_3	R_3	
I_3	S_2				5
I_4	S_2				6
I_5		R_1	S_4	R_1	
		R_2	R_2	R_2	

Now in above parsing table resolve the shift-reduce conflict and construct the error routines for the empty entries is action-table. Now the parsing table with error routines will be as follows and it contains error-recovery routines e_1 and e_2.

	Action Table				Go to Table
	id	**+**	*****	**$**	**E**
I_0	S_2	e_1	e_1	e_1	1
I_1	e_2	S_3	S_4	Accept	
I_2	R_3	R_3	R_3	R_3	
I_3	S_2	e_1	e_1	e_1	
I_4	S_2	e_1	e_1	e_1	
I_5	e_3	R_1	S_4	R_1	
I_6	e_3	R_2	R_2	R_2	

- **Routine e_1:** is called from state I_0, I_3 and I_4 which pushes and imaginary *id* on parsing stack and covers it with I_2.
- **Routine e_2:** is called from state I_1, which pushes + on the parsing stack and covers it with I_3 state.
- **Routine e_3:** is called from state I_5 and I_6.

Now let us trace the behaviour of parser on the string $W = id + * id \$$

Current Stack	Current Input	Moves
$\$ I_0$	$id + * id \$$	shift and goto state 2
$\$ I_0$ id I_2	$+ * id \$$	reduce by production 3
$\$ I_0 E I_1$	$+ * id \$$	shift and enter in state 3
$\$ I_0 E I_1 + I_3$	$* id \$$	call error routine e_1, which push *id* I_2 on stack.
$\$ I_0 E I_1 + I_3 id I_2$	$* id \$$	reduce by production number 3
$\$ I_0 E I_1 + I_3 E I_5$	$* id \$$	shift and enter into state 4
$\$ I_0 E I_1 + I_3 E I_5 * I_4$	$id \$$	shift and enter into state 4
$\$ I_0 E I_1 + I_3 E I_5 * I_4 id I_2$	\$	reduce by production 3
$\$ I_0 E I_1 + I_3 E I_5 * I_4 E I_6$	\$	reduce by production 2
$\$ I_0 E I_1 + I_3 E I_5$	\$	reduce by production 1
$\$ I_0 E I_1$	\$	accept.

9.8. RECOVERY FROM SEMANTIC ERRORS [MDU 2008]

These are the errors which can be detected both at compile time and routine.

Most common error of compile time are scope and declaration errors. For example undeclared or multiple declared identifiers. Type mismatch is another compile time error. Recovery from an undeclared name is straight forward. It is suggested to make the entry of declaration in symbol table with appropriate attribute.

❑❑❑

SECTION-D

Intermediate Code

Code Optimization

Code Generation

SYLLABUS

SECTION-D

- **Intermediate Code Generation**
- **Code Optimization and Code Generation**

SYLLABUS

Code Optimization and Code Generation : Code Generation, Forms of Objects Code, Machine Dependent Code, Optimization, Register Allocation for Temporary and User Defined Variables.

10

Chapter

INTERMEDIATE CODE GENERATION

INSIDE THIS CHAPTER

10.1. Introduction to Intermediate Code Generation **10.2.** Choosing an Intermediate Languages **10.3.** Three-Address Code **10.4.** P-Code **10.5.** Byte Code **10.6.** Intermediate Code for Control Flow.

10.1. INTRODUCTION TO INTERMEDIATE CODE GENERATION

[MDU 2009]

The final goal of a compiler is to get programs written in a high-level language to run on a computer. This means that, eventually, the program will have to be expressed as machine code which can run on the computer. This doesn't mean that we need to translate directly from the high-level abstract syntax to machine code. Many compilers use a medium-level language as a stepping-stone between the high-level language and the very low-level machine code. Such stepping-stone languages are called intermediate code. Apart from structuring the compiler into smaller jobs, using an intermediate language has other advantages:

(*i*) If the compiler needs to generate code for several different machine-architectures, only one translation to intermediate code is needed. Only the translation from intermediate code to machine language (*i.e.*, the back-end) needs to be written in several versions.

(*ii*) If several high-level languages need to be compiled, only the translation to intermediate code need to be written for each language. They can all share the back-end, *i.e.*, the translation from intermediate code to machine code.

(*iii*) Instead of translating the intermediate language to machine code, it can be interpreted by a small program written in machine code or a language for which a compiler already exists.

The advantage of using an intermediate language is most obvious if many languages are to be compiled to many machines. If translation is done directly, the number of compilers is equal to the product of the number of languages and the number of machines. If a common intermediate language is used, one front-end (*i.e.*, compiler to intermediate code) is needed for every language and one backend is needed for each machine, making the total equal to the sum of the number of languages and the number of machines. If an interpreter for the intermediate language is written in a language for which there already exist compilers on the target machines, the interpreter can be compiled on each of these. This way, there is no need to write a separate back-end for each machine.

The advantages of this approach are:

(*i*) No actual back-end needs to be written for each new machine.

(*ii*) A compiled program can be distributed in a single intermediate form for all machines, as opposed to shipping separate binaries for each machine.

(*iii*) The intermediate form may be more compact than machine code. This saves space both in distribution and on the machine that executes the programs (though the latter is somewhat offset by requiring the interpreter to be kept in memory during execution).

The disadvantage is speed: Interpreting the intermediate form will in most cases be a lot slower than executing translated code directly. Nevertheless, the approach has seen some success, *e.g.*, with Java.

Some of the speed penalty can be eliminated by translating the intermediate code to machine code immediately before or during execution of the program. This hybrid form is called just-in-time compilation and is often used for executing the intermediate code for Java.

An example of intermediate codes, we will consider three well-known examples :

(*i*) Three-address code

(*ii*) P-code, a language specific intermediate code upon which the majority of implementations of pascal are based.

(*iii*) Byte code used by the Java virtual machine.

10.2. CHOOSING AN INTERMEDIATE LANGUAGES

An intermediate language should, ideally, have the following properties:

(*i*) It should be easy to translate from a high-level language to the intermediate language. This should be the case for a wide range of different source languages.

(*ii*) It should be easy to translate from the intermediate language to machine code. This should be true for a wide range of different target architectures.

(*iii*) The intermediate format should be suitable for optimizations.

The first two of these properties can be somewhat hard to reconcile. A language that is intended as target for translation from a high-level language should be fairly close to this. However, this may be hard to achieve for more than a small number of similar languages. Furthermore, a high-level intermediate language puts more burden on the back-ends. A low-level intermediate language may make it easy to write back-ends, but puts more burden on the front-ends. A low-level intermediate language, also, may not fit all machines equally well, though this is usually less of a problem than the similar problem for front-ends, as machines typically are more similar than high-level languages.

A solution that may reduce the translation burden, though it doesn't address the other problems, is to have two intermediate levels: One, which is fairly highlevel, is used for the front-ends and the other, which is fairly low-level, is used for the back-ends. A single shared translator is then used to translate between these two intermediate formats.

(*i*) **High-Level Intermediate Representations.** Expresses high-level structure of a program.

- Closer to source language that is syntax trees.
- Easy to generate from input program.
- Code optimization may not be straight forward.

(*ii*) **Low-Level Intermediate Representations.** Expresses low-level structure of a program.

- Closer to target machine, for example 3-address code.
- Easy to generate code form.
- Generation form input program may involve some work.

10.2.1. Intermediate Languages: Design Issues

Following issues should be addressed in the designing process of intermediate language:

1. The set of operators in the intermediate language must be rich enough to allow the source language operations to be implemented.
2. A small set of operations in the intermediate language makes it easy to retarget the compiler to a new machine.
3. Intermediate code operations that are closely fied to a particular machine or architecture may make it harder to port the compiler to other architectures.
4. A small set of intermediate code operations may lead to long instruction sequences for some source language constructs. This may require more work during optimization.

10.2.2. High-Level Intermediate Representations **[MDU 2009]**

Syntax trees, and DAGs are examples of high-level intermediate representation.

(*i*) **(Abstract) syntax trees:** A compact form of a parse tree that represents the hierarchical structure of the program: nodes represent operators, the children of a node represent what it operates on.

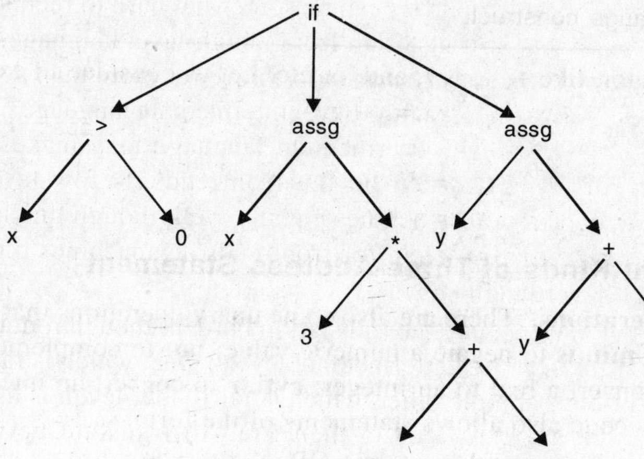

Fig. 10.1. Syntax tree.

(*ii*) **DAG:** Similar to syntax trees, except that common subexpression are represented by a single node.

For example, give syntax tree and DAG for the expression:

if ($x > 0$) then $x := 3 * (y + 1)$ else $y := y + 1$;

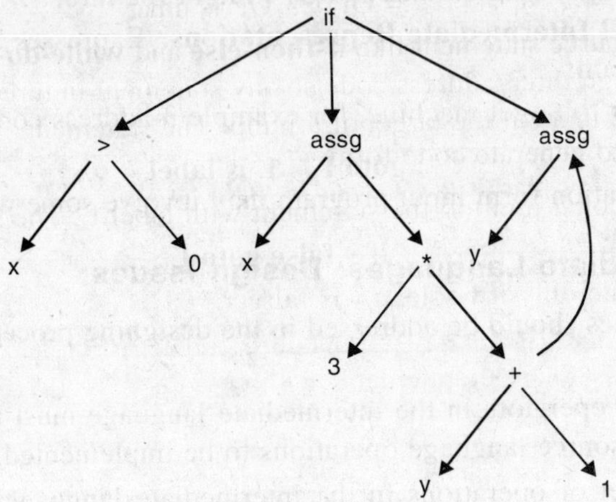

Fig. 10.2. DAG.

10.3. THREE-ADDRESS CODE [MDU 2007, 2010]

It is an example of low-level representation. Almost every arithmetic/logical operation in a computer combines two source operands to produce one result operand so **three-address code is sequence of statements of the form:**

$$x := y \text{ OP } z$$

where *x*, *y* and *z* are variables names, constants, or compiler generated variables ("temporaries").

TIPS

Only one operator is permitted on the RHS, so there is no "built-up" expression. Instead, expressions are computed using temporaries, that is source langauge construct.

OP is an operator, like +, –, *, /, and, or etc. Let us consider an example:

$$x := y + z * w$$

might translate to

$$t_1 := z * w$$
$$x := y + t_1$$

10.3.1. Different Kinds of Three-Address Statement [MDU 2007]

1. Unary Operations. There are also some unary operations that use only one source operand: **Uminus** to negate a numeric value; **not** to complement a boolean value; **cvtr2i** to convert a real to an integer; **cvti2r** to convert an integer to a real; etc. Three-address code also allows statements of the form:

$$x := \text{OP } y$$

where OP is a unary operator.

Sometimes, it is necessary to copy the value of an operand into another location so three-address code also allows copy statements of the form:

$$x := y,$$
$$x := y \text{ OP } z, \quad \text{OP is binary}$$
$$x := \text{OP } y \quad \text{OP is unary}$$

2. Jumps. Source statement like **if-then-else** and **while-do** cause jumps in the control flow through three-address code so any statement in three-address code can be given a label to make it the target of a jump. The statement:

goto L, L is label

cause an unconditional jump to the statement with label L. The statement

if y false goto L

causes a jump if and only if boolean y is false.

3. Procedure Call/Return. A procedure call like $P(x_1, x_2, ..., x_n)$ may have too many addresses for one statement in three-address code so it is shown as a sequence of $n + 1$ statements:

```
param x1
param x2
..........
..........
param xn
call p, n /* n = number of arguments */
```

Similarly, an assignment statement with a function call like $y:= f(x_1, x_2, ..., x_n)$ is shown as a sequence of $n + 1$ statement in three-address code:

```
param x1
param x2
.............
.............
param xn
y:= call f, n
```

When the source program contains a defined procedure there must be some way of delimiting the three-address code for the body of the procedure so the first statement of the body is labelled with the name of the procedure and a **return** statement marks the end of the body.

TIPS

Similarly, the three-address code for the body of a defined function is delimited with the function name as a label on the first statement of the body and a **return** y statement at the end to indicate the returned value.

4. Indexed Assignment. Reading the value of an array element is shown in three-address code with the statement:

$$x := y[i]$$

and writing a value into an array element is shown in three-address code with the statement:

$$x[i] := y$$

5. Address and Pointer Assignment. As mentioned in the text as follows, three-address code can also be enriched with statements for address and pointer assignment.

```
x:= &y (Address assignment)
x:= *y
x:= y
```

10.3.2. Implementing Three-Address Instructions as Quadruples

[MDU 2007, 2010]

A dialect of three-address code that is closer to machine assembly code reformats each statement into a structure with four fields called a quadruple:

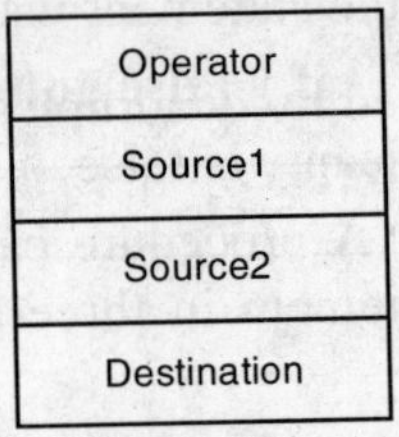

Fig. 10.3.

The field of a quadruple contain the names of the operator, the first source operand, the second source operand, and the result, respectively. Floating-point arithmatic instructions are different than fixed-point instructions in machines so quadruples also use different operator names; that is **ADDI** indicates integer addition while ADDR indicates real addition.

Example 10.1. Write the quadruple for expression:

(i) $x := y + z$

(ii) **If** $t_1 >= t_2$ **goto** L.

Solution. Let us see (*i*) $x := y + z$, the quadruple will be as follows:

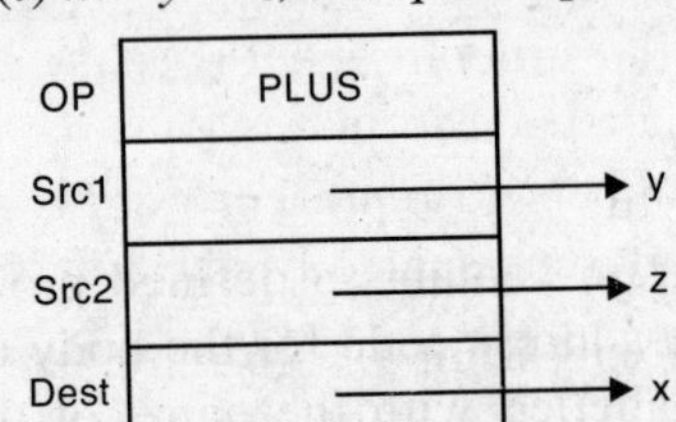

Fig. 10.4.

(*ii*) if $t_1 >= t_2$ goto L

quadruple will be as follows:

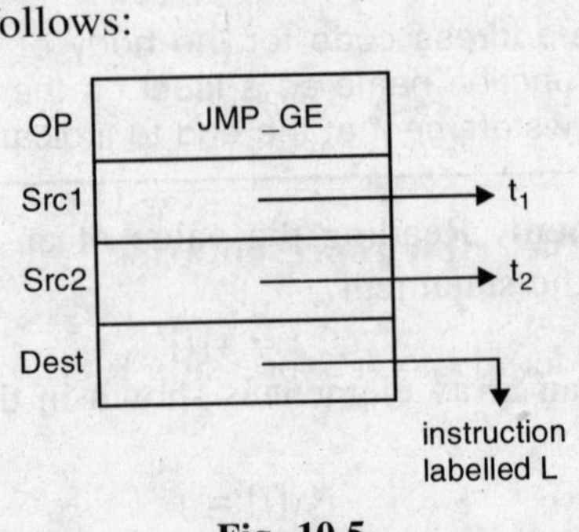

Fig. 10.5.

Example 10.2. Translate the given expression, by three-address code and also give quadruple.

$$a := -b * (c + d)$$

Solution. Given expression is:

$$a := -b * (c + d)$$

Three-address code will be as follows:

$$t_1 := -b$$
$$t_2 := c + d$$
$$t_3 := t_1 * t_2$$
$$a := t_3$$

Those statements are represented by quadruples as follows:

	Op.	Src1	Src2	Dest.
(0)	uminus	b	—	t_1
(1)	+	c	d	t_2
(2)	*	t_1	t_2	t3
(3)	:=	t_3	—	a

Fig. 10.6.

10.3.3. Implementing Three-Address Instructions as Triples

[MDU 2010]

To avoid entering temporary names into the symbol table, one can allow the statement computing a temporary value to represent the value. If we do so, three-address statements are representable by a structure with only three field OP, Src1 and Src2, where Src1 and Src2 are the arguments of OP, are either pointer to the symbol table or pointers into the structure itself (for temporary values). Since three fields are used, this intermediate code format is known as triples.

We use paranthesized numbers to represent pointers into the triple structure, while symbol-table pointers are represented by the names themselves.

Example 10.3. For the expression $a := -b * (c + d)$ write triples.

Solution. Given expression is $a := -b * (c + d)$ so triple will be as follows:

	OP	Src1	Src2
(0)	Uminus	b	—
(1)	+	c	d
(2)	*	(0)	(1)
(3)	:=	a	(2)

Example 10.4. Write the triple representation for the expressions $a[i] := b$ and $a := b$[i].

Solution. Let us consider $a[i] := b$ expression first, the triple representation is as follows:

	OP	Src1	Src2
(0)	[ ] =	*a*	i
(1)	—	*b*	—

So clearly ternary operation like $a[i] := b$ actually requires two enteries in the triple structure as shown in Fig. 10.8(*a*).

Now let us consider $a := b[i]$ expression:

	OP	Src1	Src2
(0)	= [ ]	*b*	*i*
(1)	:=	(0)	*a*

10.3.3.1. Indirect Triples Implementation **[MDU 2010]**

If we use pointers to triples, rather than listing the triples themselves, the implementation is called indirect triples implementation.

Example 10.5. Consider the expression $a := -b * (c + d)$ and show the indirect triples implementation.

Solution. We will use an **array statement** to list pointers to triples in the desired order as follows:

Statement	
(0)	(15)
(1)	(16)
(2)	(17)
(3)	(18)

	OP	Src1	Src2
(15)	Uminus	*b*	—
(16)	+	*c*	*d*
(17)	*	(15)	(16)
(18)	:=	*a*	(17)

10.4. P-CODE

We now turn to the production of another type of intermediate code designed specifically for the implementation of pascal, and widely used for this purpose. P-code instructions each have the following format:

FPQ

F is a function code and *P* or *Q* (or both) may be absent depending on the particular function code. If present, *P* can be used to specify a static block level and *Q* an offset within a frame or an immediate operand.

10.5. BYTE CODE

Byte Code is the intermediate language for the Java Virtual Machine (JVM) and, like P-code for pascal, is stack based. The JVM is designed to make Java implementation efficient, secure and portable and this is reflected in the Java run-time system, the main components of which are:

- **The execution engine,** which executes the bytecode instructions.
- **The memory manager,** which manages the heap in which all objects and arrays are stored;

- **The error and exception manager,** which is used to catch run-time failures in a planned and systematic manner;
- **The threads interface** which handles concurrency.
- **The class loader,** which loads, links and initialises classes.
- **The security manager,** which deals with attempts to run 'hostile' program.

10.6. INTERMEDIATE CODE FOR CONTROL FLOW

Code for control flow (if-then, switches and loops) consists of code to test conditions, and the use of goto instructions and labels to route execution to the correct code. Each chunk of code that is executed together (no jump into or out of it) is called a **basic block**. The basic blocks are nodes in a control flow graph, where goto instructions, as well as falling through from one basic block to another, are edges connecting basic blocks. Depending on our source language's semantic rules for things like "short circuit" evaluation for boolean operators, the operators like || and && might be similar to + and * (non-short-circuit) or they might be more like if-then code.

A general technique for implementing control flow code is to add new attributes to three nodes to hold labels that denote the possible targets of jumps. The labels in question are sort of annalogous to **First** and **Follow**, for any given list of instructions corresponding to a given tree node, we might want a **·First** attribute to hold the label for the beginning of the list, and a **·Follow** attribute to hold the label for the next instruction that comes after the list of instructions.

TIPS

The **·First** attribute can be easily synthesized. The **·Follow** attribute must be inherited from a sibling. The labels have to actually be allocated and attached to instructions at appropriate nodes in the tree corresponding to grammar production rules that govern control flow. An instruction in the middle of a basic block need neither a first nor a follow.

C code	Attribute manipulations
$S \rightarrow$ if E then S_1	E . true = new label ( ); E . false = S . follow; S_1. follow = S . follow; S . code = E . code ‖ gen (LABEL, E . true) ‖ S_1 . code
$S \rightarrow$ if E then S_1 else S_2	E . true = new label ( ); E . false = new label ( ); S_1 follow = S . follow; S_2 . follow = S . follow; S . code = E . code ‖ gen (LABEL, E . true) ‖ S_1 . code ‖ gen (GOTO, S . follow) ‖ gen (LABEL, E . false) ‖ S_2 . code

10.6.1. Syntax-Directed Translation for Intermediate Code Generation

After the above discussion, we can decide following:

(*i*) Intermediate code represented as a list of instructions. Instruction sequences are concatenated using the operator ‖.

(*ii*) Attribute for Expression E:

(*a*) E place: Denote the location that holds the value of E.

(*b*) E . code: Denote the instruction sequence that evaluates E.

(*iii*) Attribute for Statements S:

(*a*) S . begin: denotes the first instruction in the code for S.

(*b*) S . after: denotes the first instruction after the code for S.

(*c*) S . code: denotes the instruction sequence that represents S.

(*iv*) **Auxiliary functions:**

(*a*) new temp ( ): returns a new temporary each time it is called returns a pointer to the symbol table entry of a emp may take a parameter specifying the type of temp (useful if reusing temps).

(*b*) new label ( ): returns a new label name each time it is called.

(*v*) **Notation:** We write gen (*x* ':=' *y* '+' *z*) to represent the instruction *x*: *y* + *z*.

For example, let us write the intermediate code for following simple expression :

Production	Semantic Rule
$E \rightarrow$ id	E . place:= *id* . place; E . code:= ";
$E \rightarrow (E_1)$	E . place:= E_1 . place; E . code:= E_1 . code;
$E \rightarrow E_1 + E_2$	E . place:= new temp ( ); E . code:= E_1 . code ‖ E_2 . code gen (E . place ':=' E_1 . place '+' E_2 . place)
$E \rightarrow - E_1$	E . place:= newtemp ( ); E . code:= E_1 . code ‖ gen (E . place ':=' '–' E_1 . place)

Example 10.6. Consider the following grammar:

$$E \rightarrow E + T$$
$$E \rightarrow T$$
$$T \rightarrow T * F$$
$$T \rightarrow F$$
$$F \rightarrow id$$

and write the intermediate code for expression *a* + *b* * *c* as:

(*i*) Postfix representation

(*ii*) Syntax tree representation

(*iii*) Three-address code representation.

Solution. Given grammar is

$$E \rightarrow E + T$$
$$E \rightarrow T$$
$$T \rightarrow T * F$$
$$T \rightarrow F$$
$$F \rightarrow id$$

and expression is $a + b * c$, the postfix representation will be as:

$$abc * +$$

(*i*) Syntax-directed translation scheme to specify the translation of an expression into postfix notation are as follows:

Production	**Schemes**
$E \to E_1 + T$	E . code = concate (E_1 . code, T . code, "+");
$E \to T$	E . code = T . code;
$T \to T_1 * F$	T . code = concate (T_1 . code, F . code, " * ");
$T \to F$	T . code = F . code
$F \to id$	F . code = getname (id . place);

The function getname ( ) returns the name of the identifier from the symbol table record.

(*ii*) **Translation schemes for syntax tree are as follows:**

$E_1 \to E_1 + T$	E . ptr = mknode ('+', E_1 . ptr, T . ptr);
$E \to T$	E . ptr = T . ptr;
$T \to T_1 * F$	T . ptr = mknode ('*', T_1 . ptr, F . ptr);
$T \to F$	T . ptr = F . ptr;
$F \to id$	F . ptr = mkleaf (id . place);

where ptr is pointer value attribute used to link the pointer to a node in the syntax tree, and place is pointer value attribute used to link the pointer to the symbol record that contains the name of the identifier. For string $a + b * c$.

E . ptr = pointer to node +

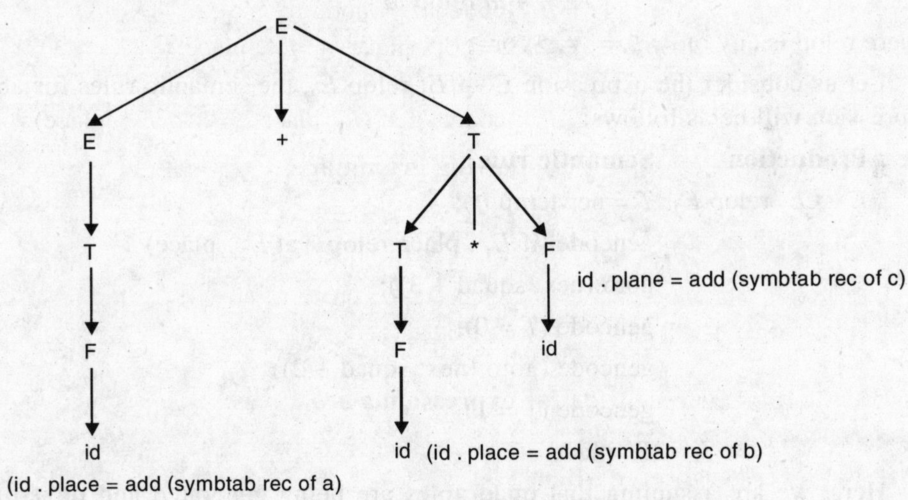

Fig. 10.7.

(*iii*) Syntax-directed translation schemes for three-address code representation are as follows:

Production	**Translation schemes**
$E \to E_1 + T$	E . place:= new temp ( );
	E . code:= E_1 . code ‖ T . code ;
	gen (E . place ':=' E_1 . place '+' T . place);

$E \rightarrow T$	E . place:= T . place;
	E . code:= T . code;
$T \rightarrow T_1 * F$	T . place:= new temp ( );
	T . code:= T_1 . code ‖ F . code
	gencode (T . place ':=' T_1 . place '+' F . place);
$T \rightarrow F$	T . place:= F . place
	T . code:= F . code
$F \rightarrow id$	F . place:= id . place
	F . code:= " ;

10.6.2. Intermediate Code Generation for Boolean Expression

We can assume expression is true and false values as the integer one and zero, respectively, we have to write the code for evaluation of the values.

Boolean expressions may be of the form

$E \rightarrow E$ or E

$E \rightarrow E$ and E

$E \rightarrow$ not E

$E \rightarrow (E)$

$E \rightarrow id$

$E \rightarrow id$ relop id

where relop is any of $<, \leq, =, \neq, >$, or $\geq$

Let us consider the expression $E \rightarrow E_1$ relop E_2, the semantic rules for above expression will be as follows:

Production	**Semantic rule**
$E \rightarrow E_1$ relop E_2	T:= newtemp ( );
	gencode (if E_1 . place relop. val E_2 . place)
	goto (next squad + 3));
	gencode (T = 0);
	gencode (goto (next squad + 2))
	gencode (T = 1)
	E . place = T

Here, we are assuming that quadruples are being generated and **nextsquad** indicates the next available entry in the quadruple array.

gencode increment **nextsquad** after producing a quadruple.

Now let see explanation of above semantic rules. Here newtemp ( ) every time generated a temporary variable whenever it is called.

Here this temporary variable is T.

gencode (if E_1 . place relop . val E_2 . place) generates the value false (0) and it is stored in temporary variable T, else true (1) is stored in T and finally E . place = T means either true or false, so clearly semantic rules will give us the true and false values of the expression $E \rightarrow E_1$ relop E_2. Let us see some for examples:

Production	Semantic rules	Explanation if any
$E \rightarrow E_1$ or E_2	t_1:= newtemp ( ); gencode (t_1 ':=' E_1 . place or . val E_2 . place) E . place ':=' t_1;	newcode ( ), generates the value of E_1 or E_2 and assign it in t_1
$E \rightarrow E_1$ and E_2	t_1:= newtemp ( ); gencode (t_1 ':=' E_1 . place and .val E_2 . place) E . place ':=' t	newcode ( ), generates the value of E_1 and E_2 and assign it to t_1
$E \rightarrow$ not E_1	t:= newtemp ( ) gencode (t ':=' not E_1 . place) E . place ':=' t;	newcode ( ) generates the value of not E_1
$E \rightarrow$ true	E . place:= newtemp ( ); gencode (E . place ':=''1')	Here true means integer value 1
$E \rightarrow$ false	E . place:= newtemp ( ); gencode (E . place ':=' '0')	Here false means integer value 0

Now let us apply the above translation scheme on expression $x < y$ and $a > b$

(0) if $x < y$ goto 3(if $x < y$ is true then jump on 3 and set $t_1 = 1$, else set $t_1 = 0$).

(1) $t_1 = 0$

(2) goto 4

(3) $t_1 = 1$

(4) if $a > b$ goto 7(if $a > b$ is true then jump on 7 and set $t_2 = 1$ else set $t_2 = 0$)

(5) $t_2 = 0$

(6) goto 8

(7) $t_2 = 1$

(8) $t_3 = t_1$ and t_2(t_3 is assign numerical value of t_1 and t_2 that is $t_3 = 1$ if both t_1 and t_2 are 1 else in any other case $t_3 = 0$).

Note. t_1 is temporary variable which stores numerical value of $x < y$ and t_2 stores numerical value of $a > b$.

10.6.3. Code Generation For Conditions

Let us analyse the following grammar, having following production:

$$S \rightarrow \text{if } E \text{ then } S_1$$
$$S \rightarrow \text{if } E \text{ then } S_1 \text{ else } S_2$$
$$S \rightarrow \text{while } E \text{ do } S_1$$

With a boolean expression E, we associate two labels, E . true, the label to which control flows if E is true, and E . false, the label to which control flows if E is false.

Let us see the block diagram of code for above productions

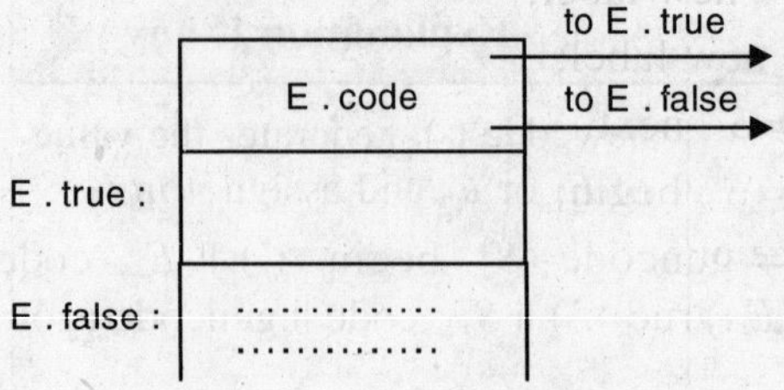

(*a*) Block for $S \rightarrow$ if E then S_1.

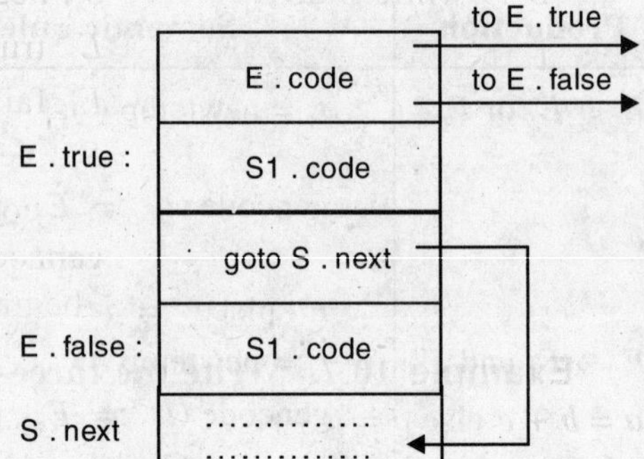

(*b*) Block for $S \rightarrow$ if E then S_1 else S_2.

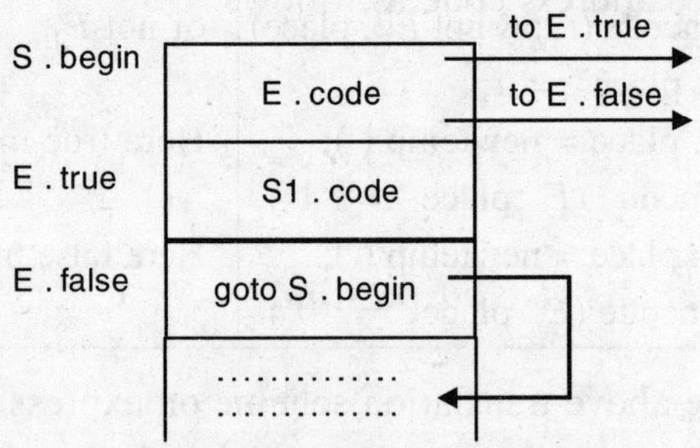

(*c*) Block for $S \rightarrow$ while E do S_1

Fig. 10.8.

After viewing the blocks, it is simple to have semantic rules:

Production	**Semantic Rules**
$S \rightarrow$ if E then S_1	E . true:= newlabel; E . false:= S . next; S_1 . next:= S . next S . code:= E . code ‖ gencode (E . true ':') ‖ S_1 . code

Explanation is simple in translating the if-then statement $S \rightarrow$ if E then S_1, a new label E . true is created and attached to the first three-address instruction generated for statement S_1. The code for E generates a jump to E . true if E is true and a jump to S . next if E is false. We therefore set E . false to S . next.

In similar fashion, we can write semantic rule for other productions:

Production	**Semantic Rules**
$S \rightarrow$ if E then S_1 else S_2	E . true:= newlabel; E . false:= newlabel; S_1 . next:= S . next; S_2 . next:= S . next; S . code:= E . code ‖ gencode (E . true ':') ‖ S_1 . code ‖ gencode (goto' S . next) ‖ gencode (E . false ':') ‖ S_2 . code

$S \rightarrow$ while E do S_1 $\quad$ S . begin:= new label;
E . true:= new label;
E . false:= S . next;
S_1 . next:= S . begin;
S . code:= gencode (S . begin ':') || E . code || gencode (E . true ':') || S_1 . code || gencode (goto S . begin)

Example 10.7. Write the three-address code for expression if $x < y$ then $a = b + c$ else $p = q + r$.

Solution. Given expression is

if $x < y$ then $a = b + c$ else $p = q + r$

We can write three address code as follows:

0. if $x = y$ goto 2
1. goto 5
2. $t_1 = b + c$
3. $a = t_1$
4. goto S . next(if S is given expression)
5. t_2 = q + r
6. p = t_2

10.6.4. Code Generation for Assignment

Let us consider the production:

$S \rightarrow$ LHS:= RHS

Semantic rule can be written as follows:

S . code = Lhs . code || Rhs . code ||
gencode (Lhs . place ':=' Rhs . place)

10.6.5. Back Patching

We know that implementation of syntax-directed definitions is possible in following two ways:

(*i*) In one pass and
(*ii*) In two passes

Definition When we generate the code for boolean expressions and flow-of-control statement in a single pass then we face the problem that, we may not know the labels that control must go to at the time the jump statements are generated. We can overcome this problem by generating a series of branching statement with the targets of the jumps temporarily left unspecified. Each such statement will be put on a list of goto statements whose labels will be filled in when the proper label can be determined. We call this subsequent filling in of **labels back patching.**

To manipulate lists of labels, we use three functions:

1. **MAKELIST (j):** creates a new list containing only j, an index into the array of quadruples being generated. MAKELIST returns a pointer to the list it has made.
2. MERGE (q_1, q_2) take the lists pointed to by q_1 and q_2 concatenates them into one list, and returns a pointer to the concatenated list.
3. BACKPATCH (q, j) insert j as the target label for each of the statements on the list pointed by q.

10.6.5.1. Code generation for the Loops

1. Let us consider the **statement production for while loop**

$$S \rightarrow \text{while } M_1 \; E \text{ do } M_2 \; S_1$$

Let us explain this. Here above while statement has two components, a boolean expression E and a statement S_1, statement S_1 executed repeatedly, till E is true. For the remembering the index of the code array where E code and S_1 code starts, we use two nullable nonterminal M_1 and M_2. We can use either $M_1 \rightarrow \in$ or $M_2 \rightarrow \in$ (since it is the property of nullable nonterminal) to get the index of E or S_1 respectively. Now let us see semantic section for above production as follows:

Production	**Semantic action**
$S \rightarrow$ while M_1 E do M_2 S_1	back patch (S_1 . next list, M_1 . quad); back patch (E . truelist, M_2 . quad); S . nextlist = E . false list gencode (goto (M_1 . quad));
$M_1 \rightarrow \in$	M_1 . quad = next quad;
$M_2 \rightarrow \in$	M_2 . quad = next quad;

Example 10.8. Write the semantic action for following productions:

$$S \rightarrow \textbf{begin } L \textbf{ end}$$
$$S \rightarrow \textbf{A}$$
$$L \rightarrow L_1; \textbf{MS}$$
$$L \rightarrow S$$

Solution.

Production	**Semantic actions**
$S \rightarrow$ begin L end	S . nextlist = L . nextlist
$S \rightarrow A$	S . nextlist = A . nextlist
$L \rightarrow L_1; MS$	{backpatch (L_1 . nextlist, M . quad); L . nextlist = S . nextlist;}
$L \rightarrow S$	L . nextlist = S . nextlist.

2. Do-while loop: Let us see the semantic action for do-while loop.

Production	**Semantic action**
$S \rightarrow$ do M_1 S_1 while M_2 E	{backpatch (S_1 . next, M_2 . quad) backpatch (E . true list, M_1 . quad) S . nextlist = E . false list gencode (goto (M_2 . quad))}

$M_1 \to \in$ $\{M_1$. quad = next quad;$\}$

$M_2 \to \in$ $\{M_2$. quad = next quad;$\}$

3. Let us consider now repeat untill loop:

Production	**Semantic action**
$S \to$ repeat M_1	{backpatch (E . falselist, M_1 . quad)
S_1 until M_2 E	backpath (S_1 . nextlist, M_2 . quad)
	S . nextlist = E . true list
	}

4. For loop can be interpreted as follows:

Production	**Semantic action**
$S \to$ for (E_1, M_1; E_2;	{backpatch (E_2 . truelist, M_3 . quad)
M_2 E_3) M_3 S_1	backpatch (M_3 . nextlist, M_1. quad)
	backpatch (S_1 . nextlist, M_2 . quad)
	gencode (goto (M_2 . quad))
	S . next = E_2 . false}

(Here M_1, M_2 and M_3 are nullable nonterminal to remember index of the code array where E_2 . code, E_3 . code and S_1 . code starts respectively)

$M_1 \to \in$ $\{M_1$. quad = next quad;$\}$

$M_2 \to \in$ $\{M_2$. quad = nextquad;$\}$

$M_3 \to \in$ $\{M_3$. next = MAKELIST (next quad)

gencode (goto ...)

M_3 . quad = next quad;

}

10.6.6. Code Generation for Conditions in Single Pass

Code for conditions can be generated in a single pass, using backpatching. Let us consider the following production for conditional statement.

$$S \to \text{if } E \text{ then } M_1\, S_1\, N \text{ else } M_2\, S_2$$

Here M_1 and M_2 are nullable nonterminal and having only significance for starting of S_1 and S_2 respectively.

In addition, an unconditional jump is required at the end of S_1 in order to transfer control to the statement that follows the if-then-else statement. To generate this unconditional jump, we add a nullable non-terminal N after S_1 to the production and associate a semantic action with the production $N \to \in$, which takes care of generating the unconditional jump.

Let us see semantic action for above statement.

```
{
    backpatch (E . truelist, M1 . quad);
    backpatch (E . falselist, M2 . quad);
  S . nextlist = merge (S1. nextlist, merge
    (N . nextlist,    S2 . nextlist));
```

```
}
For N → ∈   {N . nextlist = MAKELIST (next quad);
     gencode (goto ....)
     }
```

Now let us see another statement:

```
S → if E then MS1 having semantic action:
{
backpatch (E . truelist, M . quad);
S. nextlist = merge (E . falselist, S1 . nextlist)
}
```

10.6.7. Intermediate Code Generation for Case Statement

We need to generate code so that we can efficiently, choose one of a set of different alternatives, depending on the value of an expression.

1. Implementation Choices:

(*i*) Linear search

(*ii*) Binary search

(*iii*) Jump table

2. Implementation Consideration:

(*i*) **Execution Cost.** Linear or binary search may be cheaper if the number of cases is small.

For a large number of cases, a jump table may be cheaper.

(*ii*) **Space Cost.** A jump table may take too much space if the case values are not clustered closely together, that is:

```
switch (x)  {
        case 1: .....
        case 1000: .....
        case 1000000: .....
        }.
```

3. Implementing Jump Tables

Consider a C statement of the form

```
switch (E) {
      case C1: ...
      ....
      case Cn: ...
      }
```

Let the maximum, minimum case values be C_{max} and C_{min} respectively. The structure of the generated code is:

```
[
    t  ← value of E
    if t < Cmin goto Default-case
    if t > Cmax goto Default-case
    goto jump Table [t]
    Default - case: ...
```

Here **Jump Table** is an array of addresses:

Jump Table [*i*] is the address of the code to execute if *E* evaluates to *i*.

Note that, if there are "holes" in the set of case labels, the corresponding entries in the jump table have the address of Default-case.

Now let us consider the following switch case structure.

```
switch (E)
    {
    case X1: S1
    case X2: S2
    .............
    .............
    case Xn - 1: Sn - 1
    default: Sn
end.
```

and now we will generat translation of case statement, given above. When we see the keyword **switch**, we generate two new labels **test** and **next**, and a new temporary *t*. Then we parse the expression *E*, we generate code to evaluate *E* into *t*. After processing *E*, we generate the jump **goto test**. After this we will handle the **case** keyword, we create a new label *Li* and enter it into the symbol table. Now translation of a case statement is as follows:

```
                Code to evaluate E into t
                goto test
L1      :       code for S1
                goto next
L2      :       code for S2
                goto next
                -----------
Ln - 1  :       code for Sn - 1
                goto next
Ln      :       code for Sn
                goto next
test    :       if t = X1 goto L1
                if t = X2 goto L2
                -----------
                if t = Xn - 1 goto Ln - 1
                    goto Ln
next:
```

10.6.8. Code Generation for Function Calls

Following are the basics for code generation for function call:

(*i*) Calling Sequence: Caller:

(*a*) First, we will evaluate the actual parameters, and place actuals where the callee wants them.

Instruction: Param t

(*b*) Save machine state (current stack/or frame pointers, return address) and transfer control to callee.

Instruction: **call *p*, *n***(*n* = number of actuals)

(ii) **Calling Sequence: Callee:**

(*a*) Save registers (if necessary); update stack and frame pointers to accommodate *m* bytes of local storage.

Instruction: enter m

(iii) **Return Sequence: Callee:**

(*a*) Place return value *x* (if any) where the caller wants it; adjust stack/frame pointers (may be); jump to return address.

Instruction: return *x* or return

(iv) **Return Sequence: Caller:**

(*a*) Save the value returned by the callee (if any) into *x*.

Instruction: **retrieve *x***

Now let us consider following example:

$$x = f(0, y + 1) - 1;$$

The intermediate code for above statement can be written as follows:

```
t1 :  y + 1
param t1                /* arg 2 */
param 0                 /* arg 1 */
call f, 2
retrieve t2             /* t2:= f(0, t1) */
t3 :  =  t2 - 1
x  :  t3
```

Suppose function *f* needs 24 bytes of space for its locals and temporaries. Its code has the form

```
enter 24
----------
----------
return t17      /* suppose return value is in t17 */
```

10.6.9. Accessing Array Elements

Array elements can be accessed quickly if the elements are stored in a block of consecutive locations. Following points may be relevant to access the array elements.

(*i*) Assume, we want the *i* th element of an array *A* whose subscript ranges from **lo** to **hi**. The address of the first element of the array is **base**.

(*ii*) We can avoid address computations in the intermediate code if we have indexed "addressing modes" at the intermediate code level.

In this case, $A[i]$ is the $(i + lo)^{th}$ element of the array located at base (starting at element 0). So a reference $A[i]$ translates to the code.

```
t1  := i + lo
t2  := A [t1]
```

(*i*) Address computations can't be avoided in general, because of pointer and record types.

(*ii*) The simple approach using indexed expressions may recompute base addresses repeatedly, leading to inefficient code.

(*iii*) Assume, we want the *i*th element of an array *A* whose subscript ranges from *lo* to *hi*. The address of the first element of the array is base. Each element of *A* has width *w*.

Then, the address of $x[i]$ is

$$\text{base} + (i - lo) * w = (\text{base} - lo * w) + i * w$$
$$= C + i * w$$

where C depends on the array A and is known at compile time.

TIPS

C is memory address if A is a global, and is stack displacement if A is a local.

(*iv*) The idea extends to multidimensional array in the abvious way, need to know whether the elements are stored in row-major or column-major order.

A two-dimensional array can be stored in

- Either row-major (*row-by-row*) or
- Column-major (*column-by-column*).

Most of the programming languages use row-major method. Row-major representation of a two-dimensional array is as follows:

The location of $A[i_1,i_2]$ is:

$$\text{base}_A + ((i_1 - lo_1) * n_2 + i_2 - lo_2) * w$$

base$_A$ is the location of the array A.

***lo*$_1$** is the index of the first row

***lo*$_2$** is the index of the first column

***n*$_2$** is the number of elements in each row

w is the width of each array element

Again, this formula can be re-written as

$$((i_1 * n_2) + i_2) * \text{width} + (\text{base}_A - ((lo_1 * n_1) + lo_2) * w)$$

should be computed at run-time can be computed at compile-time

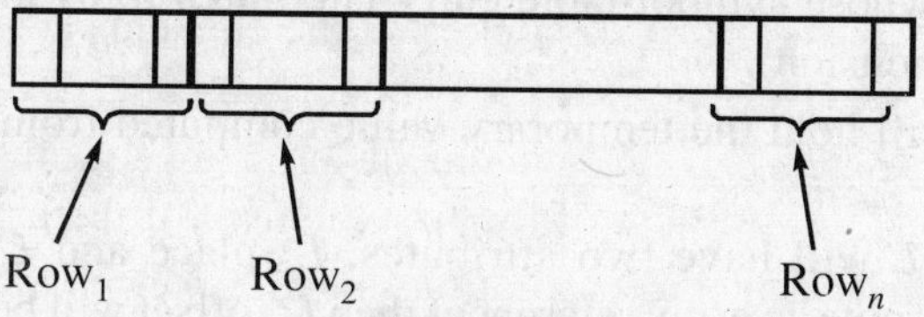

Fig. 10.9.

The location of $A[i_1, i_2]$ is $= \text{base}_A + ((i_1 - lo_1) * n_2 + i_2 - lo_2) * \text{width}$

base$_A$ is the location of the array A.

***lo*$_1$** is the index of the first row

***lo*$_2$** is the index of the first column

***n*$_2$** is the number of elements in each row

width is the width of each array element

Again, this formula can be re-written as:

$$= ((i_1 * n_2) + i_2) * \text{width} + (\text{base}_A - ((lo_1 * n_1) + lo_2) * \text{width})$$
$$= C + (\text{base}_A - ((lo_1 * n_1) + lo_2) * \text{width})$$

So clearly C can be known at compile time.

- Now we can extend the concept for ***K* dimentional array as follows:**

Let array be:

$A[i_1, i_2, ..., i_k]$ then expression can be written as follows:

$$((\ ...\ ((i_1\ n_2 + i_2)\ n_3 + i_3)\ ...\ n_x + i_x) \times w + (\text{base} - ((\ .....$$
$$((lo_1\ n_2 + lo_2)\ n_3 + lo_3)\ ....)\ n_k + lo_k) \times w)$$

Since for all j, $n_j = h_{ij} - lo_j + 1$ is fixed so above expression can be written as follows:

$$= C + ((.....\ ((i_1\ n_2 + i_2)\ n_3 + i_3)\ ....\ n_k + i_k) \times w$$

and C can be computed at compile time.

10.6.10. The Grammar for Array Reference

An array reference is an expression with l-value. Therefore, to capture its syntactic structure, we add the following productions to the grammar:

$L \rightarrow id$ [Elist]/id

Elist $\rightarrow$ Elist, E/E

But for the consideration of various dimensional limits n_j of the array, we can rewrite grammar as follows:

$L \rightarrow$ Elist]/id

Elist $\rightarrow$ Elist, E/id [E

These productions allow a pointer to the symbol-table entry for the array name to be passed as a synthesized attribute of Elist.

Now let us take the following assumption to write the translation scheme for addressing the array element.

- Elist . ndim will be used to record the number of dimensions in the Elist.
- Limit (array, i) returns n_i , the number of elements along the i th dimension of the array whose symbol-table entry is pointed to by array (that is upper bound of dimension).
- Elist place will hold the temporary value computed from index expression in Elist.
- An l . value L will have two attributes, L . place and L . offset. If L is a simple name (note an array reference) then L . offset will be null and L . place will be pointer to the symbol-table entry for that name.
- E . place plays same role (as in previous point) for nonterminal E.

10.6.11. The Translation Scheme for Addressing Array Elements

Let us consider the following grammar production for array:

(1) $L \rightarrow$ Elist]

(2) $L \rightarrow id$

(3) Elist $\rightarrow$ Elist, E

(4) Elist $\rightarrow id$ [E

Clearly above productions are representing the case of the array, without array references. So three-address code itself is produced by gencode ( ) procedure invoked in the semantic actions.

Productions	Translation Scheme
1. L → Elist]	{ L . place:= newtemp ; L . offset: newtemp ; gencode (L . place: = (Elist . array – C) gencode (L . offset:= Elist . place * width (Elist . array)) } (Note: here width (Elist . array) returns w).
2. L → id	{ L . place:= id . place ; L . offset:= null}
3. Elist → $Elist_1$, E	{ t_1: newtemp ( ) ; m: $Elist_1$. ndim + 1 gencode (t_1:= $Elist_1$. place *limit ($Elist_1$. array, m)) ; gencode (t_1:= t_1 + E . place) ; Elist . array:= $Elist_1$. array ; Elist . place:= t_1 ; Elist . ndim:= m }

(note that in above action, $Elist_1$. place has $m - 1$ components then Elist . place has m components).

4. Elist → *id* [*E* {Elist . array:= id . place ;
Elist . place:= E . place
Elist . ndim:= 1 }.

For example let us see an example of array X: B[p, q], which is 30 × 60 array with $lo_1 = lo_2 = 1$ and w to be 8. Then the three-address scheme can be as follows :

Here $n_1 = 30$ and $n_2 = 60$

$t_1 := p * 60$

$t_1 := t_1 + q$

$t_2 := C$ (here C is base address)

$t_3 := 8 * t_1$

$t_4 := t_2 [t_3]$

$X := t_4$

Example 10.9. Write the three address code for the statement while $a < b$ do $p = q + r$;

Solution. The required three address code is as follows:

```
i)      if a < b goto (i + 2)
i + 1) go to -------
i + 2) t1 = q + r
i + 3) p = t1
```

```
i + 4) goto (i)
        Statement . next → i + 1 ;
```

Example 10.10. Write the three address code for the following statement :

$$S \; : \; \text{do } x = y + z \text{ while } a < b$$

Solution. The required three-address code is as follows:

```
j)          t1 = y + z
j + 1) x = t1
j + 2) if a < b goto (i)
j + 3) goto ......
            S . next → j + 3
```

Example 10.11. Write the translation scheme for increment and decrement operators:

(i) $L \rightarrow ++\, id$

(ii) $L \rightarrow id\, ++$

(iii) $L \rightarrow id\, --$

(iv) $L \rightarrow --\, id.$

Solution.

Production	Translation Scheme
(i) L → ++ id	{ t_1 = gentemp ( ) ; gencode (t_1 = id . place +1) ; gencode (id . place = t_1) ; L . place = t_1 ; }
(ii) L → id ++	{ t_1 = gentemp ( ) ; t_2 = gentemp ( ) ; gencode (t_1 = id . place) ; gencode (t_2 = id . place + 1) ; gencode (id . place = t_2) ; L . place = t_1 ; }

(iii) L → id – –	{
	t_1 = gentemp ( ) ;
	t_2 = gentemp ( ) ;
	gencode (t_1 = id . place) ;
	gencode (t_2 = id . place . 1) ;
	gencode (id . place = t_1) ;
	L . place = t_1 ;
	}

(iv) L → – – id	{
	t_1 = gentemp ( ) ;
	gencode (t_1 = id . place – 1) ;
	gencode (id . place = t_1) ;
	L . place = t_1 ;
	}

Example 10.12. Write the three-address code for the following C program :

```
main ( )
{
    int x = 1 ;
    int y [20] ;
    while (x < = 20)
    a[x] = ;
}
```

Solution. The three-address code for the above C program is:

```
(1)   x = 1
(2)   if x < = 20 goto (4)
(3)   goto next
(4)   t1 = x * w
(5)   t2 = add (y) - w
(6)   t2 [t1] = 0
(7)   goto (2)
```

where w is the number of bytes required for each element.

Example 10.13. Write the three-address code for the following program fragment:

```
while (p < q and r > s) do
    if p = 1 then q = q + 1
    else
    while (p < = s) do
        p = A + 3
```

Solution. The three-address code is:

```
(1) if (p < q) goto (3)
(2) goto next
(3) if (r > s) goto (5)
(4) goto next
(5) if p = 1 goto (7)
(6) goto (10)
(7) t1 = q + 1
(8) q = t1
(9) goto (1)
(10)   if p < = S goto
(11)   goto (1)
(12)   t2 = p + 3
(13)   p = t2
(14)   goto (10)
(15)   goto (1)
```

Example 10.14. For the following program fragment:

```
sum = 0
for (j = 1 ; j < = 10 ; j ++)
sum = sum + a[j] + b[j] ;
```

Generate the three-address code. There are 8-bytes per word.

Solution. The three address code will be as follows:

```
(1)   sum = 0
(2)   j = 1
```

```
(3)  if j <= 10 goto (8)
(4)  goto next
(5)  t1 = j + 1
(6)  j = t1
(7)  goto (3)
(8)  t2 = j * 8
(9)  t3 = add (9) - 8
(10) t4 = t3 [t2]
(11) t5 = j * 8
(12) t6 = add (b) - 8
(13) t7 = t6 [t5]
(14) t8 = sum + t4
(15) t8 = t8 + t7
(16) sum = t8
(17) goto (5)
```

Example 10.15. Consider the following switch statement:

```
switch (i + j)
  {
  case 1: x = y + z
  default: p = q + r
  case 2: u = v + w
  }
```

and write the translation scheme to translate the above program fragment into three-address code.

Solution. The translation scheme for translating the given fragment into three-address code is as follows:

```
(0) t1 = i + j
(1) goto (11)
(2) t1 = y + z
(3) x  = t1
(4) goto next
(5) t2 = q + r
(6) p  = t2
(7) goto next
(8) t3 = u + w
(9) u  = t3
(10)    goto next
(11)    if t1 = 1 goto (2)
(12)    if t1 = 2 goto (8)
(13)    goto (5).
```

EXERCISE

1. Consider the following code fragment. Generate the three-address code for it.

```
for (j = 1; j <= 10; j++)
    if x = y then a = b + c
```

2. Consider the following code fragment. Generate the three-address code for it.

```
if x < y then
    while z > d do
a = a + b
else
```

```
do p = p + q
while e <= f
```

3. Consider the following code fragment. Generate the three-address code for it.

```
switch (a + b)
{
    case 1: x = x + 1
    case 2: y = y + 2
    case 3: z = z + 3
    default: C = C - 1
}
```

4. Write quadruples from the expression:

$$(a + b) * (c + d) - (a + b + c)$$

5. Write triples for the expression: $(a + b) * (c + d) - (a + b + c)$.
6. Translate the expression:

```
A:= -B * (C + D)/E
```

into quadruples and triples representation.

7. Describe the various representation of three-address codes. Translate the expression :

$$-(a + b) * (c + d) + (a + b + c)$$

into quadruples and triples.

8. Explain the back patching. Also generate three-address code for the following program segment:

```
while (a < c and b > d) do
    if a = 1 then c = c + 1;
    else
    while a <= d do
        a = a + 3;
```

9. Consider the following grammar for array references. Give syntax directed translation scheme to generate three address codes for addressing array elements. Translate the statement $X = A[i, j]$ where lower bounds of i and c are 1 and 1, upper bounds are 10 and 20 respectively.
Grammar is as follows:

$$S \rightarrow L := E$$
$$E \rightarrow E + E \,/\, (E) \,/L$$
$$L \rightarrow \text{Elist }] \,/\, id$$
$$\text{Elist} \rightarrow \text{Elist}, E \,/\, id\, [\, E$$

10. Translate the following statement into three-address statement:

```
switch (a + b)
{
    case 2: {x = y: break;}
    case 5: {switch (x)
        {
            case 0: {a = b + 1; break;}
            case 1: {a = b + 3; break;}
            default: {a = 2:}
        }
        break;
    case 9: {x = y - 1; break;}
    default: {a = 2;}
}
```

❑❑❑

11

Chapter

CODE OPTIMIZATION AND CODE GENERATION

INSIDE THIS CHAPTER

11.1. Introduction to Code Optimization **11.2.** Optimization Strategy **11.3.** Loop Optimization **11.4.** Global Data-Flow Analysis **11.5.** Elimination of Induction Variables **11.6.** Elimination of Common Subexpression **11.7.** Loop Unrolling **11.8.** Introduction to Code Generation **11.9.** Machine Architecture Currently in Use **11.10.** Major Issues of Code Generation **11.11.** A Machine Model and Code Generation **11.12.** A Simple Code Generator **11.13.** Code Generation From DAG **11.14.** Register Allocation **11.15.** Algebraic Properties **11.16.** Peephole Optimization

11.1. INTRODUCTION TO CODE OPTIMIZATION

Due to the cost factor of the software, now-a-days emphasis has shifted from the construction of efficient programs to that of more natural or more 'convenient' programs.

The process of optimization is aimed at improving the execution efficiency of a program. **Optimization aims mainly at rearranging the computation in a program so as to gain the advantage of execution speed, without changing the meaning of a program.**

Definition

"So finally we can say that code optimization refers to the technique used by the compiler to improve the execution efficiency of the generated object code."

Optimization may be machine dependent or may be machine independent. Machine-independent optimization can be performed independently of the target machine for which the compiler is generating the code. On the other hand, machine-dependent optimization requires knowledge of target machine architecture. We should know the target machine architecture so that better utilization of registers and other architecture characteristics may be possible.

A good optimizer has to exploit:

(*i*) Characteristics of the target machine in terms of the addressing structure and instruction repertoire.

(*ii*) Characteristics of the particular program in terms of program structure and nature of computation involved.

Optimization may be local and global:

In case of local optimization, we consider only small portion of program code. In this type of optimization only restricted amount of analysis required, but gain of optimization is restricted in this type of optimization. On the other hand in global optimization the scope is not restricted to short code sequences. Thus optimization of large program sections-including entire loops can be performed. While this requires considerable analysis of the program structure, its gains are also higher than those of local optimization.

11.2. OPTIMIZATION STRATEGY [MDU 2007, 2009]

Some common steps in optimization may be as follows:

- Examine the program to find out certain properties of interest, it is said to be **Data Slow Analysis**.
- Use this information to change the code in a way that improves performance. ("code optimization"). Examples of optimization include:

(*a*) **Code Motion out of Loops.** If a computation inside a loop produces the same result for all iterations (that is computing the base address of a local array), it may be possible to move the computation outside the loop. Complete process, we will study under loop optimization.

(*b*) **Common Sub-expression Elimination.** If the same expression is computed in many places (that is array address computations; result of macro expansion), compute it once and reuse the result.

(*c*) **Copy Propagation.** If we have an intermediate code "copy" instruction '*x* : *y*', replace subsequent uses of *xbyy* (where possible).

(*d*) **Dead Code Elimination.** Delete instruction whose results are not used.

11.3. LOOP OPTIMIZATION

Loop optimization is most valuable machine-independent optimization because a program's inner loop are good candidate for improvement.

The execution efficiency of a loop can be optimized by moving out some code which is loop-invariant. Such code would now be executed just once prior to loop entry, and not repeatedly during execution of the loop. For example, in

```
while (i < j)
{
a = x * 2 - z * 2 ;
-----------
}
```

The evaluation of $(x * 2 - z * 2)$ can be moved out of the loop provided it is invariant. **This kind of code movement also called Frequency Reduction** transformation.

The major tasks in loop optimization are:

(*i*) Detecting loops in a program.

(*ii*) Deducing the feasibility of moving certain code out of the loop.

11.3.1. Analysis of Program for Loop Optimization

For program analysis and loop optimization, it is usually necessary, to know control flow relationships between different pieces of code.

For this we:

- group three-address instruction in to **Basic Blocks.**
- represent control flow relationships between basic blocks using a **Control flow graph**.

For example:

```
L1: if x > y goto L0
        t1 = x + 1
        x = t1
L0 :    y = 0
        goto L1
```

can be represent as follows:

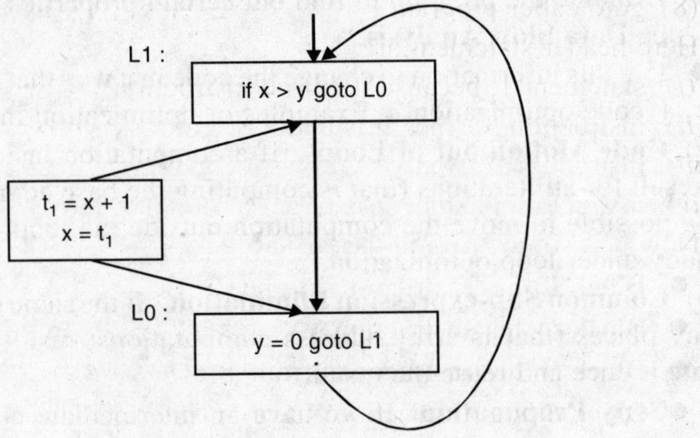

Fig. 11.1.

1. Representation of Three-address Code in to Basic Blocks. A **basic block** is a sequence of consecutive instruction such that:

1. Control enters at the beginning.
2. Control leaves at the end; and
3. Control cannot halt or branch except at the end.

Identification of Basic Block can done by following steps:

Step 1. Determine the set of **Leaders**, that is the first instruction of each basic block :

(*a*) The first instruction of the function is a header.

(*b*) Any instruction that is the target of a branch is a header (target of conditional and unconditional goto)

(*c*) Any instruction immediately following a (conditional or unconditional goto) a branch is a header.

Step 2. For each header, its basic block consists of itself and all instruction upto, but not including, the next header (or end of function).

For example let us consider the following program fragment:

```
exp (x)
{
int P = 1;
for (i = 2 ; i < = x ; i++)
```

```
p = p * i
return (p);
}
```

The three-address-code for above code fragment is as follows:

(1) $p = 1$;
(2) $i = 2$
(3) if $i <= x$ goto (8)
(4) $p = p * i$
(5) $t_1 = i + 1$
(6) $i = t_1$
(7) goto (3)
(8) goto caller program

Here header statement are:

(*i*) statement 1, because it is first instruction.
(*ii*) instruction 3, since it is target of goto.
(*iii*) instruction 4, since it follows a conditional goto.
(*iv*) instruction 8, since it is target of conditional goto statement.

Now it is simple to find the basic blocks:

- Block B_1 has instructions 1 and 2.
- Block B_2 has instruction 3.
- Block B_3 has instructions from 4 to 7.
- Block B_4 has instruction 8.

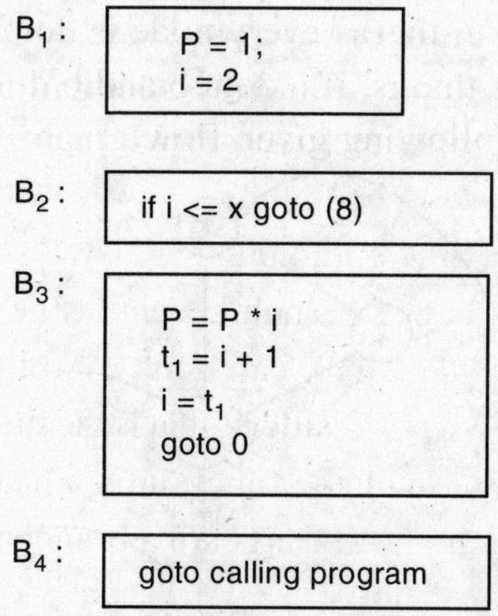

Fig. 11.2.

2. Finding Control Flow Graph from Basic-Blocks: [MDU 2009, 2010]

Definition A flow graph for a function is a directed graph $G = (V, E)$ whose nodes are the basic blocks of the function, and where $a \rightarrow b \in E$ iff control can leave '*a*' and immediately enter *b*.

The distinguished **initial** node if a flow graph is the basic block whose header is the first instruction of the function.

Flow graph can be construct as follows:

Step 1. Identify the basic block of the function.

Step 2. There is a directed edge from block B_1 to block B_2 if

(*a*) There is a (conditional or unconditional) jump from the last instruction of B_1 to the first instruction of B_2; or

(*b*) B_2 immediately follows B_1 in the textual order of the program, and B_1 does not end in an unconditional jump.

For example the control flow graph for the basic blocks in Fig. 11.3 is as follows.

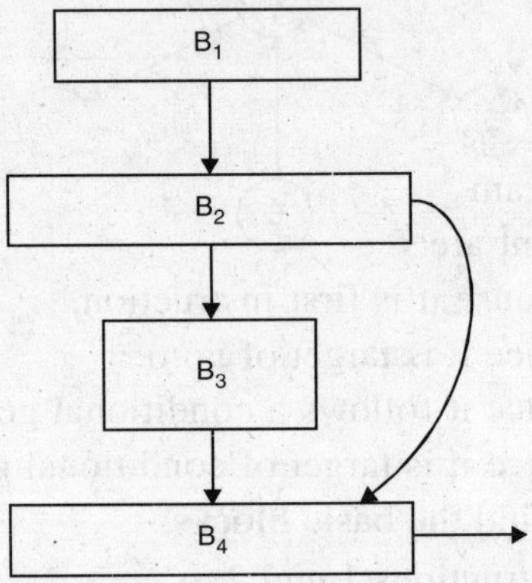

Fig. 11.3.

3. Identifying the Dominators. A node **a** is a dominator of node **b** if every path from the initial node of the flow graph to **b** goes through **a**.

Dominators (dom) relationship have the following properties:

(*i*) There are reflexive; that is every node is dominators of itself.

(*ii*) Transitivity holds that is, if a dom b and b dom c this implies a dom c.

For example for the following given flow graph:

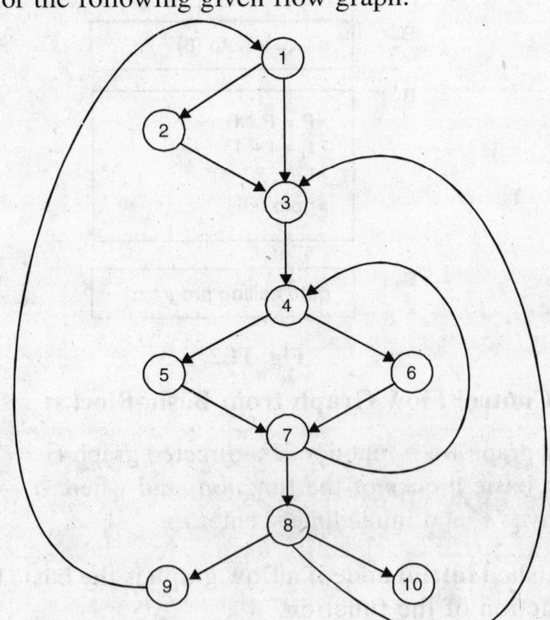

Fig. 11.4.

Dominators tree can be shown as follows:

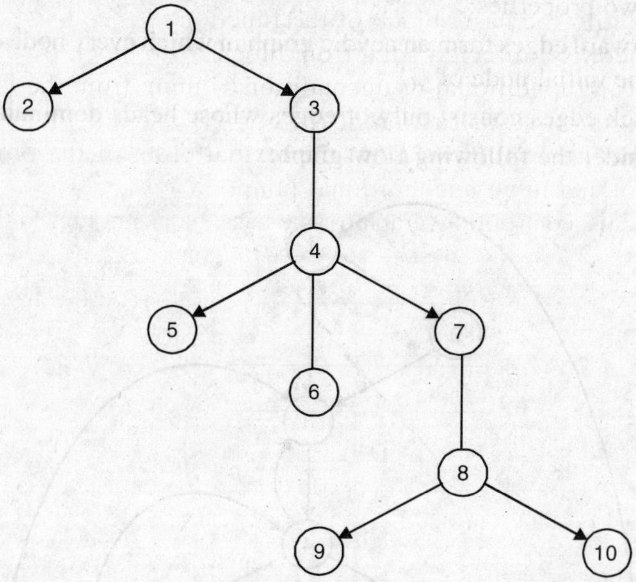

Fig. 11.5.

4. Back Edges Identification. All those edges are called **Back Edges**, whose heads dominate their tails.

11.3.2. Loop Detection

Now after previous analysis, it is to detect a loop in a flow graph. **A loop is a cycle in the flow graph that satisfies two properties:**

1. A loop must have a single entry point, called the "header". This entry point dominates all nodes in the loop, it makes possible to move all of the loop invariant computations to a unique place, called a "preheader", which is a block/node placed outside the loop, just infront of the header.
2. There must be at least one way to iterate the loop, that is one path back to the header.

For example consider the flow graph given in Fig. 11.4. In which

4 dom 7, 7 dom 10, 3 dom 4, 3 dom 8 and 1 dom 9.

Now it is easy to detect the loop. "If a given back edge is $\boldsymbol{a \rightarrow d}$, we define the natural loop of the edge to be **d** plus the set of nodes that can reach **a** without going through *d*. Node *d* is the header of the loop".

So clearly the natural loop of the edge $10 \rightarrow 7$ consists of nodes 7, 8 and 10, since 8 and 10 are all those nodes that can reach 10 without going through 7. The natural loop of $9 \rightarrow 1$ is the entire flow graph.

11.3.3. Reducible Flow Graph

It is assumed that code optimization operations can be easily performed by using reducible flow graph. A flow graph *G* is reducible if we can partition the edges into two disjoint groups, often called the forward edges and back edges, with the following two properties:

(*i*) The forward edges form an acydic graph in which every node can be reached from the initial node of *G*.

(*ii*) The back edges consist only of edges whose heads dominates their tails.

Let us consider the following flow graph:

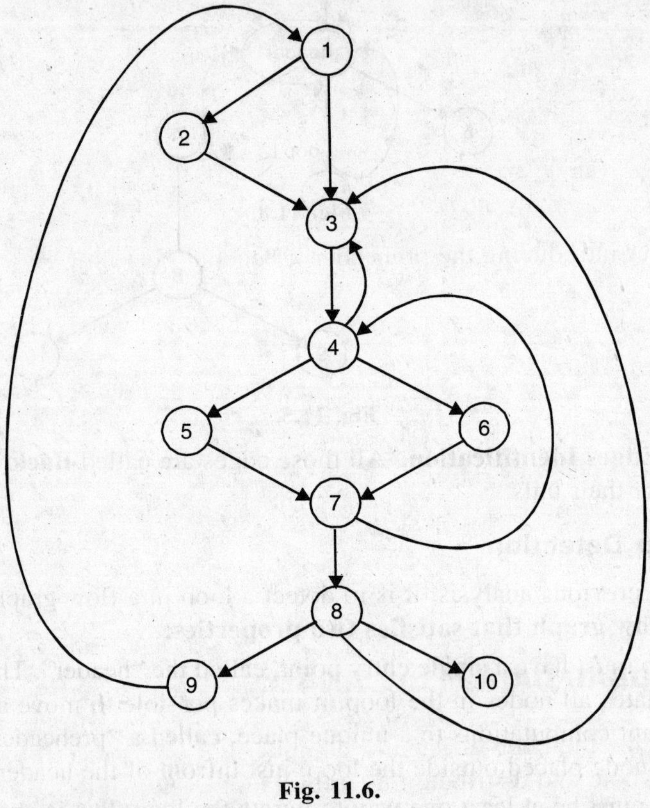

Fig. 11.6.

Here 4 → 3, 7 → 4, 8 → 3, 9 → 1 and 10 → 7 are back edges, whose heads dominates their tail.

Now let us remove the back edges from the flow graph shown in Fig. 11.6, it gives us reduced flow graph.

Now new flow graph is as follows:

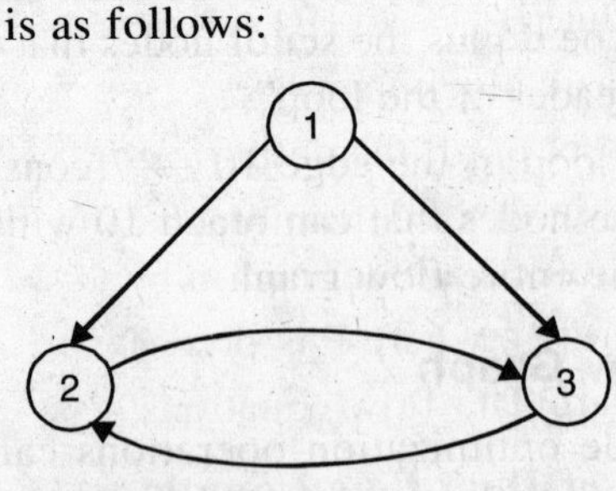

Fig. 11.7 A non-reducible flow graph.

11.3.4. Pre-Headers

To more statements "before the header node" we define a new header called preheader. The preheader has only the header as successor, all the edges enters in header through preheader.

For example Fig. 11.8 shows a flow graph with loop:

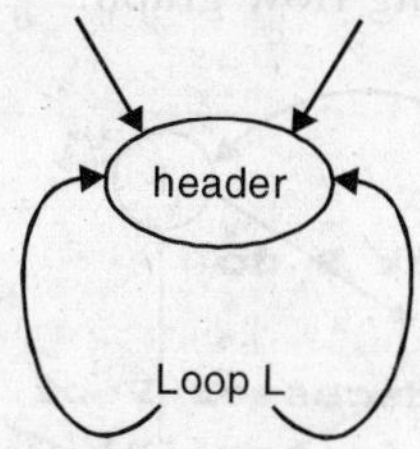

Fig. 11.8.

Now after introducing the preheader it becomes as follows:

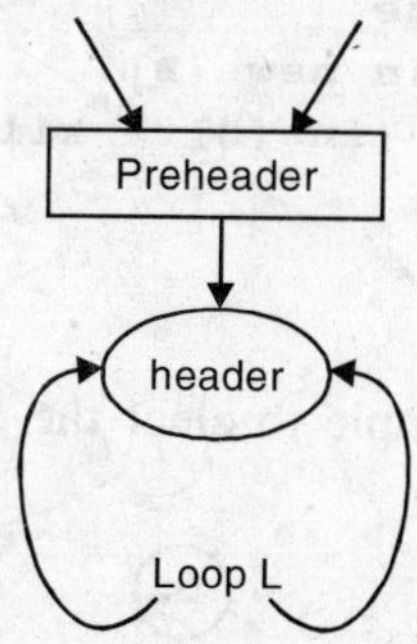

Fig. 11.9.

11.4. GLOBAL DATA-FLOW ANALYSIS [MDU 2008]

To optimize the code efficiently compiler collects the information about the program as whole and distribute this information to each block of the flow graph. This process is known as data-flow analysis.

To get the data-flow information we define certain data-flow equations and solution of those equation gives the data-flow information.

- **Data-flow-equations.** A typical equation is

$$\text{Out } [B] = (\text{in } [B] - \text{Kill } [B] \vee \text{gen } [B] \quad \ldots(1)$$

where: B : is basic block

gen [B] : The set of all the definitions generated in block B.

Kill [B] : The set of all definitions outside block B that define the same variable as are defined in block B.

in [B] : $\vee$ out [P], where P is predecessor of B. ...(2)

Equation (1) and (2) are data-flow equations.

- **Itrative solution of Data-flow Equations.** We can define the following algorithm to find out the solution to equations as follows:

```
/* initialize out on the assumption in [B] = ϕ for all B */
for each block B do
    {
      out [B]: = gen [B] ;
    }
    flag: = true
    while (flag) do
    {
        flag = false
        for each block B do
        {
        for each predecessor P of B
        In new [B] = in new [B] ∨ out [P] ;
        if in new [B] ≠ in [B] then
        {
            flag = true
            in [B] = in new [B]
            out [B] = (in [B] - kill [B]) ∨ gen [B] ;
            }
        }
    }.
```

Now let us consider an example to clear the concept. Consider the following flow graph:

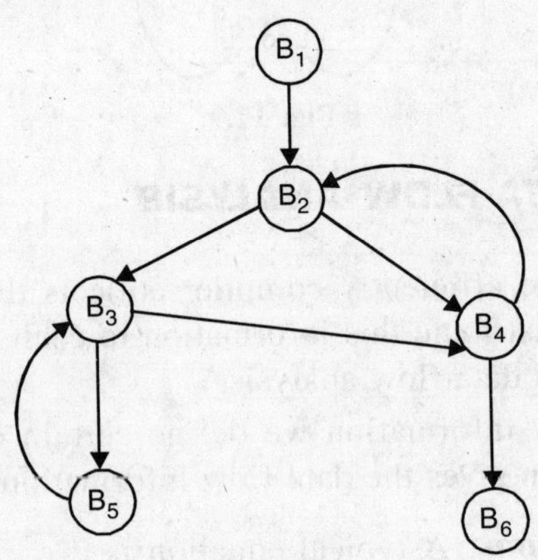

Fig. 11.10.

where B_1: (1) $b = 1$
(2) $c = 2$
B_2 : (3) $a = b + c$
(4) $d = a - b$
B_3 : (5) $d = c * d$
B_4 : (6) $c = b + c$
(7) $e = a - b$
B_5 : (8) $d = b + c$
(9) $e = e + 1$
B_6 : (10) $b = c * d$
(11) $c = b - d$

Step 1. Calculate the gen and kill as follows:

Block	gen	kill
B_1	{1, 2}	{6, 10, 11}
B_2	{3, 4}	{5, 8}
B_3	{5}	{4, 8}
B_4	{6, 7}	{2, 9, 11}
B_5	{8, 9}	{4, 5, 7}
B_6	{10, 11}	{1, 2, 6}

Step 2. Now let us apply the iterative method to solve the data-flow equations.

Initially In [B] = ϕ and out [B] = gen [B]

So:

Block	In	Out
B_1	ϕ	{1, 2}
B_2	ϕ	{3, 4}
B_3	ϕ	{5}
B_4	ϕ	{6, 7}
B_5	ϕ	{8, 9}
B_6	ϕ	{10, 11}

TIPS

Initially, we take In [B] for every block that is to be an empty set, and we take out [B] to be gen [B], and we compute In new [B]. If it is different from In [B], we compute a new out [B] and more to next iteration, according to algorithm discussed in previous section.

So after first iteration In and out values are:

Block	In	Out
B_1	ϕ	{1, 2}
B_2	{1, 2, 6, 7}	{1, 2, 3, 4, 6, 7}
B_3	{3, 4, 8, 9}	{3, 5, 9}
B_4	{3, 4, 5}	{3, 4, 5, 6, 7}
B_5	{5}	{8, 9}
B_6	{6, 7}	{7, 10, 11}

After second iteration In and Out values are:

Block	In	Out
B_1	ϕ	{1, 2}
B_2	{1, 2, 3, 4, 5, 6, 7}	{1, 2, 3, 4, 6, 7}
B_3	{1, 2, 3, 4, 6, 7, 8, 9}	{1, 2, 3, 5, 6, 7, 9}
B_4	{1, 2, 3, 4, 5, 6, 7, 9}	{1, 3, 4, 5, 6, 7}
B_5	{3, 5, 9}	{3, 8, 9}
B_6	{3, 4, 5, 6, 7}	{3, 4, 5, 7, 10, 11}

After third iteration In and out values are as follows:

Block	In	Out
B_1	ϕ	{1, 2}
B_2	{1, 2, 3, 4, 5, 6, 7}	{1, 2, 3, 4, 6, 7}
B_3	{1, 2, 3, 4, 5, 6, 7, 8, 9}	{1, 2, 3, 5, 6, 7, 9}
B_4	{1, 2, 3, 4, 5, 6, 7, 9}	{1, 3, 4, 5, 6, 7}
B_5	{1, 2, 3, 5, 6, 7, 9}	{1, 2, 3, 6, 8, 9}
B_6	{1, 3, 4, 5, 6, 7}	{1, 3, 4, 5, 7, 10, 11}

After Fourth Iteration In and Out values are as follows:

Block	In	Out
B_1	ϕ	{1, 2}
B_2	{1, 2, 3, 4, 5, 6, 7}	{1, 2, 3, 4, 6, 7}
B_3	{1, 2, 3, 4, 6, 7, 8, 9}	{1, 2, 3, 5, 6, 7, 9}
B_4	{1, 2, 3, 4, 5, 6, 7, 9}	{1, 3, 4, 5, 6, 7}
B_5	{1, 2, 3, 5, 6, 7, 9}	{1, 2, 3, 6, 8, 9}
B_6	{1, 3, 4, 5, 6, 7}	{1, 3, 4, 5, 7, 10, 11}

Here In [B] becomes to Out [B], so we must stop iteration here.

Step 3. Computing of Use-Definition Chains:

It is easy to store the information as "Use-definition chains" or "ud-chains", which are lists, for each use of a variable, of all the definition that reach that use.

We can calculate ud-chain as follows:

(*i*) If a use of variable ***a*** in block *B* is preceded by no unambiguous definition of ***a***, then the ud-chain for that use of ***a*** is the set of definition in In[B] that are definition of ***a***.

(*ii*) If the use of ***a*** in block B is proceeded by its definition, then the ud-chain of ***a*** contains only the definition prior to this use of *a*.

For example, consider flow graph is Fig. 11.10.

The use of **a** in definition 4, block B_2 is preceded by definition 3, which is the definition of **a**. Hence ud-chain for this use of **a** contains only definition 3. But the use of **b** in B_2 is not preceded by any definition of **b** in B_2, so ud-chain for this use of **b** will be {13, which is the only definition of *b* in In[B]}.

Step 4. Detection of Loop-Invariant Computations:

By the use of ud-chain, we can detect the loop-invariant.

Suppose $a := b + c$ is at a position in the loop where all possible definition of *b* and *c* are outside the loop (including the special case where *b* and/or *c* are constant), then the calculation of $b + c$ will be the same each time encountered, **so it is loop-invariant in the loop.**

Step 5. Code Motion (Moving loop-invariant code out of the loop): We can move loop-invariant outside the loop if following conditions are met:

(*a*) The block in which a loop invariant statement occurs should be a dominator of all exists of the loop, or the name assigned to the block should not be used outside the loop.

(*b*) We cannot move a loop-invariant statement assigned to ***a*** into preheader if there is another statement in the loop that assign to ***a***.

For example, consider the following flow graph:

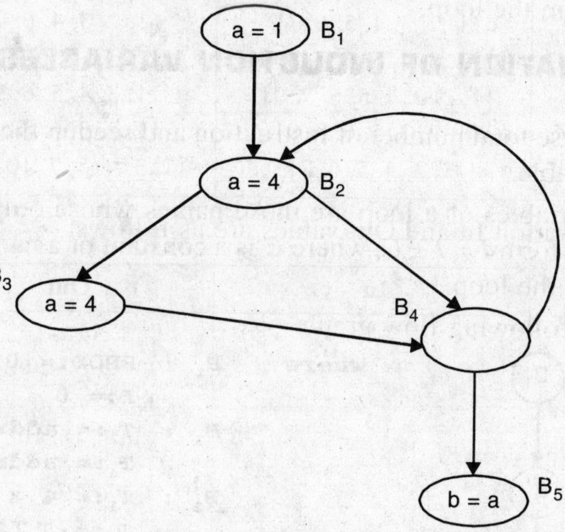

Fig. 11.11.

Here $a = 4$ in B_2 cannot be moved outside the loop inspite of that B_2 is dominator of other block of loop, since moving it to the preheader will change the meaning of the program. Because if $a = 4$ is moved out of the loop, then the value that will be assigned to y in B_5 will be 3 if the execution path is $B_1 - B_2 - B_3 - B_4 - B_2 - B_4 - B_5$ whereas for same execution path, the original program assign x to y in B_5.

(*i*) The move is illegal if ***a*** is used in the loop and ***a*** is reached by any definition of ***a*** other than the statement to be moved.

For example, consider the following flow-graph.

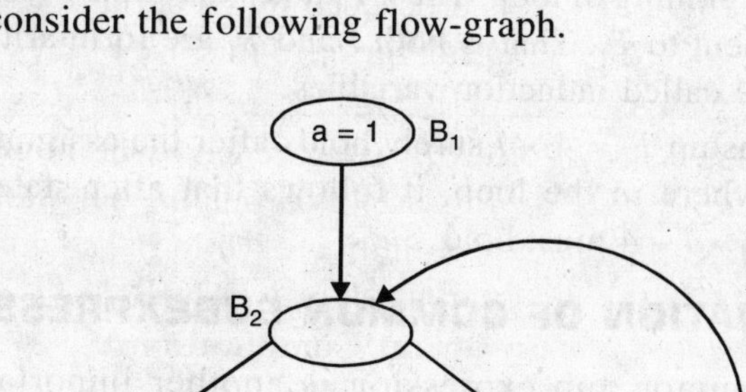

Fig. 11.12.

Here $a = 2$ (from B_3) cannot be moved out since use of a in B_4 is also reached by the definition $a = 1$ in B_1.

Note that, if all above discussed condition are met then we can move out loop-invariant code, from the loop.

11.5. ELIMINATION OF INDUCTION VARIABLES [MDU 2007, 2010]

We can decrease total number of instruction and seedup the loop by eliminating the induction variable.

"Induction variables of a loop are those names whose only assignment within the loop are of the form $I = I \pm C$, where C is a constant or a name whose value does not change within the loop."

Consider the following flow graph.

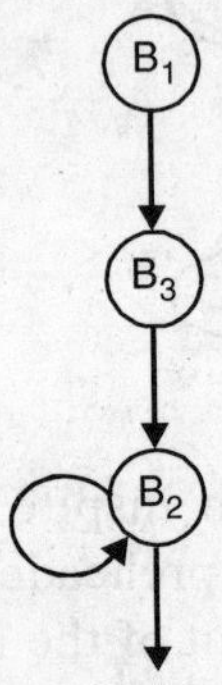

```
where   B1 : PROD:= 0
             I:= 0
        B3 : T2:= addr (A) - 4
             T4:= addr (B) - 4
        B2 : T1:= 4 × I
             T3:= T2[T1]
             T5:= T4 [T1]
             T6:= T3 * T5
             PROD:= PROD + T6
             I:= I + 1
          if I ≤ 20 goto B2
```

Fig. 11.13.

We note that the purpose of I is to count from 1 to 20 in the loop, while the purpose of T_1 is to step through the arrays, four bytes at a time, since we are assuming four bytes/word. At the assignment $T_1 := 4 * I$, I takes on the values 1, 2, ..., 20 each time through the beginning of loop. Thus, T_1 takes the values 4, 8, ..., 80 immediately after each assignment to T_1. That is both I and T_1 are form arithmetic progression, such identifiers are called induction variables.

As the relationship $T_1 = 4 * I$ surely holds after the assignment to T_1, and T_1 is not changed elsewhere in the loop, it follows that after statement $I := I + 1$ the relationship $T_1 = 4 * I - 4$ must hold.

11.6. ELIMINATION OF COMMON SUBEXPRESSION

Removing common sub expression is another important feature of code optimization. Common subexpression elimination implies the elimination of an expression evaluation from a place in the program if an "equivalent" value has already been computed and can be used. For example,

```
A = B ** C
.......
X  =  B ** C + 5
```

can be write as follows:

$$T = B ** C$$

$$A = T$$

....

$$X = T + 5$$

Above optimization is possible only when, none of the statements intervening between the evaluation of $B ** C$ and its use in the computation of X changes the value of $B ** C$.

11.6.1. Elimination of Local Common Subexpression

Local common subexpression can be eliminating by Dag Construction.

Dag Construction

Let us define a function node (id), which return the most recently created node associated with its id. For every three address statement $x = y$ OP z, $x =$ OP y, or $x = y$ in the block, we have to follow the following steps:

Step 1. If node (y) is undefined, create a leaf labelled y, and let node (y) be this node. If node (z) is undefined, create a leaf labelled z, and let that leaf be node (z). If the statement is of the form $x =$ OP y or $x = y$, then if node (y) is undefined, creates a leaf labelled y, and let node (y) be the node.

Step 2. If a node exists that is labelled OP whose left child is node (y) and whose right child is node (z) (to catch common subexpression), then returns this node. Otherwise, create such a node, and return it. If the statement is of the form $x =$ OP y, then check if a node exist that is labelled OP whose only child is node (y). Return this node. Otherwise, create such a node and return. Let the returned node be n.

Step 3. Append x to the list of identifiers for the node n returned in Step 2. Delete x from the list of attached identifiers for node (x), and set node (x) to be node n.

Let us consider an example to make the concept clear, consider the following the block of code:

(1) S_1:= 4 ∗ I
(2) S_2:= addr (A) – 4
(3) S_3:= S_2 [S_1]
(4) S_4:= 4 ∗ I
(5) S_5:= addr (B) – 4
(6) S_6:= S_5 [S_4]
(7) S_7:= S_3 ∗ S_6
(8) S_8:= PROD + S_7
(9) PROD:= S_8
(10) S_9:= I + 1
(11) $I = S_9$
(12) if I <= 20 got 0(1)

Now let us construct the Dag for above code step-by-step:

1. Very first instruction (1) is S_1:= 4 ∗ I, we will create leaves n_1 and n_2 labelled 4 an I_0 then a node n_3 labelled ∗ and we also attach an identifier S_1 to it:

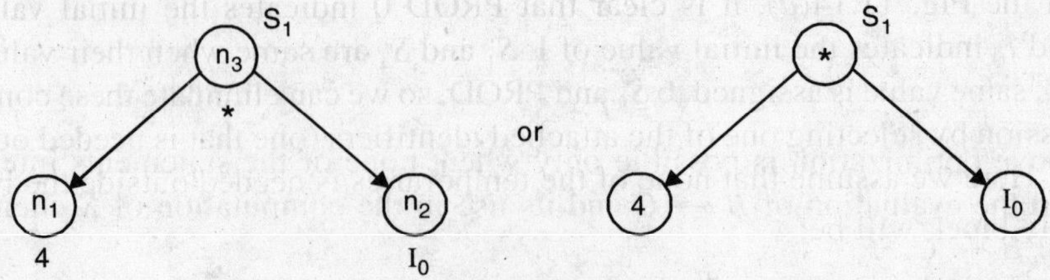

Fig. 11.14(*a*)

2. Now for S_2:= add (A) – 4 in same manner.

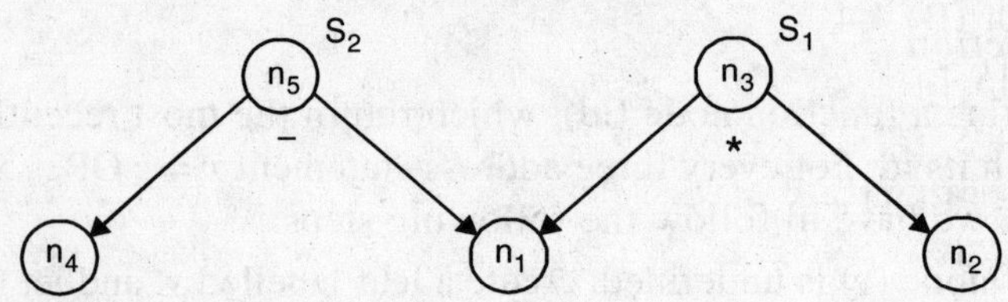

Fig. 11.14(*b*).

3. Figure 11.14(*c*) represents the Dag for S_3:= $S_2[S_1]$

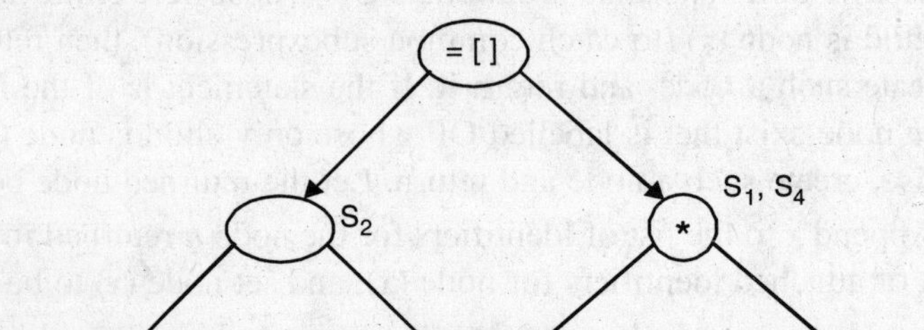

Fig. 11.14(*c*).

4. Finally in same way, we can find the Dag for complete code, as follows:

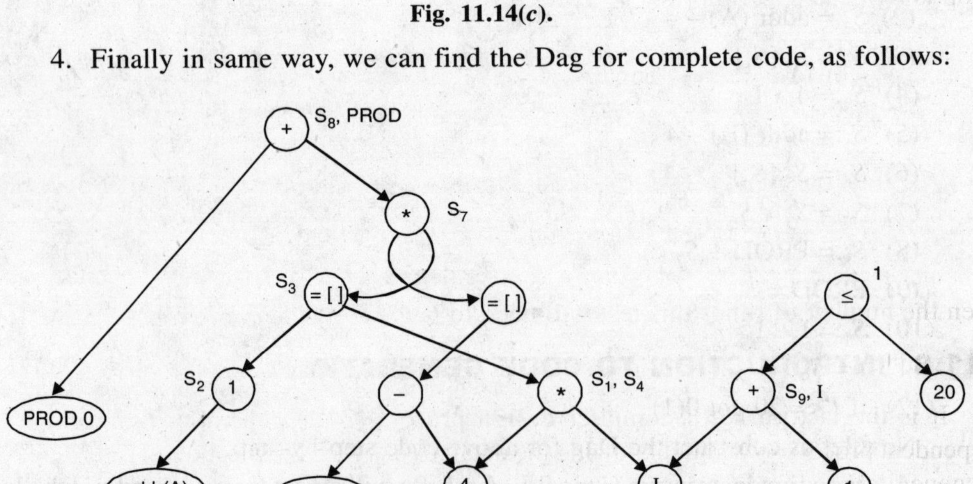

Fig. 11.14(*d*).

From the Fig. 11.14(*d*), it is clear that PROD 0 indicates the initial value of PROD and I_0 indicates the initial value of I. S_1 and S_4 are same when their value are computed, same value is assigned to S_8 and PROD, so we can eliminate these common subexpression by selecting one of the attached identifiers (one that is needed outside the loop). Here we assume that none of the temporaries is needed outside the block.

Finally block will be:

(1) $S_1 := 4 * I$
(2) $S_2 :=$ addr (A) – 4
(3) $S_3 := S_2\ [S_1]$
(4) $S_5 :=$ addr (B) – 4
(5) $S_6 := S_5\ [S_1]$
(6) $S_7 := S_3 * S_6$
(7) PROD:= PROD + S_7
(8) $I := I + 1$
(9) if $I \le 20$ goto (1)

11.7. LOOP UNROLLING

If number of iterations are constant then we can reduce the replicating the body of the loop to reduce the required number of test, concept is called loop unrolling.

For example, consider the following loop:

```
I = 1
while (I <= 100)
{
    x[I]  = 0;
    I:= I + 1;
}
```

In this case the test I <= 100 will be performed 100 times. But if the body of the loop is replicated as follows:

```
I = 1
while (I <= 100)
{
    x[I]  = 0
    I:= I + 1;
    x[I]  = 0;
    I:= I + 1
{
```

then the number of times this test will need to be performed will be 50.

11.8. INTRODUCTION TO CODE GENERATION

It is the last phase of compiler design process. Code generation is machine dependent process so without knowing the detail of machine architecture, it is difficult to generate good code. We must be very carefull about code generation algorithm selection since a efficient algorithm may be twice as fast as code generated by an rough algorithm.

Let us see pictures view of code generation phase:

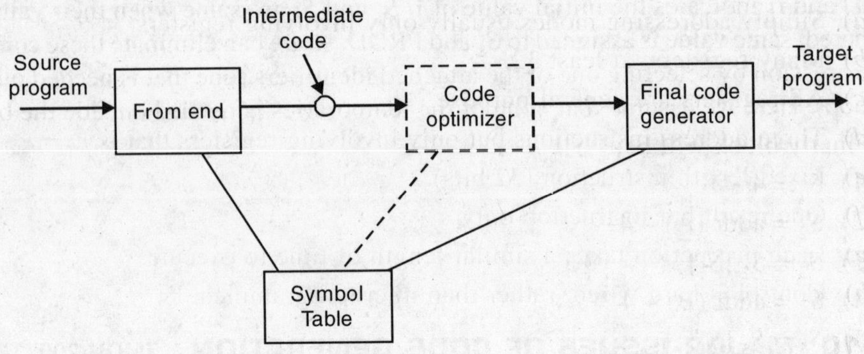

Fig. 11.15.

A good generation must do following things:

(*i*) Produce correct code

(*ii*) Make use of machine architecture

(*iii*) Run efficiently.

- **Input to code generator:**

(*i*) Intermediate code program

(*ii*) Symbol table

- **Output of code generator:** is target program

This can be any of:

(*i*) Assembly language program

(*ii*) Absolute machine-language program.

(*iii*) Relocatable machine-language program.

11.9. MACHINE ARCHITECTURE CURRENTLY IN USE

Before considering other issues, we should look at the two main types of machine architecture currently in use

1. CISC (Complex Instruction Set Computer): CISCs are designed with the implementation of high-level languages in mind and would therefore seem, the ideal candidate as target for compilers. Typical characteristics are:

(*a*) A wide range of addressing modes to support access to arrays, records, lists, stack frames etc.

(*b*) A small number of registers (usually 16 or less).

(*c*) Many special purpose registers, for example registers may be reserved for indexation.

(*d*) Two-address instructions such as $A + B - A$, where A, B may be complex addresses.

(*e*) Variable length instruction.

(*f*) Instructions with side effects, for example auto-increment instructions.

(*g*) Widely different execution times for instructions.

(*h*) Control implemented by microprogram.

2. RISC (Reduced Instruction Set Computer): RISCs have the following characteristics.

(*a*) Simply addressing modes usually only involving registers.

(*b*) Many registers, at least 32.

(*c*) All registers are general purpose.

(*d*) Three-address instructions but only involving registers that is $r_3 = r_1 + r_2$.

(*e*) Fixed length instruction (32 bits).

(*f*) One result per instruction only.

(*g*) Each instruction takes a similar length of time to execute.

(*h*) Control is hard wired, rather than microprogrammed.

11.10. MAJOR ISSUES OF CODE GENERATION [MDU 2007, 2009]

The code generator mainly concern with

- Memory management
- Instruction selection
- Register Utilization (Allocation)
- Storage allocation.

1. Memory Management. During code generation process references to symbol-table entries have to be mapped to actual P addresses and levels have to be mapped to instruction addresses.

Mapping names in the source program to address of data is co-operating done in pass 1 (Front end) and pass 2 (code generator).

Local variables (local to functions or procedures) are stack-allocation in the activation record while global variables are in static area.

2. Instruction Selection. The nature of instruction set of the target machine determines selection. Peculiarities in the target instruction set have to be taken into account. For example:

(*a*) The MC68K treats address and data registers differently.

(*b*) Some instructions (that is MUL, DIV) may need pairs of registers and take several cycles to complete (that is can't use those registers for other purposes).

(*c*) Addressing mode (PC-relative vs. absolute), span (short vs. long).

Instruction selection is "easy" if instruction set is regular that is uniform (all triple addresses all stack single addresses) and complete (use of register for any operation).

If we don't care about efficiency of target program, instruction selection is straight forward.

For example, the three address code is:

$$a := b + c$$
$$d := a + e$$

Inefficient assembly code is

1. MOV b, R_0 $\quad R_0 \leftarrow b$
2. ADD c, R_0 $\quad R_0 \leftarrow c + R_0$

3. MOV R_0, a $\qquad a \leftarrow R_0$
4. MOV a, R_0 $\qquad R_0 \leftarrow a$
5. ADD e, R_0 $\qquad R_0 \leftarrow e + R_0$
6. MOV R_0, d $\qquad d \leftarrow R_0$

3. Register Allocation: Register can be accessed faster than memory words. Frequently accessed variables should reside in registers (register allocation). Register assignment is picking a specific register for each such variable.

Formally, there are two steps in register allocation.

(*a*) Register allocation (what register ?): This is a register selection process in which we select the set of variables that will reside in register.

(*b*) Register assignment (what variable ?): Here we pick the register that contain variable. Note that this is a NP-complete problem.

Some of the issues that complicate register allocation

(*i*) Special use of hardware for example, some instruction require specific register.

(*ii*) Convention for software:

For example:

- Register R_6 (say) always return address.
- Register R_5 (say) for stack pointer.
- Similarly, we assigned registers for branch and link, frames, heaps etc.

(*iii*) Choice of evaluation order, may produce more efficient code.

This is NP-complete problem but we can by pass this hindrance by generating code for quadruples in the order in which they have been produced by intermediate code generator.

```
ADD x, y, T1
ADD a, b, T2
```

is legal because x, y and a, b are different.

4. Storage Allocation: Delay decision about storage allocation until final code generation, when the width of each type is known:

- Initialize location field of each identifier/temporary to a special value UNDEF.
- Maintain a counter that gives the displacement of the next available slot in the stack frame.
- During code generation, if there is a reference to a symbol table entry whose location is UNDEF, then set the value using the counter before generating any code. Update the location counter using type info.

There are following advantages for this kind of delay decision about storage allocation.

(*i*) We can push machine dependencies to the back end;

(*ii*) Variables that are optimized away need not take up space;

(*iii*) Variables that can be kept in registers do not take up storage unless necessary.

11.11. A MACHINE MODEL AND CODE GENERATION

It is very clear now that code generation required very deep knowledge of machine architecture. By knowing the machine specification, we can generate a code for complete language efficiently.

For example, let us see two type of machine specification:

1. **Target Machine First** having following specification:
   (*i*) We have byte addressable machine (that is 2^{15} 16-bit words) of memory.
   (*ii*) 4 byte per word.
   (*iii*) We have 16 to 32 (or *n*) general purpose registers.
   (*iv*) Machine has two addressable instruction of the form:

```
OP source, destination
that is, MOV A, B
ADD A, D
```

2. **Target Machine Second** having following specification:
   (*i*) Machine has bit addressing (factor of 1)
   (*ii*) Three address instruction of forms:

```
OP source1, source2, destination
```

For example:

```
ADD A, B, C
```

Now let us see two address instruction of the form **mnemonic source, destination** with their addressing mode.

Mode	Form	Address	Example	Added-cost
Absolute	M	M	Add R_0, R_1	1
Register	R	R	Add temp, R_1	0
Index	C(R)	C + contents (R)	ADD 100 (R_2), R_1	1
Indirect register	*R	Content (R)	ADD * R_2, R_1	0
Indirect Index	*C(R)	Content (C+) Content (R)	ADD * 100 (R_2), R_1	1
Literal	#C	Constant C	ADD #3, R_1	1

Note, here cost 1 means that it occupies only one word of memory.

"Each instruction has a cost of 1 plus added costs for the source and destination".

Cost of instruction = 1 + Cost of source and destination mode

This cost corresponds to the length (in words) of instruction. Let us analyse the some of example:

1. **Move register to memory $R_0 \leftarrow M$**

MOV R_0, *M* cost = 1 + 1 = 2(since address of memory location *M* is in word following the instruction)

2. Indirect indexed mode:

MOV $* 4 (R_0), M$

cost = 1 + 1 + 1 (since one word for memory location M, one word for result of $* 4(R_0)$ and one for instruction)

3. Indexed Mode:

MOV $4(R_0), M$

cost = 1 + 1 + 1 = 3

4. Literal Mode:

MOV #1, R_0

cost = 1 + 1 = 2 (one word for constant 1 and one for instruction)

5. MOVE memory to memory:

MOV m, m cost = 1 + 1 + 1 = 3 (two for memory locations and one for instruction)

11.12. A SIMPLE CODE GENERATOR

Give a sequence of three-address statement partitioned into basic blocks, straight forward code generation involves generating code for each three-address statement in turn by taking the advantages of any of the operands of the three address statement that are in the register as long as possible. We store it only if the register is needed for another computation or just before a procedure call, jump or labelled statement, such as at the end of a basic block is encountered. The reason for this that after leaving a basic block, we may go to several different blocks, or we may go to one particular block that can be reached from several others.

To avoid the error, our code-generation strategy stores everything across the basic block boundaries. Note that to make this strategy successful we should keep track of what in currently in each register. For this we maintain the **"register descriptor"** which is simply a pointer to a list that contains information about what is currently in each of the register. Initially all registers are empty.

We also maintain the **"address descriptor"** for each name in black to keep track for the location for each name where the current value of the name can be found at run time.

A space is required to perform the computation specified by each of the three-address statement. We will define a function getreg ( ) to find that location.

After the execution of the statement a = b OP c, those register will no longer contain b and/or C.

```
}
Store all the results
End
```

Mechanism of Getreg ( ) Function: We can have the view of mechanism of getreg ( ) function by following points.

When the index of statement $a = b$ OP c is passed to the getreg ( ) as argument then it performs followings jobs.

(*a*) It searches for a register already containing the name *b*. If such a register exists, and if *b* has no further use after the execution of $a = b$ OP c and if it is not live at the end of the block and holds the value of no other name, then return the register for L.

(*b*) Otherwise, getreg ( ) searches for an empty register, and if an empty register is available, then it returns it for *L*.

(*c*) If no empty register exist and if ***a*** has further use in the block or OP is an operator such as indexing that requires a register, then getreg ( ) finds a suitable, occupied register. The register is emptied by storing its value in the proper memory location *M*, the address descriptor is updated, the register is returned to *L*. The least-recently used strategy can be used to find a suitable, occupied register to be obtained.

Code Generation Algorithm: Let $a = b$ OP c is to be performed, then when we call getreg ( ) it returns a location ***L,*** where the computation of ***b OP c*** take place. Basically getreg ( ) returns a register.

Let us discuss the algorithm for code generation.

```
Begin
    For every three-address statement of the form a =b
OP c in any basic block
    {
```

Step 1. Call function getreg ( ) to find out the location (register) L to performed the computation for ***b OP c***; getreg ( ) function takes index of the statement $a = b$ OP c as argument.

Step 2. Try to find out the location of the operand ***b*** by consulting its **address descriptor** and if the value of b is currently both in the memory location as well as in the integer, then prefer the register. If the value of *b* is currently not available in *L* then generate an instruction **MOV *b, L*** update the address descriptor for show that now is in *L*.

Step 3. Do the same for *c* (Now *b* and *c* are in *L*)

Step 4. If the current values of *b* and/or *c* are in the register, and we have no further use for them, and they are not live at the end of the block, then after the register descriptor to indicate that.

(*d*) If *a* is not used in block or not suitable, occupied register can be found, getreg ( ) selects a memory location and returns it for *L*.

Example 11.1. Consider the following expression, with its three address code:

$$\boldsymbol{a = (p + q) - ((r + s) - t)}$$

$$\boldsymbol{t_1 = p + q}$$

$$\boldsymbol{t_2 = r + s}$$

$$\boldsymbol{t_3 = t_2 - t}$$

$$\boldsymbol{\mathrm{a} = t_1 - t_3}$$

Apply the algorithm of code generation on it.

Solution. Given expression is

$$a = (p + q) - ((r + s) - t)$$

Following table will generates the instruction

Statement	Location L	Instruction generated	Register descriptor	Address descriptor
			All registers empty initially	
$t_1 = p + q$	R_0	MOV p, R_0 ADD q, R_0	R_0 will hold t_1	t_1 is in R_0
$t_2 = r + s$	R_1	MOV r, R_1 ADD s, R_1	R_1 will hold t_2	t_2 is in R_1
$t_3 = t_2 - t$	R_1	SUB t, R_1	R_1 will hold t_3	t_3 is in R_1
$a = t_1 - t_3$	R_0	SUB R_1, R_0 MOV R_0, a	R_0 will hold a	a is in R_0 a is in R_0 and memory

11.13. CODE GENERATION FROM DAG [MDU 2007, 2010]

From the DAG we can move easily rearrange the order of the final computation sequence than we can starting from a linear sequence of quadruples. So first we obtain the DAG representation of the basic block, and then we order the nodes of the DAG using heuristics. Heuristics attempt to order the nodes of DAG so that, if possible, a node immediately follows the evaluation of its left-most operand.

Algorithm for Ordering Nodes of DAG: This algorithm lists the nodes of a DAG such that the node's reverse listing results in the computation order.

```
Begin
  while there exists an unlisted interior node do
      {
      select an unlisted node n whose parent have been listed
      list n
      while there exists a left-most child m of n that
  has no parent listed and m is not a leaf do
      {
          list m
             m = n
          }
  }
      order = reverse of the order of listing of nodes
      End
```

Let us see an example, consider the given DAG in Fig. 11.16.

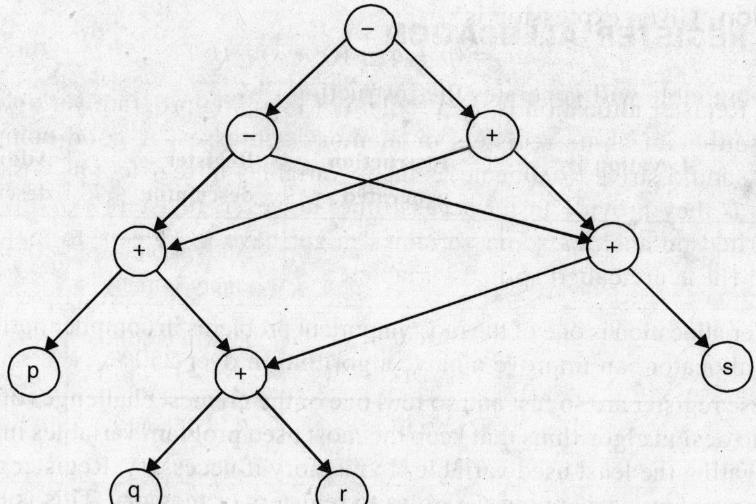

Fig. 11.16.

Now let us apply the above discussed algorithm, which gives following DAG with heuristic ordering.

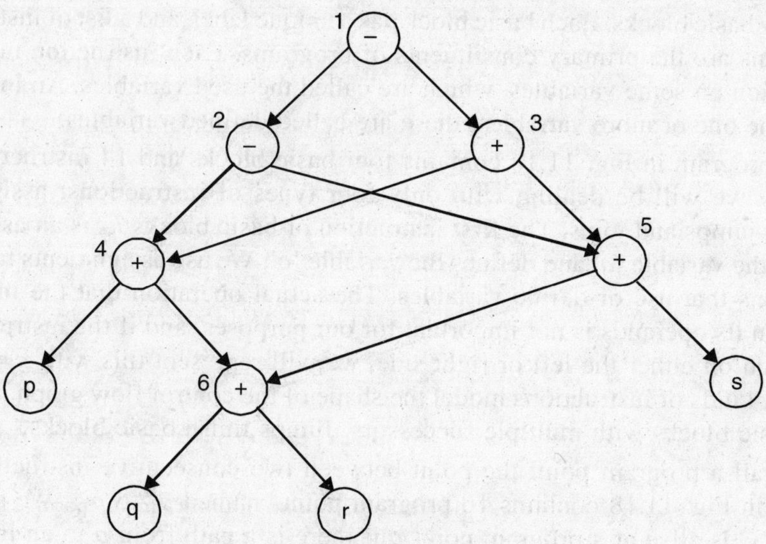

Fig. 11.17.

So the computation order will be as follows:

$$t_6 = q + r$$
$$t_5 = t_6 + s$$
$$t_4 = p + t_6$$
$$t_3 = t_4 + t_5$$
$$t_2 = t_4 - t_5$$
$$t_1 = t_2 - t_3$$

11.14. REGISTER ALLOCATION

Definition Register allocation is the problem of mapping program variables to either machine registers or memory addresses. A good compiler should strive to allocate as many variables in registers as possible, as they provide faster access time; however, registers exist in limited number, as some variables might have to be sent to memory. These are called spilled variables.

Register allocation is one of the most important problems in compiler optimization, as a good allocator can improve a naive algorithm in over 250%.

Because register are so fast and so few, one of the greatest challenges of compiler writers is to design algorithms that keep the most used program variables in registers, while relegating the least used variables to memory if necessary. Register allocation is thus the problem of mapping variables to registers or memory. This is one of the most important compiler optimizations. As an example, experiments show that an optimal allocator can produce code that is over 250% faster than the code produced by a naive algorithm that maps all the variables to memory.

A program can be described by its control flow graph. The control flow graph is formed by basic blocks. Each basic block has a unique label, and a list of instructions. Instructions are the primary constituents of programs. Each instruction may apply an operation on some variables, which are called the used variables. An instruction may define one or more variables; these are called defined variables.

The program in Fig. 11.18 contains four basic blocks and 14 instructions. For simplicity we will be dealing with only four types of instructions: assignments, branches, jumps and joins. The first instruction of basic blocks L_2 is an assignment that uses the variable '*a*' and defines the variable '*c*'. We use assignments to abstract instructions that use or define variables. The actual operation that the instruction applied on its operands is not important for our purposes, and if the instruction has no operand on either the left or right side, we will represent this with a • symbol. The other kinds of instructions model the shape of the control flow graph. Branches finish basic blocks with multiple successors. Jumps finish basic blocks.

We call a program point the point between two consecutive instructions. The program in Fig. 11.18 contains 16 program points named p_1 to p_{16}. We say that a variable '*v*' is alive at a program point *p* if there is a path from *p* to an instruction that uses '*v*' where '*v*' is not re-defined by any instruction. The collection of program points where a variable is alive is called its live range. For instance, the live range of variable *B* is $\{p_2, p_3, p_4, p_9\}$, whereas the live range of variable a includes all the program points but p_{11}, p_{12} and p_{16}. Notice that '*a*' is not alive at program points p_{11} and p_{12} because this variable is redefined by the instruction $a = f$, and its old value is not necessary after the instruction $f = a$. A simple algorithm to compute liveness information is given by Appel and Palsberg.

We say that two variables interfere if the intersection between their live ranges is non-empty. In this case. we also say that their live ranges overlap. For instance, variables *c* and *d* interfere, because their live ranges overlap at program point p_5; however, variables *c* and *E* do not interfere.

TIPS

This concept is very important for register allocation, because two variables that do not interfere can be stored in the same register.

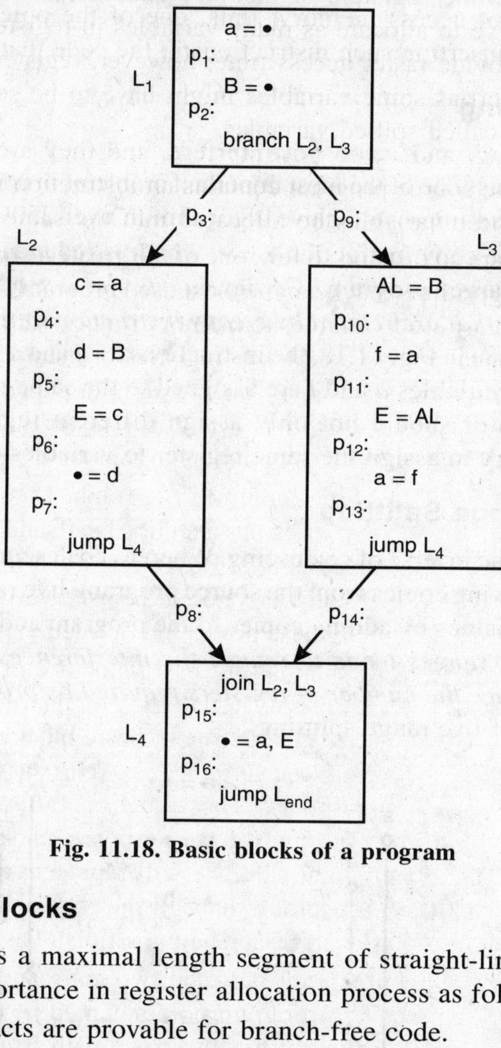

Fig. 11.18. Basic blocks of a program

11.14.1. Basic Blocks

A basic block is a maximal length segment of straight-line (*i.e.*, branch free) code. It is very importance in register allocation process as follows:

- Strongest facts are provable for branch-free code.
- If any statement executes, they all execute.
- Execution is totally ordered.

A local transformation operates on basic blocks and many optimizations are done locally. It produces decent register use inside a block but inefficiencies can arise at boundaries between blocks.

11.14.2. Spilling

Because registers exist in limited number, they may not be enough to store all the variables in the source program. If that is the case, then some variables must be mapped to memory. *The act of storing a variable into memory is called spilling. Spill is normally undesirable because it forces the register allocator to insert special instructions, that we will call spill code, in the target program to access values stored in memory.* An instruction that copies a value from a register to a memory address is called a store. The opposite instruction, which copies a value from memory into a register, is called a load. These instructions tend to be slow compared to operations that do not access memory; thus, one of the objectives of a register allocator is to avoid inserting such instructions in the code that it produces.

11.14.3. Coalescing

If two variables v_1 and v_2 do not interfere, and they are related by a copy instruction, that is, the source program contains an instruction such as $v_1 = v_2$, then it is desirable that these variables be allocated into the same register '*r*'. In this case, we will have the copy instruction $r = r$, which is redundant and can safely be removed from the target program. *Coalescing is the act of mapping two non-interfering variables that are related by a copy instruction to the same register.* For instance, in the program in Fig. 11.18, the instructions $f = a$ and $a = f$ can be eliminated from the program if variables *a* and *f* are assigned to the same register. Therefore, a good register allocator should not only assign different registers to interfering variables, but also, try to assign the same register to variables related by copies.

11.14.4. Live Range Splitting

This concept is the inverse of coalescing. Whereas coalescing join the live ranges of variables by removing copies from the source program, live range splitting divides the live range of variables by adding copies to the program and renaming variables. *The splitting of live ranges tends to reduce the interferences between variables; thus, it might minimize the number of registers required by programs.* Figure 11.19 shows an example of live range splitting.

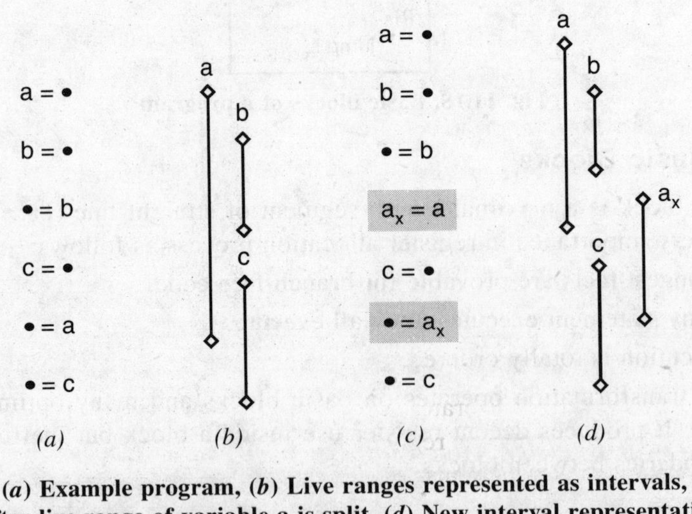

Fig. 11.19. (*a*) Example program, (*b*) Live ranges represented as intervals, (*c*) Program after live range of variable a is split, (*d*) New interval representation.

11.14.5. Interference Graph

The Interference Graph of a program is the intersection graph of the live ranges of the variables in the program. That is, given a program *P*, its interference graph

$G = (V, E)$ contains a vertex for each variable v of P. An edge (u, v) is in E if the intersection of the live ranges of variables v and u is non-empty.

Let us consider following code sequence:

```
        A = ............
        B = ............
        ...........B............
        C = ............
        ...........A............
        D = ............
        ...........D............
        ...........C............
```

In above code sequence there will be an edge between Node '*A*' and '*B*', an edge between Node '*A*' and '*C*', and an edge between Node '*C*' and '*D*'. Since intersection of their live ranges are non-empty. The Interference Graph of above code sequence will be as follows:

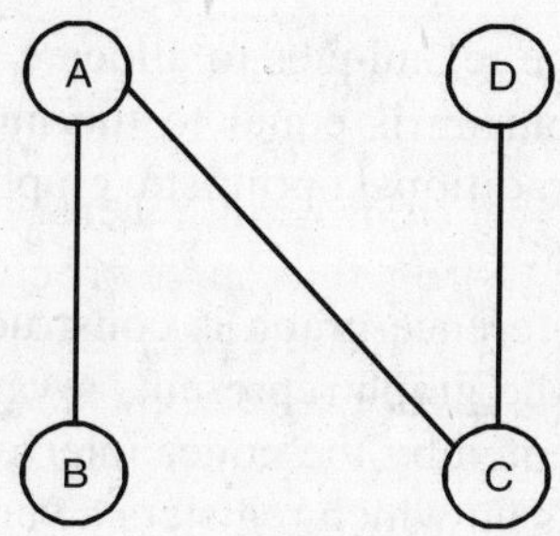

Fig. 11.20. Interference Graph

11.14.6. Register Allocation Approaches

There are two basic approaches are there for register allocation as follows:

1. **Local Allocators:** Use Instruction-level knowledge
   - Top-Down: Use frequency of variables use for allocation
   - Bottom-Up: Evaluate instructions needs and reuse registers
2. **Global Allocators:** Use a graph-colouring paradigm
   - Build a k-conflict graph or k-interference graph
   - Find a k-colouring for the graph, or change the code to a nearby problem that it can k-colour.

11.14.6.1. Local Register Allocation

Allocating registers for a single basic block with following assumptions:

- Assumes register-to-register memory model
- Input program assumes infinite number of registers
- Assume all registers on target machine are equivalent.

There are following two approaches are there for local register allocation.

(*a*) **Top-Down:** Here we assign registers to the most heavily used variables:

- Traverse the block
- Count uses
- Use count as a priority function
- Assign registers to higher priority variables first

Advantage is that heavily used values reside in register and disadvantage is that it does not consider non-uniform distribution of uses.

(*b*) **Bottom-up:** Here we spill the value that is needed the latest

- For each variable use, compute the distance of its next use.
- Process each instruction in evaluation order; when running out of registers, spill the value whose next use is farthest in the future.

It produces excellent result in many cases but not optimal. All spilling may not take the same number of cycles.

11.14.6.2. Register Allocation by Graph Colouring

The graph colouring is the techniques to allocate the register in which number of colours available for colouring is equal to the number of registers which are available for allocation. The traditional optimistic graph colouring algorithm consists of six main phases as follows:

Step 1: Build: An interference graph is constructed using the results of data flow analysis. Each node in the graph represents a variable. An edge connects two nodes if the variables represented by the nodes interfere and cannot be allocated to the same register. Restrictions on which registers a variable may be allocated to can be implemented by adding precoloured nodes to the graph.

Step 2: Coalesce: After building the graph, unneeded copies (register-to-register move instructions) are eliminated in the coalesce phase. This might sound unnecessary, but it is actually an extremely important step, partially because copies are often introduced during optimization, but more importantly because coalescing allows the allocator to achieve targeting—where specific registers are used for specific purposes such as argument passing. With luck, those values can be computed directly into the appropriate registers in the final code. Coalescing also allows better handling of self-updating instructions, such as auto-increment addressing, as well as the two address instructions.

Step 3: Potential Spill: If only nodes with degree greater than *k* are left, we mark a node as a potential spill node, remove it from the graph, and optimistically push it onto the stack in the hope that we might be able to assign it a colour. We repeat this process until there exists nodes in the graph with degree less than k, at which point we return to the Simplify phase.

Step 4: Simplify: The core of the colouring process itself starts with the *simplify* phase, sometimes called *pruning*. Here, the graph is repeatedly examined and nodes with fewer than *k* neighbours are removed (where *k* is the number of colours we have to offer). As each node is removed it is placed on a stack and its edges are removed from the graph, thereby decreasing the degree of interference of its neighbours. If a point is reached where every node has k neighbours (or more), a node is chosen as a possible spill candidate (just because a node has *k* neighbours doesn't necessarily mean it will spill—the neighbours may not all be different colours). Choosing *which* node to spill is, of course, the hard part—often a variant of the ratio of spill cost to degree of interference is used, or sometimes a combination of several different metrics. Once a spill candidate is chosen, the node is then removed from the graph and pushed onto the stack, just like the others, and the algorithm continues on to the next node.

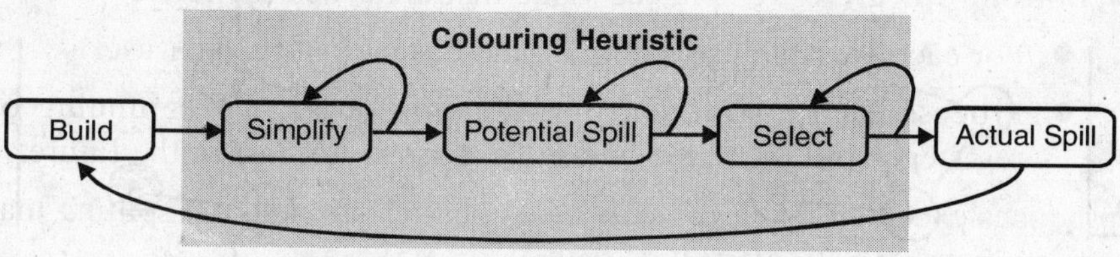

Fig. 11.21. Colouring heuristic

Step 5: Select: In this phase all of the nodes have been removed from the graph. We now assign colours to nodes in the order we pop them off the stack. If the node was not marked as a potential spill node then there must be a colour we can assign this node that does not conflict with any colours already assigned to this node's neighbours. If it is a potential spill node, then it still may be possible to assign it a colour; if it is not possible to colour the potential spill node, we mark it as an actual spill and leave it uncoloured.

Step 6: Actual Spill: If any nodes are marked as actual spills, we generate spill code which loads and stores the variable represented by the node into new, short lived, temporary variables everywhere the variable is used and defined. Because new variables are created, it is necessary to rebuild the interference graph.

Note that the Simplify, Potential Spill, and Select phases together form a heuristic for graph colouring. If this heuristic is successful, there will be no actual spills. Otherwise, the graph is modified so that it is easier to colour by spilling variables and the entire process is repeated.

Note: This book is not using colours other than black and white so here we are taking following assumption to make concept more clear:

(*a*) Three registers are available so we can use only three colours for graph colouring.

(*b*) Let us assume that red colour node is represented by , green colours node is represented by  and yellow coloured node is represented by .

An example of the core simplify/select process is shown below:

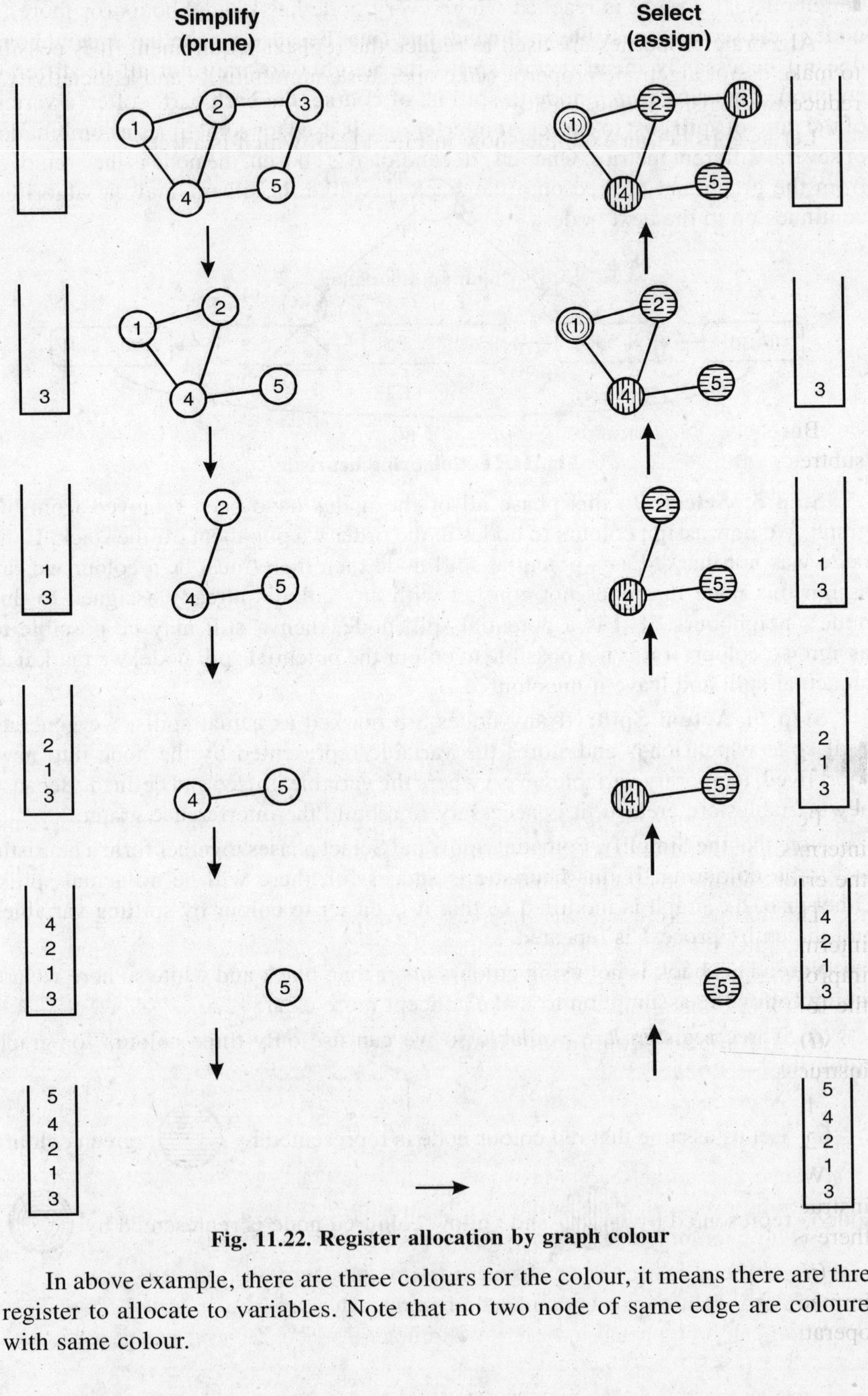

Fig. 11.22. Register allocation by graph colour

In above example, there are three colours for the colour, it means there are three register to allocate to variables. Note that no two node of same edge are coloured with same colour.

11.15. ALGEBRAIC PROPERTIES

Algebraic properties are used to reduce the register requirement. It is possible to make use of algebraic properties like operator commutativity and associativity to reduce the register requirement.

Let us consider an example show in Fig. 11.23, which is a tree.

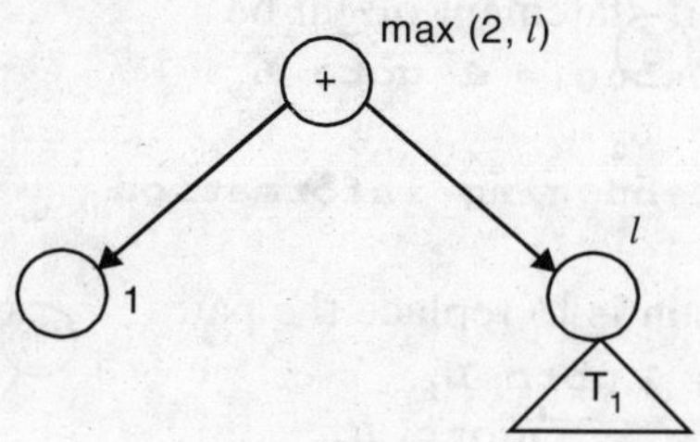

Fig. 11.23

But since + is commutative operator so we can interchange the left and right subtree and this brings the register requirement of the tree down.

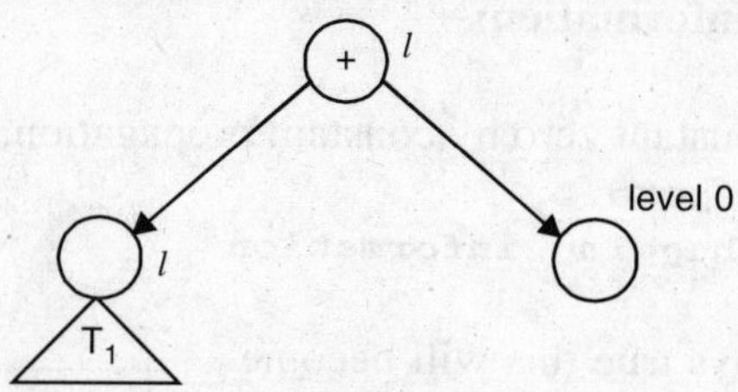

Fig. 11.24.

11.16. PEEPHOLE OPTIMIZATION **[MDU 2007, 2009]**

Peephole optimization technique is used by many compilers to optimize their intermediate or object code. By performing peephole optimization we can control the errors during syntax-directed translation.

This optimization is called peephole since in it we look for small segment of intermediate code at time. Peephole optimization is an effective technique for locally improving the target code. During the peephole optimization generally we perform the following operations:

***(i)* Eliminating Redundant loads and stores:** If the target code contains the instruction sequence:

1. MOV R, *x*
2. MOV *x*, R

We can delete the second instruction if it is unlabelled. This is because the first instruction ensures that the value of a is already in the register R. If it is labelled, there is no guarantee that step 1 will always be executed before step 2.

***(ii)* Unreachable code Elimination:** An unlabelled instruction that immediately follows an unconditional jump can possibly be removed, and this operation can be repeated in order to eliminate a sequence of instructions.

For example, consider the following source code:

```
Debug:= 0;
  ..........
  ..........
  if Debug = 1 then
     print debugging information
```

In intermediate code if-statement might be

```
        if Debug = 1 goto L1
        goto L2
  L1: Print Debugging information
  L2:
```

One of the optimization is to replace the pair:

```
  if Debug = 1 goto L1
               goto L2
```

Within the statements with a single conditional goto statement by regarding the condition and changing its target, as shown below:

```
  if Debug ≠ 1 goto L2
```

Print Debugging information:

```
  L2:
```

Since debug is a constant zero by constant propagation, this code will become :

```
  if 0 ≠ 1 goto L2
  print debugging information
  L2:
```

Since 0 ≠ 1 is always true this will become

```
  goto L2
  print debugging information
  L2:
```

Therefore, the statements that print the debugging information unreachable and can be eliminated, one at a time.

***(iii)* Eliminating Multiple Jumps:** If we have jump to other jumps, then the unnecessary jumps can be eliminated in either intermediate code or the target code. If we have a jump sequence:

```
  goto L1
  ...........
  ...........
  L1: goto L2
  then this can be replaced by
  goto L2
  ...........
  ...........
  L1: goto L2
```

EXERCISE

1. Construct DAG for the following code:

$$a = a + b$$
$$e = a + d + e$$

2. In following flow graph, find out the dominators of the node C.

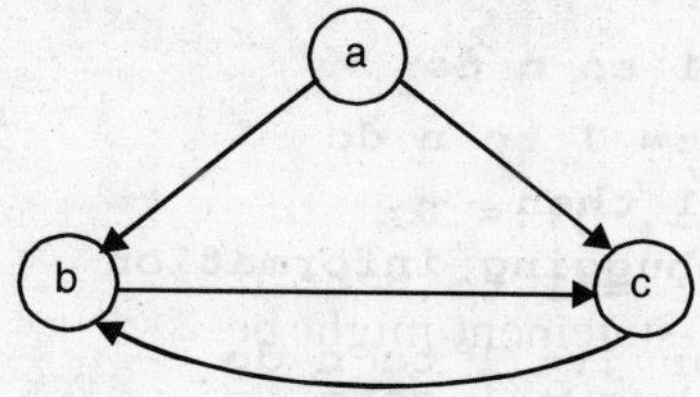

Fig. 11.25.

3. Explain the need of code optimization. With example, illustrate loop optimization.
4. What is global data flow analysis ? What is its use in code optimization ?
5. What is DAG ? What are its advantages in content of optimization ? How does it help in elimination of common subexpression ?
6. Construct DAG for the following code sequence.

A[I] := B
*P := C
D := A[J]
E := *P
*P := A[I]

Assume that:

(*i*) P can point anywhere

(*ii*) P point to only B or D.

7. Consider the following sequence of three address code:

(1) PROD:= 0
(2) I:= 1
(3) T_1:= 4 * I
(4) T_2:= addr (A) – 4
(5) T_3:= T_2 [T_1]
(6) T_4:= addr (B) – 4
(7) T_5:= T_4 [T_1]
(8) T_6:= T_3 * T_5
(9) PROD:= PROD + T_6
(10) I:= I + 1
(11) If I <= 20 goto (3)

(*i*) Find the basic blocks and construct a flow graph.

(*ii*) Eliminate common subexpression.

(*iii*) Move the loop-invariant computation out of the loop.

(*iv*) Find the induction variables and eliminate then where possible.

8. Discuss the algorithm for the elimination of induction variables with the help of example.
9. Show the DAG for the following statement:

$$Z = X - Y + X * Y * U - V/W + X + V$$

10. Consider the following matrix multiplication program:

```
begin
    for i:= 1 to n do
        for j:= 1 to n do
            c[i, j]:= 0;
        for i: = 1 to n do
            for j:= 1 to n do
                for k:= 1 to n do
    c[i, j]:= c[i, j] + a[i, k] * b[k, j]
end
```

(*i*) Assuming *a*, *b* and *c* are allocating static storage and there are four bytes per word in a byte-addressed memory, produce three address statement for the program.

(*ii*) Construct a flow graph from the three-address statements.

(*iii*) Eliminate the common subexpression from each basic block.

(*iv*) Find the loop in the flow graph.

(*v*) Move the loop-invariant computation out of the loops.

(*vi*) Find the induction variables of each loop and eliminate them where possible.

❑❑❑

Appendix

QUICK REVIEW OF COMPILER DESIGN

1. What is a Complier?

Ans. A complier is a program that reads a program written in one language-the source language-and translates it in to an equivalent program in another language-the target language. As an important part of this translation process, the compiler reports to its user the presence of errors in the source program.

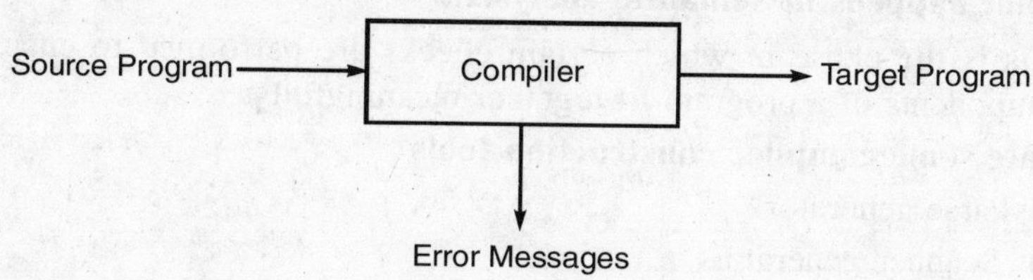

2. State some software tools that manipulate source program?

Ans. (*i*) Structure editors (*ii*) Pretty printers

(*iii*) Static checkers (*iv*) Interpreters.

3. What are the cousins of compiler?

Ans. The following are the cousins of compilers

(*i*) Preprocessors (*ii*) Assemblers

(*iii*) Loaders (*iv*) Link editors.

4. What are the main two parts of compilation? What are they performing?

Ans. The two main parts are

- **Analysis part** breaks up the source program into constituent pieces and creates an intermediate representation of the source program.
- **Synthesis** part constructs the desired target program from the intermediate representation

5. What is a Structure editor?

Ans. A structure editor takes as input a sequence of commands to build a source program. The structure editor not only performs the text creation and modification functions of an ordinary text editor but it also analyzes the program text putting an appropriate hierarchical structure on the source program.

6. What are a Pretty Printer and Static Checker?

Ans.
- A Pretty printer analyses a program and prints it in such a way that the structure of the program becomes clearly visible.
- A static checker reads a program, analyses it and attempts to discover potential bugs with out running the program.

7. How many phases does analysis consists?

Ans. Analysis consists of three phases

(*i*) Linear analysis (*ii*) Hierarchical analysis

(*iii*) Semantic analysis

8. What happens in lexical analysis?

Ans. This is the phase in which the stream of characters making up the source program is read from left to right and grouped in to tokens that are sequences of characters having collective meaning.

9. What happens in Hierarchical analysis?

Ans. This is the phase in which characters or tokens are grouped hierarchically in to nested collections with collective meaning.

10. What happens in Semantic analysis?

Ans. This is the phase in which certain checks are performed to ensure that the components of a program fit together meaningfully.

11. State some compiler construction tools?

Ans. (*i*) Parse generator

(*ii*) Scanner generators

(*iii*) Syntax-directed translation engines

(*iv*) Automatic code generator

(*v*) Data flow engines.

12. What is a Loader? What does the loading process do?

Ans. A Loader is a program that performs the two functions

(*i*) Loading (*ii*) Link editing

The process of loading consists of taking relocatable machine code, altering the relocatable address and placing the altered instructions and data in memory at the proper locations.

13. What does the Link Editing does?

Ans. **Link editing:** This allows us to make a single program from several files of relocatable machine code. These files may have been the result of several compilations, and one or more may be library files of routines provided by the system and available to any program that needs them.

14. What is a preprocessor?

Ans. A preprocessor is one, which produces input to compilers. A source program may be divided into modules stored in separate files. The task of collecting the source program is sometimes entrusted to a distinct program called a preprocessor.

The preprocessor may also expand macros into source language statements.

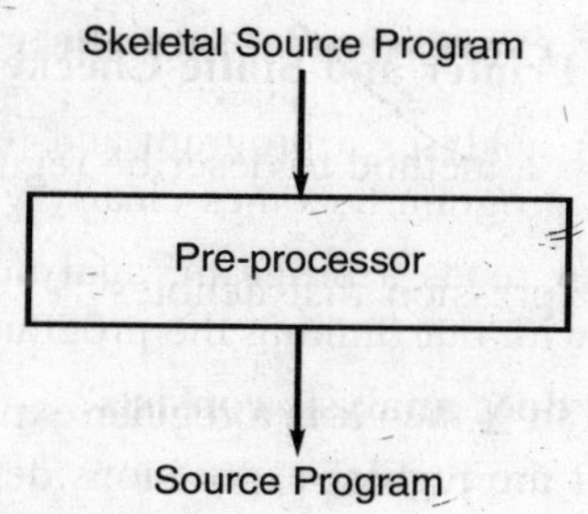

15. State some functions of Preprocessors

Ans. (*i*) Macro processing (*ii*) File inclusion
(*iii*) Relational Preprocessors (*iv*) Language extensions

16. What is a Symbol table?

Ans. A Symbol table is a data structure containing a record for each identifier, with fields for the attributes of the identifier. The data structure allows us to find the record for each identifier quickly and to store or retrieve data from that record quickly.

17. State the general phases of a compiler.

Ans. (*i*) Lexical analysis (*ii*) Syntax analysis
(*iii*) Semantic analysis (*iv*) Intermediate code generation
(*v*) Code optimization (*vi*) Code generation

18. What is an assembler?

Ans. Assembler is a program, which converts the source language in to assembly language.

19. What is the need for separating the analysis phase into lexical analysis and parsing?

(Or)

What are the issues of lexical analyzer?

Ans.
- Simpler design is perhaps the most important consideration. The separation of lexical analysis from syntax analysis often allows us to simplify one or the other of these phases.
- Compiler efficiency is improved.
- Compiler portability is enhanced.

20. What is Lexical Analysis?

Ans. The first phase of compiler is Lexical Analysis. This is also known as linear analysis in which the stream of characters making up the source program is read from left-to-right and grouped into tokens that are sequences of characters having a collective meaning.

21. What is a lexeme? Define a regular set.

Ans.
- A Lexeme is a sequence of characters in the source program that is matched by the pattern for a token.
- A language denoted by a regular expression is said to be a regular set

22. What is a sentinel? What is its usage?

Ans. A Sentinel is a special character that cannot be part of the source program. Normally we use 'eof' as the sentinel. This is used for speeding-up the lexical analyzer.

23. What is a regular expression? State the rules, which define regular expression?

Ans. Regular expression is a method to describe regular language

Rules :

(1) e-is a regular expression that denotes $\{e\}$ that is the set containing the empty string

(2) If a is a symbol in Σ,then a is a regular expression that denotes $\{a\}$

(3) Suppose r and s are regular expressions denoting the languages $L(r)$ and $L(s)$. Then,

(a) $(r)/(s)$ is a regular expression denoting $L(r) \cup L(s)$.

(b) $(r)(s)$ is a regular expression denoting $L(r)L(s)$

(c) $(r)^*$ is a regular expression denoting $L(r)^*$.

(d) (r) is a regular expression denoting $L(r)$.

24. What are the Error-recovery actions in a lexical analyzer?

Ans. 1. Deleting an extraneous character

2. Inserting a missing character

3. Replacing an incorrect character by a correct character

4. Transposing two adjacent characters

25. Construct Regular expression for the language

$L = \{w\ e\{a, b\}/w$ ends in $abb\}$

Ans. $\{a/b\}^*abb$.

26. What is recognizer?

Ans. Recognizers are machines. These are the machines which accept the strings belonging to certain language. If the valid strings of such language are accepted by the machine then it is said that the corresponding language is accepted by that machine, otherwise it is rejected.

27. What is the output of syntax analysis phase? What are the three general types of parsers for grammars?

Ans. Parser (or) parse tree is the output of syntax analysis phase.

General types of parsers:

(1) Universal parsing

(2) Top-down

(3) Bottom-up

28. What are the different strategies that a parser can employ to recover from a syntactic error?

Ans.
- Panic mode
- Phrase level
- Error productions
- Global correction

29. What are the goals of error handler in a parser?

Ans. The error handler in a parser has simple-to-state goals :

- It should report the presence of errors clearly and accurately.
- It should recover from each error quickly enough to be able to detect subsequent errors.
- It should not significantly slow down the processing of correct programs.

30. What is phrase level error recovery?

Ans. On discovering an error, a parser may perform local correction on the remaining input; that is, it may replace a prefix of the remaining input by some string that allows the parser to continue. This is known as phrase level error recovery.

31. Define context free language. When will you say that two CFGs are equal?

Ans.
- A language that can be generated by a grammar is said to be a context free language.
- If two grammars generate the same language, the grammars are said to be equivalent.

32. Differentiate sentence and sentential form.

Ans.

Sentence	Sentential form
• If $S \Rightarrow w$ then the string w is called Sentence of G. • Sentence is a string of terminals. Sentence is a sentential form with no nonterminals.	• If $S \Rightarrow a$ then a is a sentential form of G. • Sentential form may contain non terminals.

33. Give the definition for leftmost and canonical derivations.

Ans.
- Derivations in which only the leftmost nonterminal in any sentential form is replaced at each step are termed leftmost derivations
- Derivations in which the rightmost nonterminal is replaced at each step are termed canonical derivations.

34. What is a parse tree?

Ans. A parse tree may be viewed as a graphical representation for a derivation that filters out the choice regarding replacement order. Each interior node of a parse tree is labeled by some nonterminal A and that the children of the node are labeled from left to right by symbols in the right side of the production by which this A was replaced in the derivation. The leaves of the parse tree are terminal symbols.

35. What is an ambiguous grammar? Give an example.

Ans.
- A grammar that produces more than one parse tree for some sentence is said to be ambiguous
- An ambiguous grammar is one that produces more than one leftmost or rightmost derivation for the same sentence.

Example :

$$E \rightarrow E + E \,/\, E{*}E \,/\, id$$

36. Why do we use regular expressions to define the lexical syntax of a language?

(*i*) The lexical rules of a language are frequently quite simple, and to describe them we do not need a notation as powerful as grammars.

(*ii*) Regular expressions generally provide a more concise and easier to understand notation for tokens than grammars.

(*iii*) More efficient lexical analyzers can be constructed automatically from regular expressions than from arbitrary grammars.

(*iv*) Separating the syntactic structure of a language into lexical and non lexical parts provides a convenient way of modularizing the front end of a compiler into two manageable-sized components.

37. When will you call a grammar as the left recursive one?

Ans. A grammar is a left recursive if it has a nonterminal A such that there is a derivation $A \Rightarrow A\alpha$ for some string.

38. Define left factoring.

Ans. Left factoring is a grammar transformation that is useful for producing a grammar suitable for predictive parsing. The basic idea is that when it is not clear which of two alternative productions to use to expand a nonterminal "*A*", we may be able to rewrite the "*A*" productions to refer the decision until we have seen enough of the input to make the right choice.

39. Left factor the following grammar:

$$S \rightarrow iEtS \mid iEtSeS \mid a$$
$$E \rightarrow b.$$

Ans. The left factored grammar is,

$$S \rightarrow iEtSS' \mid a$$
$$S' \rightarrow eS \mid \in$$
$$E \rightarrow b$$

40. What is parsing?

Ans. Parsing is the process of determining if a string of tokens can be generated by a grammar.

41. What is Top Down parsing?

Ans. Starting with the root, labeled, does the top-down construction of a parse tree with the starting nonterminal, repeatedly performing the following steps.

(*i*) At node *n*, labeled with non terminal "*A*", select one of the productions for "*A*" and construct children at *n* for the symbols on the right side of the production.

(*ii*) Find the next node at which a sub tree is to be constructed.

42. What do you mean by Recursive Descent Parsing?

Ans. Recursive Descent Parsing is top down method of syntax analysis in which we execute a set of recursive procedures to process the input. A procedure is associated with each nonterminal of a grammar.

43. What is meant by Predictive parsing?

Ans. A special form of Recursive Descent parsing, in which the look-ahead symbol unambiguously determines the procedure selected for each nonterminal, where no backtracking is required.

44. Define Bottom Up Parsing.

Ans. Parsing method in which construction starts at the leaves and proceeds towards the root is called as Bottom Up Parsing.

45. What is Shift-Reduce parsing?

Ans. A general style of bottom-up syntax analysis, which attempts to construct a parse tree for an input string beginning at the leaves and working up towards the root.

46. Define handle. What do you mean by handle pruning?

Ans. (*i*) An Handle of a string is a sub string that matches the right side of production and whose reduction to the nonterminal on the left side of the production represents one step along the reverse of a rightmost derivation.

(*ii*) The process of obtaining rightmost derivation in reverse is known as Handle Pruning.

47. Define LR (0) items.

Ans. An LR (0) item of a grammar *G* is a production of *G* with a dot at some position of the right side. Thus the production $A \rightarrow XYZ$ yields the following four items,

$$A \rightarrow .XYZ$$
$$A \rightarrow X.YZ$$
$$A \rightarrow XY.Z$$
$$A \rightarrow XYZ.$$

48. What do you mean by viable prefixes?

Ans.
- The set of prefixes of right sentential forms that can appear on the stack of a shift-reduce parser are called viable prefixes.
- A viable prefix is that it is a prefix of a right sentential form that does not continue the past the right end of the rightmost handle of that sentential form.

49. What is meant by an operator grammar? Give an example.

Ans. A grammar is operator grammar if,

(*i*) No production rule involves "e" on the right side.

(*ii*) No production has two adjacent nonterminals on the right side..

Example :

$$E \rightarrow E + E \mid E - E \mid E*E \mid E/E \mid E - E \mid (E) \mid -E \mid id$$

50. What are the disadvantages of operator precedence parsing?

(*i*) It is hard to handle tokens like the minus sign, which has two different precedences.

(*ii*) Since the relationship between a grammar for the language being parsed and the operator – precedence parser itself is tenuous, one cannot always be sure the parser accepts exactly the desired language.

(*iii*) Only a small class of grammars can be parsed using operator precedence techniques.

51. State error recovery in operator-Precedence Parsing.

Ans. There are two points in the parsing process at which an operator-precedence parser can discover the syntactic errors:

(*i*) If no precedence relation holds between the terminal on top of the stack and the current input.

(*ii*) If a handle has been found, but there is no production with this handle as a right side.

52. LR (k) parsing stands for what?

Ans. The "L" is for left-to-right scanning of the input, the "R" for constructing a rightmost derivation in reverse, and the k for the number of input symbols of lookahead that are used in making parsing decisions.

53. Why LR parsing is attractive one?

Ans.
- LR parsers can be constructed to recognize virtually all programming language constructs for which context free grammars can be written.
- The LR parsing method is the, most general nonbacktracking shift-reduce parsing method known, yet it can be implemented as efficiently as other shift reduce methods.
- The class of grammars that can be parsed using LR methods is a proper superset of the class of grammars that can be parsed with predictive parsers.
- An LR parser can detect a syntactic error as soon as it is possible to do so on a left-to-right scan of the input.

54. What is meant by *goto* function in LR parser? Give an example.

Ans.
- The function *goto* takes a state and grammar symbol as arguments and produces a state.
- The *goto* function of a parsing table constructed from a grammar G is the transition function of a DFA that recognizes the viable prefixes of G.

Example :

***goto*(I,X)**

Where I is a set of items and *X* is a grammar symbol to be the closure of the set of all items $[A \rightarrow \alpha X.b]$ such that $[A \rightarrow \alpha.X\beta]$ is in I.

55. Write the configuration of an LR parser?

Ans. A configuration of an LR parser is a pair whose first component is the stack contents and whose second component is the unexpended input:

$$(s_0 \, X_1 \, s_1 \, X_2 \, s_2 \ldots X_m \, s_m, \, a_i \, a_{i+1} \ldots a_n \, \$)$$

56. Define LR grammar.

Ans. A grammar for which we can construct a parsing table is said to be an LR grammar.

57. What are kernel and non kernel items?

(*i*) The set of items which include the initial item, S′→.S, and all items whose dots are not at the left end are known as kernel items.

(*ii*) The set of items, which have their dots at the left end, are known as non kernel items.

58. Why SLR and LALR are more economical to construct than canonical LR?

Ans. For a comparison of parser size, the SLR and LALR tables for a grammar always have the same number of states, and this number is typically several hundred states for a language like Pascal. The canonical LR table would typically have several thousand states for the same size language. Thus, it is much easier and more economical to construct SLR and LALR tables than the canonical LR tables.

59. What is ambiguous grammar? Give an example.

Ans. A grammar *G* is said to be ambiguous if it generates more than one parse trees for sentence of language *L*(*G*).

Example: $E \rightarrow E + E \mid E{*}E \mid \text{id}$

60. What are the benefits of using machine-independent intermediate form?

Ans. (*a*) Retargeting is facilitated; a compiler for a different machine can be created by attaching a back end for the new machine to an existing front end.

(*b*) A machine-independent code optimizer can be applied to the intermediate representation.

61. List the three kinds of intermediate representation.

Ans. The three kinds of intermediate representations are :

(*i*) Syntax trees

(*ii*) Postfix notation

(*iii*) Three address code

62. How can you generate three-address code?

Ans. The three-address code is generated using semantic rules that are similar to those for constructing syntax trees for generating postfix notation.

63. What is a syntax tree? Draw the syntax tree for the assignment statement

a := b * -c + b * -c.

Ans. • A syntax tree depicts the natural hierarchical structure of a source program.

• **Syntax tree :**

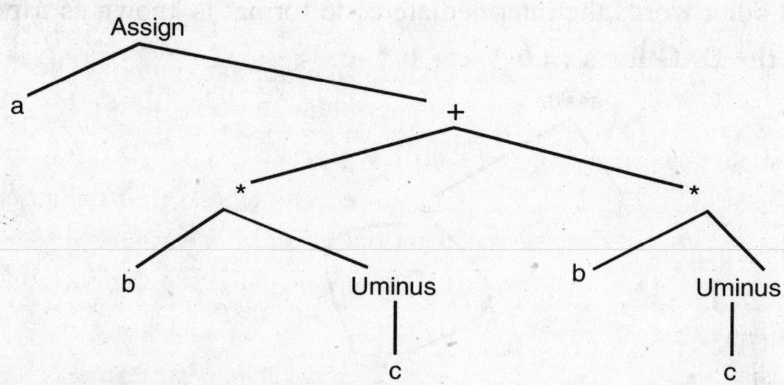

64. What is postfix notation?

Ans. A Postfix notation is a linearized representation of a syntax tree. It is a list of nodes of the tree in which a node appears immediately after its children.

65. What is the usage of syntax directed definition.

Ans. Syntax trees for assignment statement are produced by the syntax directed definition.

66. Why "Three address code" is named so?

Ans. The reason for the term "Three address code" is that each usually contains three addresses, two for operands and one for the result.

67. Define three-address code.

Ans. • Three-address code is a sequence of statements of the general form

x := y *op* z

where x, y and z are names, constants, or compiler-generated temporaries; op stands for any operator, such as fixed or floating-point arithmetic operator, or a logical operator on boolean-valued data.

• Three-address code is a linearized representation of a syntax tree or a dag in which explicit names correspond to the interior nodes of the graph.

68. State quadruple.

Ans. A quadruple is a record structure with four fields, which we call op, arg1, arg2 and result.

69. What is called an abstract or syntax tree?

Ans. A tree in which each leaf represents an operand and each interior node an operator is called as abstract or syntax tree.

70. Construct Three address code for the following

```
position := initial + rate * 60
```

Ans.

```
temp1 := inttoreal(60)
temp2 := id3 * temp1
temp3 := id2 + temp2
   id1 := temp3
```

71. What are triples?

Ans.
- The fields arg1,and arg2 for the arguments of op, are either pointers to the symbol table or pointers into the triple structure then the three fields used in the intermediate code format are called triples.
- In other words the intermediate code format is known as triples.

72. Draw the DAG for a := b * -c + b * -c

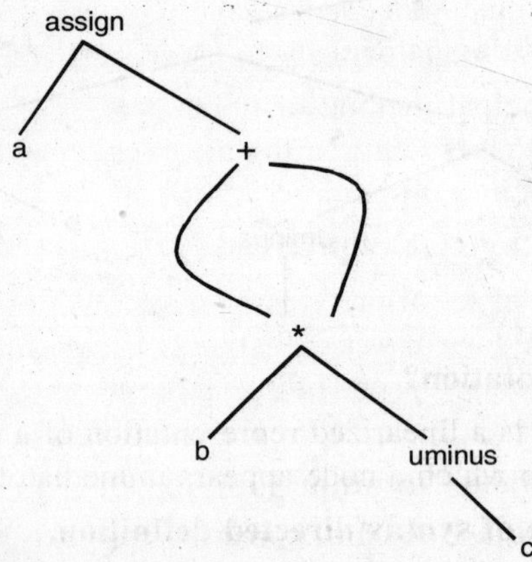

73. List the types of three address statements.

Ans. The types of three address statements are :

(*a*) Assignment statements
(*b*) Assignment Instructions
(*c*) Copy statements
(*d*) Unconditional jumps
(*e*) Conditional jumps
(*f*) Indexed assignments
(*g*) Address and pointer assignments
(*h*) Procedure calls and return

74. What are the various methods of implementing three-address statements?

Ans. (*i*) Quadruples (*ii*) Triples
(*iii*) Indirect triples

75. What is meant by declaration?

Ans. The process of declaring keywords, procedures, functions, variables, and statements with proper syntax is called declaration.

76. How semantic rules are defined?

Ans. The semantic rules are defined by the following ways :

(*a*) mktable(previous)

(*b*) enter(table,name,type,offset)

(*c*) addwith(table,width)

(*d*) enterproc(table,name,newtable)

77. What are the two primary purposes of Boolean Expressions?

Ans.
- They are used to compute logical values
- They are used as conditional expressions in statements that alter the flow of control, such as if-then, if-then-else, or while-do statements.

78. Define Boolean Expression.

Ans. Expressions which are composed of the Boolean operators (and, or, and not) applied to elements that are Boolean variables or relational expressions are known as Boolean expressions.

79. What are the two methods to represent the value of a Boolean expression?

(*i*) The first method is to encode true and false numerically and to evaluate a Boolean expression analogously to an arithmetic expression.

(*ii*) The second principal method of implementing Boolean expression is by flow of control that is representing the value of a Boolean expression by a position reached in a program.

80. What do you mean by viable prefixes.

Ans. Viable prefixes are the set of prefixes of right sentinels forms that can appear on the stack of shift/reduce parser are called viable prefixes. It is always possible to add terminal symbols to the end of the viable prefix to obtain a right sentential form.

81. What is meant by Shot-Circuit or jumping code?

Ans. We can also translate a Boolean expression into three-address code without generating code for any of the Boolean operators and without having the code necessarily evaluate the entire expression. This style of evaluation is sometimes called "short-circuit" or "jumping" code.

82. What is known as calling sequence?

Ans. A sequence of actions taken on entry to and exit from each procedure is known as calling sequence.

83. What is the intermediate code representation for the expression a or b and not c? (Or) Translate a or b and not c into three address code.

Ans. Three-address sequence is

t_1 := not c

t_2 := b and t_1

t_3 := a or t_2

84. Translate the conditional statement if *a<b then 1 else 0* into three address code.

Ans. Three-address sequence is

```
100: if a < b goto 103
101: t := 0
102: goto 104
103: t := 1
104:
```

85. Explain the following functions:

(*i*) makelist(*i*) (*ii*) merge(p_1, p_2) (*iii*) backpatch(*p*, *i*)

Ans. (*i*) makelist(*i*) creates a new list containing only I, an index into the array of quadruples; makelist returns a pointer to the list it has made.

(*ii*) merge(p_1, p_2) concatenates the lists pointed to by p_1 *and* p_2, and returns a pointer to the concatenated list.

(*iii*) backpatch(*p*, *i*) inserts *i* as the target label for each of the statements on the list pointed to by *p*.

86. Define back patching.

Ans. Back patching is the activity of filling up unspecified information of labels using appropriate semantic actions in during the code generation process.

87. What are the methods of representing a syntax tree?

(*i*) Each node is represented as a record with a field for its operator and additional fields for pointers to its children.

(*ii*) Nodes are allocated from an array of records and the index or position of the node serves as the pointer to the node.

88. Give the syntax directed definition for if-else statement.

Ans.

Production	Semantic rule
S → if E then S_1 else S_2	E.true := newlabel; E.false := newlabel; S_1.next := S.next S_2.next := S.next S.code := E.code ‖ gen(E.true ':') ‖ S_1.code ‖ gen('goto' S.next) ‖ gen(E.false ':') ‖ S_2.code

89. What are basic blocks?

Ans. A sequence of consecutive statements which may be entered only at the beginning and when entered are executed in sequence without halt or possibility of branch , are called basic blocks.

90. What is a flow graph?

Ans.
- The basic block and their successor relationships shown by a directed graph is called a flow graph.
- The nodes of a flow graph are the basic blocks.

91. Mention the applications of DAGs.

Ans. (*a*) We can automatically detect common sub expressions.

(*b*) We can determine the statements that compute the values, which could be used outside the block.

(*c*) We can determine which identifiers have their values used in the block.

92. What are the advantages and disadvantages of register allocation and assignments?

Advantages:

(*i*) It simplifies the design of a compiler

Disadvantages:

(*i*) It is applied too strictly.

(*ii*) It uses registers in efficiently. Certain registers may go unused over substantial portions of the code, while unnecessary load and stores are generated.

93. What is meant by virtual machine?

Ans. An intermediate language as a model assembly language, optimized for a non-existent but ideal computer called a virtual machine.

94. Discuss back-end and front end?

Ans. **Back-end :** Intermediate to binary translation is usually done by a separate compilation pass called back end.

Front end : There are several back ends for different target machines, all of which use the same parser and code generator called front end.

95. Define relocatable object module.

Ans. The unpatched binary image is usually called a relocatable object module.

96. What is meant by multiregister operations?

Ans. We can modify our labeling algorithm to handle operations like multiplication, division, or function calls which normally requires more than one register to perform. Hence this operation is called multiregister operations.

97. What is meant by peephole optimization?

Ans. Peephole optimization is a technique used in many compliers, in connection with the optimization of either intermediate or object code. It is really an attempt to overcome the difficulties encountered in syntax directed generation of code.

98. List the types of addressing modes.

Ans.
1. Intermediate mode
2. Direct mode
3. Indirect mode
4. Effective address mode

99. What is input to code generator?

Ans. The input to code generator consists of the intermediate representation of the source program produced by the front end together with information in the symbol table that is used to determine the run time addresses of the data objects denoted by the names in the intermediate representation.

100. How the use of registers is subdivided into 2 sub-problems?

(*a*) During register allocation we select the set of variables that will reside in registers at a point in the program.

(*b*) During a subsequent register assignment phase, we pick the specific register that a variable will reside in.

101. How would you calculate the cost of an instruction?

(*a*) The cost of an instruction to be one plus the costs associated with the source and destination address modes. This cost corresponds to the length of the instruction.

(*b*) Address modes involving registers have cost zero, while those with a memory location or literal in them have cost one.

102. What are the primary structure preserving transformations on basic blocks?

Ans. (*a*) Common sub-expression elimination

(*b*) Dead-code elimination

(*c*) Renaming of temporary variable

(*d*) Interchange of 2 independent adjacent statements.

103. Give some examples for 3 address statements.

Ans. (*a*) Call (*b*) Return

(*c*) Halt (*d*) Action

104. What are the characteristics of peephole optimization?

Ans. (*a*) Redundant –instruction elimination

(*b*) Flow-of control optimizations

(*c*) Algebraic simplifications

(*d*) Use of machine idioms

105. What is a recursive procedure?

Ans. A procedure is recursive a new activation can begin before an earlier activation of the same procedure has ended.

106. What are the common methods for associating actual and formal parameters?

Ans. (*a*) Call-by-value (*b*) Call-by-reference

(*c*) Copy-restore (*d*) Call-by-name

(*e*) Macro-expansion

107. Define DAG.

Ans. A DAG for a basic block is a directed acyclic graph with the following labels on nodes:

1. Leaves are labeled by unique identifiers, either variable names or constants.
2. Interior nodes are labeled by an operator symbol.
3. Nodes are also optionally given a sequence of identifiers for labels.

108. What are the issues in the design of code generators?

Ans. 1. Input to the code generator 2. Target programs

3. Memory management 4. Instruction selection

5. Register allocation 6. Choice of evaluation order

7. Approaches to code generation

109. What are the various forms of target programs?

Ans. 1. Absolute machine language

2. Relocatable machine language

3. Assembly language

110. What is memory management?

Ans. Mapping names in the source program to addresses of data object in run time memory done comparatively by the front end and the code generator is called memory management

111. What is backpatching?

Ans. Process of leaving a blank slot for missing information and fill in the slot when the information becomes available is known as backpatching.

112. What are the rules to determine the leaders of basic blocks?

Ans. 1. The first statement is a leader

2. Any statement that is the target of a conditional or unconditional goto is a leader

3. Any statement that immediately follows a goto or conditional goto statement is a leader.

113. What is the use of algebraic transformation?

Ans. Algebraic transformation can be used to change the set of expressions computed by a basic blocks into an algebraically equivalent set.

114. What is meant by loop?

Ans. A loop is a collection of nodes in a flow graph such that

1. All nodes in the collection are strongly connected *i.e.*, from any node in the loop to any other, there is a path of length one or more, wholly within the loop.

2. The collection of nodes has a unique entry, *i.e.*, a node in the loop such that the only way to reach a node of the loop from a node outside the loop is to first go through the entry.

115. What is register descriptor and address descriptor?

Ans.
- A register descriptor keeps track of what is currently in each register.
- An address descriptor keeps track of the location where the current value of the name can be found at run time.

116. What are the characteristics of peephole optimization?

Ans. 1. Redundant – instruction elimination

2. Flow – of – control optimizations

3. Algebraic simplifications

4. Use of machine idioms

117. How the quality of object program is measured?

Ans. The quality of an object program is measured by its Size or its running time. For large computation running time is particularly important. For small computations size may be as important or even more.

118. What is the more accurate term for code optimization?

Ans. The more accurate term for code optimization would be "code improvement"

119. Explain the principle sources of optimization.

Ans. Code optimization techniques are generally applied after syntax analysis, usually both before and during code generation. The techniques consist of detecting patterns in the program and replacing these patterns by equivalent and more efficient constructs.

120. What are the patterns used for code optimization?

Ans. The patterns may be local or global and replacement strategy may be a machine dependent or independent

121. What are the 3 areas of code optimization?

Ans.
- Local optimization
- Loop optimization
- Data flow analysis

122. Define local optimization.

Ans. The optimization performed within a block of code is called a local optimization.

123. Define constant folding.

Ans. Deducing at compile time that the value of an expression is a constant and using the constant instead is known as constant folding.

124. What do you mean by inner loops?

Ans. The most heavily traveled parts of a program, the inner loops, are an obvious target for optimization. Typical loop optimizations are the removal of loop invariant computations and the elimination of induction variables.

125. What is code motion?

Ans. Code motion is an important modification that decreases the amount of code in a loop.

126. What are the properties of optimizing compilers?

Ans. (*a*) Transformation must preserve the meaning of programs.

(*b*) Transformation must, on the average, speed up the programs by a measurable amount.

(*c*) A Transformation must be worth the effort.

127. Give the block diagram of organization of code optimizer.

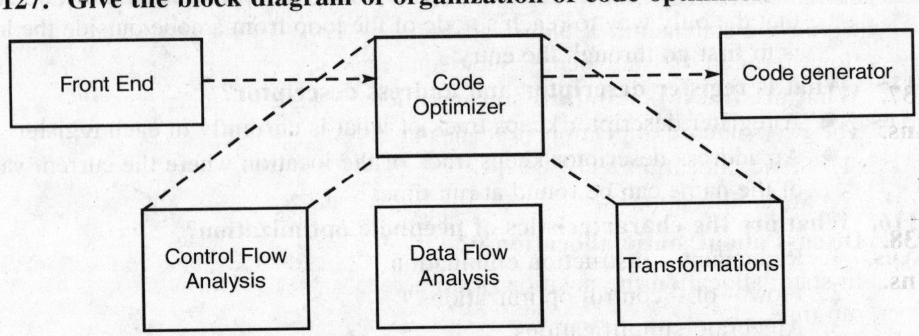

128. What are the advantages of the organization of code optimizer?

Ans. (*a*) The operations needed to implement high level constructs are made explicit in the intermediate code, so it is possible to optimize them.

(*b*) The intermediate code can be independent of the target machine, so the optimizer does not have to change much if the code generator is replaced by one for a different machine

129. Define Local transformation & Global Transformation.

Ans. A transformation of a program is called Local, if it can be performed by looking only at the statements in a basic block otherwise it is called global.

130. Give examples for function preserving transformations.

(*a*) Common subexpression elimination

(*b*) Copy propagation

(*c*) Dead – code elimination

(*d*) Constant folding

131. What is meant by Common Subexpressions?

Ans. An occurrence of an expression E is called a common subexpression, if E was previously computed, and the values of variables in E have not changed since the previous computation.

132. What is meant by Dead Code?

Ans. A variable is live at a point in a program if its value can be used subsequently otherwise, it is dead at that point. The statement that computes values that never get used is known Dead code or useless code.

133. What are the techniques used for loop optimization?

(*i*) Code motion

(*ii*) Induction variable elimination

(*iii*) Reduction in strength

134. What is meant by Reduction in strength?

Ans. Reduction in strength is the one which replaces an expensive operation by a cheaper one such as a multiplication by an addition.

135. What is meant by loop invariant computation?

Ans. An expression that yields the same result independent of the number of times the loop is executed is known as loop invariant computation.

136. Define data flow equations.

Ans. A typical equation has the form

```
Out[S] = gen[S] U (In[S] - kill[S])
```

and can be read as, "the information at the end of a statement is either generated within the statement, or enters at the beginning and is not killed as control flows through the statement". Such equations are called data flow equations.

137. What are the two standard storage allocation strategies?

Ans. The two standard allocation strategies are:

1. Static allocation.
2. Stack allocation

138. Discuss about static allocation.

Ans. In static allocation the position of an activation record in memory is fixed at run time.

139. Write short notes on activation tree.

Ans. (*a*) A tree which depicts the way of control enters and leaves activations.

(*b*) In an activation tree

(*i*) Each node represents an activation of an procedure

(*ii*) The root represents the activation of the main program.

(*iii*) The node for *a* is the parent of the node for *b*, if and only if control flows from activation *a* to *b*.

(*iv*) Node for *a* is to the left of the node for *b*, if and only if the lifetime of *a* occurs before the lifetime of *b*.

140. Define control stack.

Ans. A stack which is used to keep track of live procedure actions is known as control stack.

141. Define heap.

Ans. A separate area of run-time memory which holds all other information is called a heap.

142. Give the structure of general activation record

Ans.

Returned Value
Actual Parameters
Optional Control Link
Optional Access Link
Saved Machine Status
Local Data
Temporaries

143. Discuss about stack allocation.

Ans. In stack allocation a new activation record is pushed on to the stack for each execution of a procedure. The record is popped when the activation ends.

144. What are the 2 approaches to implement dynamic scope?

(*a*) Deep access

(*b*) Shallow access

145. What is padding?

Ans. Space left unused due to alignment consideration is referred to as padding.

146. What are the 3 areas used by storage allocation strategies?

Ans. (*a*) Static allocation

(*b*) Stack allocation

(*c*) Heap allocation

147. What are the limitations of using static allocation?

Ans. (*a*) The size of a data object and constraints on its position in memory must be known at compile time.

(*b*) Recursive procedure are restricted, because all activations of a procedure use the same bindings for local name.

(*c*) Data structures cannot be created dynamically since there is no mechanism for storage allocation at run time.

148. Define calling sequence and return sequence.

Ans. (*a*) A call sequence allocates an activation record and enters information into its fields.

(*b*) A return sequence restores the state of the machine so that calling procedure can continue execution.

149. When dangling reference occurs?

Ans. (*a*) A dangling reference occurs when there is storage that has been deallocated.

(*b*) It is logical error to use dangling references, since the value of deallocated storage is undefined according to the semantics of most languages.

150. Define static scope rule and dynamic rule.

Ans. (*a*) Lexical or static scope rule determines the declaration that applies to a name by a examining the program text alone.

(*b*) Dynamic scope rule determines the declaration applicable to name at runtime, by considering the current activations.

151. What is block? Give its syntax.

Ans. (*a*) A block is a statement containing its own data declaration.

(*b*) **Syntax:**

```
{
    Declaration statements
}
```

152. What is access link?

Ans. An access link is a pointer to each activation record which obtains a direct implementation of lexical scope for nested procedure.

153. What is known as environment and state?

Ans. (*a*) The term environment refers to a function that maps a name to a storage location.

(*b*) The term state refers to a function that maps a storage location to the value held there.

154. How the run-time memory is sub-divided?

Ans. (*a*) Generated target code

(*b*) Data objects

(*c*) A counterpart of the control stack to keep track of procedure activations.

❑❑❑

EXAMINATION PAPERS

B.TECH.

SEVENTH SEMESTER EXAMINATION, MAY 2008

COMPILER DESIGN

Time : 3 hours *Total Marks : 100*

1. (*a*) What do you mean by a translator? Explain any three categories of translators.

(*b*) What do you mean by cross compilation? Explain with suitable diagram.

2. (*a*) Write a SCANNER for FORTRAN that uses look ahead to determine the next token. Use an input buffer.

(*b*) What sets to strings do the following R. E.'s describe.

(*i*) {*a b*} {*b*/*a*}

(*ii*) Digit {digit}/∈ (0/2/4/6/8) where digit represents/0/1/....9.

3. What do you mean by ambiguous grammar? State whether the following grammar is ambiguous or not? If yes then remove the ambiguity.

E → E + E / E – E / E * E / E / E/

E → E / (E) / - E / id

4. Following is an LL(1) grammar for R. E.'s over alphabet {*a*, *b*}.

E → TE′, E′ → +E/∈

T → FT′, T′ → +T/∈

F → PF′, F' → +F′/∈

P → (E)/*a*/*b*/∈

(*a*) Compute FIRST and Follow for each non-terminal of the grammar.

(*b*) Show that the grammar is LL(1).

(*c*) Construct predictive parsing table for grammar.

5. State and explain the syntax directed translation scheme for the desic calculator? Give the parse tree and translations for the expression (4*7 + 19)*2.

6. (*a*) What are the contents of a symbol table? Give a brief description for each.

(*b*) How symbol table space can be re-used. Give some example.

7. Write the algorithm for construction of SLR parsing table. Construct the SLR parsing table for the grammar.

E → E+T/T

T → T*F/F

F → (E)/id

8. (*a*) What do you mean by DAG? What are its applications?

(*b*) What do you mean by instruction variables how they are eliminated?

B.TECH.

SEVENTH SEMESTER EXAMINATION, DECEMBER 2008

THEORY OF AUTOMATA COMPUTATION

Time : 3 hours *Total Marks : 100*

1. (*a*) Describe the structure of compiler with its phases with the help of an example.

(*b*) Which translator is better single pass or multi pass and why.

2. (*a*) Define following term Lex, Lexical analyzer and token. Describe the role of a lexical analyzer.

(*b*) Convert the regular expression (a+b)* abb into ∈-NFA and then corresponding DFA.

3. (*a*) What is a parser? Write the predictive passing algorithm with example.

(*b*) What is left recursion and left factoring? How these can be removed?

4. (*a*) What is ambiguity? How a ambiguous grammar is converted to unambiguous grammar? Explain with the help of an example.

(*b*) What is handle pruning? In which passer it is used and how.

5. (*a*) Write the 3-address code, quadruple, triple and indirect triple for the expression :

$$-(a + b) * ((-d) * e)$$

(*b*) How syntax directed translation is implemented?

6. (*a*) What is a symbol table? Explain the use of hash for the symbol table.

(*b*) What are semantic errors? Explain with the help of an example.

7. (*a*) What is code optimization? What are its various type and stages? Explain in detail.

(*b*) Describe the concept of register allocation.

8. Write short note on the following:

(*a*) Labeling algorithm

(*b*) Global data flow analysis

(*c*) Back patching

(*d*) Canonical LR Parser.

SOLVED EXAMINATION PAPER

B.E.

SEVENTH SEMESTER EXAMINATION, MAY 2010

COMPILER DESIGN

Time : 3 hours *Total Marks : 100*

1. (*a*) What is translator? Why we need translator?

(*b*) Explain Bootstrapping?

(*c*) What is a cross-compiler? Explain in detail.

Answer:

1. (*a*) See Section 1.4 for the answer.

(*b*) See Section 1.10.2 for the answer.

(c) Cross Compiler.

A cross-compiler is a compiler that runs on one machine and produces object code for another machine. The cross-compiler is used to implement the compiler, which is characterized by three languages:

1. The source language,
2. The object language, and
3. The language in which it is written.

If a compiler has been implemented in its own language, then this arrangement is called a "bootstrap" arrangement. The implementation of a compiler in its own language can be done as follows.

Working of Cross Compiler : Suppose we have a new language, *L*, that we want to make available on machines *A* and *B*. As a first step, we can write a small compiler: ${}^{S}C_{A}{}^{A}$, which will translate an *S* subset of *L* to the object code for machine *A*, written in a language available on *A*. We then write a compiler ${}^{S}C_{S}{}^{A}$, which is compiled in language *L* and generates object code written in an *S* subset of *L* for machine *A*. But this will not be able to execute unless and until it is translated by ${}^{S}C_{A}{}^{A}$; therefore, ${}^{S}C_{S}{}^{A}$ is an input into ${}^{S}C_{A}{}^{A}$, as shown below, producing a compiler for *L* that will run on machine *A* and self-generate code for machine *A*: ${}^{S}C_{A}{}^{A}$.

$$ {}^{S}C_{S}{}^{A} \rightarrow {}^{S}C_{A}{}^{A} \rightarrow {}^{L}C_{A}{}^{A} $$

Now, if we want to produce another compiler to run on and produce code for machine *B*, the compiler can be written, itself, in *L* and made available on machine *B* by using the following steps :

$$ {}^{L}C_{L}{}^{B} \rightarrow {}^{L}C_{A}{}^{A} \rightarrow {}^{L}C_{A}{}^{B} $$

$$ {}^{L}C_{L}{}^{B} \rightarrow {}^{L}C_{A}{}^{B} \rightarrow {}^{L}C_{B}{}^{B} $$

2. (*a*) Convert the regular expression $(a + b)^*\ aba\ (a + b)^*$ into the corresponding minimized finite state automata.

(*b*) What is lexical analyzer? How it is designed?

Answer:

2. (*a*) Given regular expression is $(a + b)^*\ aba(a + b)^*$. The FA for this given regular expression will be as followed :

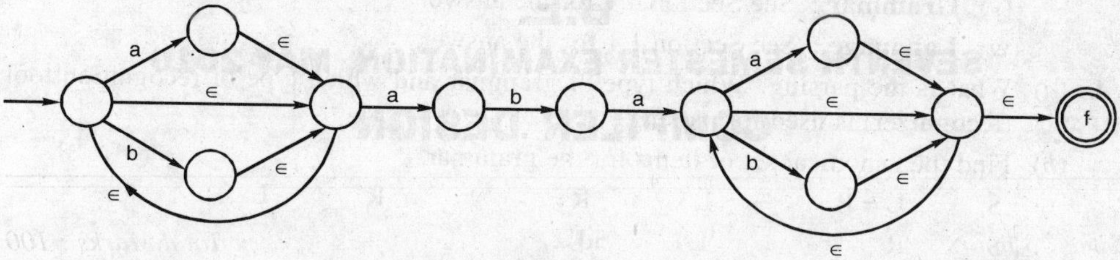

Fig. 1.

(*b*) See Section 3.1 and 3.4 for the answer.

3. (*a*) What are various parsing techniques? Explain the top down parsing with example.

(*b*) Explain the following terms: Syntax tree, right most derivation, terminal, non-terminal, grammar, language, handle.

Answer:

3. (*a*) See Section 4.10 and 5.1 for the answer.

(*b*) (*i*) **Syntax tree:** See Section 7.5 for the answer

(*ii*) **Right most derivation:** See Section 4.3 for the answer

(*iii*) **Terminal :** A **Terminal Symbol** is a Symbol (from a Terminal Symbol Set) that can be used to generate a Formal Sentence.

Terminal symbols, or just terminals, are the elementary symbols of the language defined by a formal grammar. That is, they are literal strings that can appear in the inputs to or outputs from the production rules of a formal grammar and that cannot be broken down into "smaller" units. To be precise, terminal symbols cannot be changed using the rules of the grammar; that is, they are the end of the line, or terminal, as far as the production rules are concerned.

For example, if the grammar rules are that *X* can become *Xa* and *X* can become *aX*, then a is a terminal symbol because it cannot become something else (but *X* can, so it is nonterminal). A formal language defined (or generated) by a particular grammar is the set of strings that can be produced by the grammar and that consist only of terminal symbols; nonterminals that do not consist entirely of terminals may not appear in the lexemes that are said to belong to the language.

In the context of syntax analysis, as opposed to the theory of programming languages and compilers, the terms "terminal symbol" and "token" are often treated as synonymous.

(*iv*) **Non Terminal :** Nonterminal symbols, or just nonterminals, are the symbols which can be replaced; thus they are composed of some combination of terminal and nonterminal symbols. They may also be called simply syntactic variables. A formal grammar includes a start symbol, a designated member of the set of nonterminals from which all the strings in the language may be derived by successive applications of the production rules. In fact, the language defined by a grammar is precisely the set of terminal strings that can be so derived.

Context-free grammars are those grammars in which the left-hand side of each production rule consists of only a single nonterminal symbol. This restriction is non-trivial; not all languages can be generated by context-free grammars. Those that can are called context-free languages. These are exactly the languages that can be recognized by a non-deterministic pushdown automaton. Context-free languages are the theoretical basis for the syntax of most programming languages.

(*v*) **Grammar :** See Section 4.1 for the answer

(*vi*) **Language :** See Section 1.1 for the answer.

4. (*a*) What is the parsing? Which type of grammar and which type of recognition tool (recognizer) is used for parsing ?

(*b*) Find the canonical set of items for the grammar.

S	L = R	L	*R	R	L
S	R	L	id		

also construct SLR parsing table.

Answer:

(*a*) The basic connection between a sentence and the grammar it derives from is the parse tre, which describes how the grammar was used to produce the sentence. For the reconstruction of this connection we ned a parsing technique.

"The Parsing is the process of determining if a string of tokens can be generated by a grammar", we cannot always construct a parser for a Type 0 grammar, but for a Type 1 grammar we always can. All known parsing algorithms for Type 0, Type 1 have exponential time dependency.

A recognizer will tell whether the string belongs to the language defined by the grammar. A parser will try to build a tree corresponding to the string, according to the rules of the grammar.

A context-free language can be recognized in $O(n^3)$ time (*see* Big O notation) by an algorithm such as Earley's algorithm. That is, for every context-free language, a machine can be built that takes a string as input and determines in $O(n^3)$ time whether the string is a member of the language, where n is the length of the string. One way of building a recognizer from a grammar is called recursive descent.

All languages generated by a **regular grammar** can be recognized in linear time by a finite state machine. Although, in practice, regular grammars are commonly expressed using regular expressions, some forms of regular expression used in practice do not strictly generate the regular languages and do not show linear recognition performance due to those deviations.

(*b*) See Section 6.5.4 for the answer

5. (*a*) Write the triple, quadruple, 3-addrus code and indirect triple for the expression.

$$(a - b)^* (c + d) - e$$

(*b*) Explain the construction of syntax tree with the help of an example.

Answer:

(*a*) The given expression is $(a - b) * (c + d) - e$

$$A = (a - b) * (c + d) - e$$

(*i*) **Three-address code is :**

$$T_1: = (a - b)$$
$$T_2: = (c + d)$$
$$T_3: = e$$
$$T_4: = T_1 * T_2$$
$$T_5: = T_4 - T_3$$
$$A: = T_5$$

(*ii*) **Quardrople is :**

Operator (op)	Source 1 (Src-1)	Source 2 (Src-2)	Destination (Dest)
–	a	b	T_1
+	c	d	T_2
=	e	–	T_3
*	T_1	T_2	T_4
–	T_4	T_3	T_5
=	T_5	–	A

(*iii*) **The Triple is :**

	Op	Src-1	Src-2
(0)	–	a	b
(1)	+	c	d
(2)	=	e	–
(3)	*	(0)	(1)
(4)	–	(3)	(2)
(5)	=	A	(4)

(*iv*) **The Indirect Triple is :**

Statement	
(0)	(14)
(1)	(15)
(2)	(16)
(3)	(17)
(4)	(18)
(5)	(19)

	Op	Src-1	Src-2
(14)	–	a	b
(15)	+	c	d
(16)	=	e	–
(17)	*	(14)	(15)
(18)	–	(16)	(17)
(19)	=	A	—

(*b*) See Section 7.5.2 for the answer in addition of following text:

The construction of a syntax tree for an expression is similar to the translation of the expression into postfix form. We construct subtrees for the sub expressions by creating a node for each operator and operand. The children of an operator node are the roots of the nodes representing the subexpresions constituting the operands of that operator.

Each node in a syntax tree can be implemented as a record with several fields. In the node for an operator, one field identifies the operator and the remaining fields contain pointers to the nodes for the operands. The operator is often called the *label* of the node. When used for translation, the nodes in a syntax tree may have additional fields to hold the values of attributes attached to the node. Usually there are a number of functions defined to create the nodes of syntax trees. Each function returns a pointer to a newly created node.

Consider for example the expression $a - 4 + c$. Here, we make use of the following functions to create the nodes of syntax trees for expressions with binary operators.

(*a*) mknode(*op* , *left* , *right*) creates an operator node with label *op* and two fields containing pointers to *left* and *right*.

(*b*) mkleaf (id, *entry*) creates a identifier node with label id and a field containing *entry*, a pointer to the symbol-table entry for the identifier.

(*c*) mkleaf(num, *val*) creates a number node with label num and a field containing *val*, the value of the number.

The following sequence of function calls creates the syntax tree for the expression a– 4 + c.In this sequence, *p1* , *p2* , *p3* , *p4* , *p5* are pointers to nodes, and *entrya* and *entyrc* are pointers to the symbol-table entries for identifiers *a* and *c*,respectively.

(*i*) p1 := mkleaf(id , *entrya*);

(*ii*) p2 := mkleaf(num , 4);

(*iii*) p3 := mknode(' - ', *p1* , *p2*);

(*iv*) p4 := mkleaf(id , *entryc*);

(*v*) p5 := mknode('+' , *p3* , *p4*);

The tree is constructed bottom up.The function calls mkleaf(id,*entrya*) and mkleaf(num, 4) construct the leaves for a and 4;the pointers to these nodes are saved using *p1* and *p2*. The call mknode (' - ', *p1*, *p2*) then constructs the interior node with the leaves for a and 4 as children. After two mor steps, *p5* is left pointing to the root.

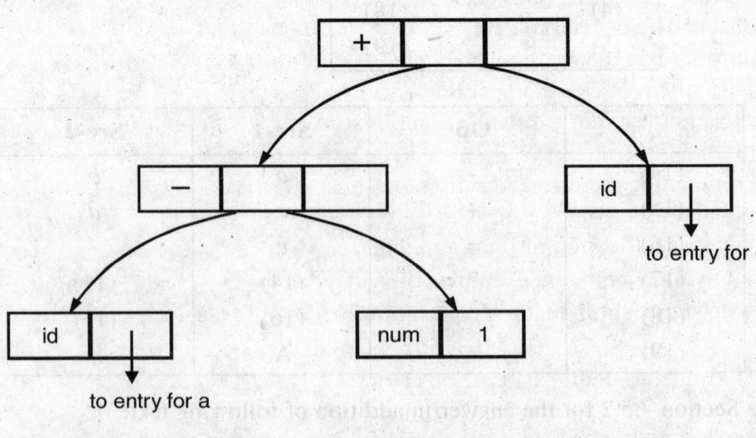

6. (*a*) Give linked list implementation of symbol table with an example.

(*b*) What is an error? Explain various syntactic phase errors in detail.

Answer:

(*a*) A *symbol table* A *symbol table* is a data structure that associates a *value* with a *key*. The simplest and easiest to implement data structure for symbol table is a linear list of records. We use single array or collection of several arrays for this purpose to store name and their associated information. Now names are added to end of array.

Complexity : If any symbol table has 'n' names then for inserting any new name we must search list 'n' times in worst case. So cost of searching is $O(n)$ and if we want to insert 'n' name then cost of this insert is $O(n\char94 2)$ in worst case.

To reduce the time of searching we can add an addition field 'linker' to each record field or each array index. When a name is inserted then it will insert at 'space' and manage all linkers to other existing name.

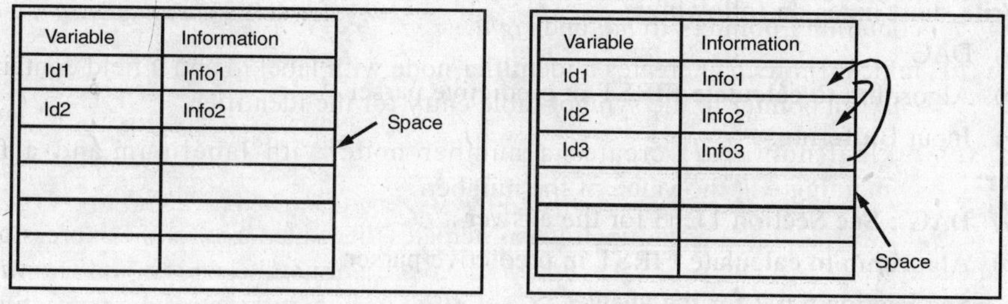

(*a*) (*b*)

Fig. 2

In above Fig. 2(*a*) represent the simple list and Fig. 2(*b*) represent self organizing list in which Id1 is related to Id2 and Id3 is related to Id1

(*b*) A software error is present for when the program does not do what its end user reasonability expects to do. In programming, a development error is one that can be prevented. Such an error can occur in syntax or logic. Syntax errors, which are typographical mistakes or improper use of special characters, are handled by rigorous proofreading. Logic errors, also called bugs, occur when executed code does not produce the expected or desired result. Logic errors are best handled by meticulous program debugging. This can be an ongoing process that involves, in addition to the traditional debugging routine, beta testing prior to official release and customer feedback after official release.

A run-time error takes place during the execution of a program, and usually happens because of adverse system parameters or invalid input data. An example is the lack of sufficient memory to run an application or a memory conflict with another program.

Often much of the error detection and recovery in a compiler is centered around syntax analysis. One reason is the high degree of precision we can achieve in the syntactic specification of programming languages using context free grammars. Here are a few common examples of syntactic errors:

(*i*) Missing right parenthesis (Deletion Error)

MIN(A, 2 * (3 + B)

(*ii*) Extraneous comma (Insertion error)

DO 10, I = 1, 100

(*iii*) Colon in place of semicolon (Replacement error)

I=1:

J=2;

(*iv*) Misspelled Keyword : (Transposition error)

F: PORCEDURE OPTIONS (MAIN)

v) Extra blank (Insertion error)

/* COMMENT * /

See Section 11.1 and 11.3 for answer.

7. (*a*) What is data flow analysis?

(*b*) What are induction variables? How these are removed?

Answer:

(*a*) See Section 12.4 for answer.

(*b*) See Section 12.5 for answer.

8. Write short notes on following:

(*a*) DAG

(*b*) Algorithm to calculate FIRST in predictive parser

(*c*) Input Buffering.

Answer:

(*a*) **DAG :** See Section 11.13 for the answer

(*b*) Algorithm to calculate FIRST in predictive parser

See Section 5.6.2 for the answer.

(*c*) **Input Buffering :** The lexical analyzer scans the characters of the source program one a t a time to discover tokens. Often, however, many characters beyond the next token many have to be examined before the next token itself can be determined. For this and other reasons, it is desirable for the lexical analyzer to read its input from an input buffer. Figure shows a buffer divided into two halves of, say 100 characters each. One pointer marks the beginning of the token being discovered. A look ahead pointer scans ahead of the beginning point, until the token is discovered .we view the position of each pointer as being between the character last read and the character next to be read. In practice each buffering scheme adopts one convention either a pointer is at the symbol last read or the symbol it is ready to read.

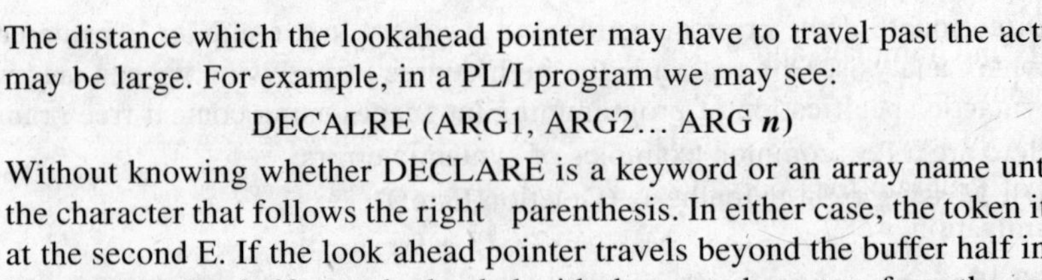

The distance which the lookahead pointer may have to travel past the actual token may be large. For example, in a PL/I program we may see:

DECALRE (ARG1, ARG2... ARG ***n***)

Without knowing whether DECLARE is a keyword or an array name until we see the character that follows the right parenthesis. In either case, the token itself ends at the second E. If the look ahead pointer travels beyond the buffer half in which it began, the other half must be loaded with the next characters from the source file.

Since the buffer shown in above figure is of limited size there is an implied constraint on how much look ahead can be used before the next token is discovered. In the above example, if the look ahead traveled to the left half and all the way through the left half to the middle, we could not reload the right half, because we would lose characters that had not yet been grouped into tokens. While we can make the buffer larger if we chose or use another buffering scheme, we cannot ignore the fact that overhead is limited.

❑❑❑

INDEX

A

A Comparison of Predictive Parser with Shift-Reduce Parsers 208
A Machine Model And Code Generation 307
A Simple Code Generator 308
A Syntax Directed Definition For Constructing Syntax Trees 224
Acceptance of A string by NFA with ∈-transitions 59
Accessing Array Elements 280
Advantages and Disadvantages for Both Single and Multipass Compilers 26
Advantages and Disadvantages of LL(1) Parsing 154
Advantages and Disadvantages of Operator Precedence Parsing 177
Algebraic Laws for Regular Expressions 77
Algebraic Properties 319
Algorithm to Calculate the Follow (X) for a Non-terminal X 150
Alphabets 38
Ambiguity in Grammars and Languages 125
Ambiguity 139
Analysis of Program for Loop Optimization 289
Analysis 14
Annotated Parse Tree 223
Assembler 10
Associativity and Commutativity 78
Attribute Grammars 219
Augmented Grammar 181

B

Back Patching 275
Backtracking a Problem of Top-down Parsing 137
Basic Definition of Attributes and Semantic Rules 215
Bootstrapping (Compiling Compilers) 28
Bottom-Up Evaluation of S-Attributed Definitions 233
Bottom-Up Parsing Techniques 162
Building Regular Expression 68
Byte Code 268

C

Categories and Goals of Error Handling 251
Chomsky Hierarchy 99
Choosing an Intermediate Languages 262
Classic BNF Notation for Productions 99
Coalescing 314
Code Generation for Assignment 275
Code generation for Conditions in Single Pass 277
Code Generation For Conditions 273
Code Generation for Function Calls 279
Code Generation 22
— for the Loops 276
— From DAG 310
Code Optimization 21
— and Code Generation 288
Comparative study of regular expression, regular sets and finite automata 74
Compiled Languages 12

Compilers 10
— Requirements 14
Concatenation of Languages 40
Conflicts in Shift-Reduce Parsing 208
Constructing M-table/Predictive Parsing Table 151
Constructing the LALR Parsing Table 205
Constructing the SLR Parsing Tables (or Tables for LR(0) Parsers) 181
Construction of FA for Regular Expression 74
Construction of LR(1) Parsing Table 197
Construction of Syntax Tree 222
Context Sensitive Grammar 102
Context-Free Grammars 102
Contextual Structure 9
Conversion of NFA with ∈-transitions to NFA without ∈-transitions 59
Cycle-free Grammar 124

D

Dag Construction 301
Definition of Regular Expression 68
Derivations and Their Variations 98
Deterministic Finite Automata 41
Developing Some Other Operator-Precedence Relation 171
Difference Between Single Pass Compiler and Multi-Pass Compiler 26
Different Kinds of Three-Address Statement 264
Directed Acyclic Graph for Expression 224
Distributed Law 78

E

Eliminating Use Less Symbols 117
Elimination of Common Subexpression 300
Elimination of Induction Variables 300
Elimination of Local Common Subexpression 301
Error Detection and Recovery 251
Error Handler 23
Error Recovery in LR parsing 255
Error Recovery in Predictive Parsing 254
Error Recovery Strategies 252
Extended Transition Function for NFA 53
Facts in Designing Procedure of FA 44
First and Follow 149

G

Global Data-flow Analysis 295
Grammars 97

H

Handle of a right sentential form 163
High Level Programming Languages 5
High-Level Intermediate Representations 263

I

Identifier 7
Identities for Regular Expression 79
Identities 78
Implementation Issues of Syntax-directed Translation 229
Implementation of Syntax-Directed Translation 235
Implementing Three-Address Instructions as Quadruples 266
Implementing Three-Address Instructions as Triples 267
Indirect Triples Implementation 268
Information Caching 241
Inherent Ambiguity 130
Inherently Ambiguous Context-Free Grammar 130
Inherited Attributes 218
Interference Graph 314
Intermediate Code for Control Flow 269
Intermediate Code Generation for Boolean Expression 272
Intermediate Code Generation for Case Statement 278
Intermediate Code Generation 261
Intermediate Code Generator 21
Intermediate Languages: Design Issues 263
Interpreted Languages 12

Interpreter 10
Introduction to Symbol Table 240
Issues in Lexical Analyzer 88
Issues of CFG for the
  Programming Languages 139

K

Keywords 7

L

LALR Parsing 204
Language Categories 5
Languages 39
Languages in Set Forms 40
L-Attributed Definitions 226
Laws for Closure 79
Left most and Right Most Derivations 104
Left Recursion and its Removal 141
Left-Factoring 143
Left-Factoring–Algorithm 143
LEX 30
LEX and YAAC As Team 35
LEX File Format 30
LEX Predefined Variables 31
LEX Source Program 31
Lexical Analysis 85
Lexical Analyzer (Scanner) 17
Lexical Structure 7
Limitation of SLR Parser 195
Linkers 11
Literals 8
Live Range Splitting 314
LL (1) Grammars 153
Loaders 11
Local Register Allocation 315
Loop Detection 293
Loop Optimization 289
Loop Unrolling 303
LR Grammars 180
LR Parsers 178

M

Machine Architecture
  Currently in Use 304
Major issues of Code Generation 305
Managing Attributes in a LL(1)
  Parser 234
Minimization Algorithm for DFA 63
Minimization of DFA's 62
Modern Approach of Compiler Design 16
Multi-pass Compilers 25

N

Need of Compilers 11
Need of Translators 11
NFA with ∈-transitions 58
Nondeterminism 49
Notations 27

O

One-Pass Compilers 25
Operator Precedence Parsers 168, 173
Operator-Precedence Relation 169
Operator-Precedence Parser 168
Operators 7
Optimization Strategy 289
Other Sources of Ambiguity 140

P

Panic-Mode Error Recovery
  in LL(1) Parsing 254
Parse Tree 104
P-Code 268
Peephole Optimization 23, 319
Phrase Structure and
  Lexical Structure 115
Porting 29
Pre-Headers 295
Procedure For Converting NFA to
  Equivalent DFA 55
Process of Bootstrapping 28
Processing of Strings by DFA 41
Proving Inherent Ambiguity 130

R

Reasons for Decouple the scanner
  from the parser 86
Recovery from Lexical Phase Errors 252
Recovery from Semantic Errors 257

Recovery from Syntax Error 253
Recursive-Decent Parsing 145
Reduce/Reduce Conflict 204
Reducible Flow Graph 293
Reduction of Context Free Grammar 117
Register Allocation Approaches 315
Register Allocation by Graph Colouring 316
Register Allocation 312
Regular Expression 68
Regular Grammars 103
Removal of ∈-productions and Nullable Non-terminal 122
Removal of Unit Production 119
Removing Ambiguity from Grammars 127
Role of Parsers 132

S

S-Attributed Definitions 226
S-Attributed Grammars and L-Attributed Grammars 219
Scope-by-Location 249
Scope-by-Number 248
Scoping 247
Semantic Actions 236
Semantic Analyzer 19
Separators 7
Shift Reduce Parsers 165
Shift/Reduce Conflict 204
Shift-Reduce Parsing 163
Shortcomings of Shift-reduce Parsing 167
Simpler Notations for DFA's 42
Storage-Allocation 22
Strategies in lexical analyzer construction 87
Strings 38
Structure of Syntax Trees 221
Symbol Table Management 240
Symbol Table Manager 20
Symbol Table Organization 244
Syntactic Structure 8
Syntax Analysis 131
Syntax Analysis 97
Syntax Analyzer (Parser) 18
Syntax and Semantics of the Languages 6
Syntax Diagrams 116
Syntax tree 220
Syntax-Directed Definition 217
Syntax-Directed Translation for Intermediate Code Generation 269
Syntax-Directed Translation 213
Synthesis 15
Synthesized Attributes 217

T

The components of a Compiler 14
The Grammar for Array Reference 282
The History of Computer Programming Languages 3
The Idempotent Law 79
The Language of a DFA 43
The Language of NFA 54
The LR Parsing Algorithm 178
The Number of Compiler Passes 24
The String Table 87
The Structure of the Program 6
The Translation Scheme for Addressing Array Elements 282
The Yield of Parse Tree 104
Three-Address Code 264
Top-Down Parsing Techniques 135
Top-down Predictive Parsers 148
Top-down Translation 229
Top-down-parsing 135
Transformation of NFA to DFA 54
Translation Scheme with Inherited Attributes 228
Translation Scheme with Synthesized Attributes 228
Triggering Semantic Actions in a LL(1) Parser 234
Two-pass Vs One Pass Compilation and Syntax Directed Translation 216
Types of Shift-Reduce Parsers 168

U

Unrestricted (Type-0) Grammars 101

Y

YAAC 34
YACC File Format 34

□□□

Win Prizes

ATTENTION READERS

Win Prizes

We request for your frank assessment regarding various aspects of the book as given below :

ELEMENTS OF COMPILER DESIGN

By : ADESH KUMAR PANDEY (MDU)

Please Fill up the Sheet in Neat and Clean Writing

(i) What topics of your syllabus, important for your examination point of view, are not covered in this book ?

....

....

....

....

(ii) What are the chapters wherein the treatment of the subject matter is not systematic or organised or updated ?

....

....

....

....

(iii) Have you come across misprints/mistakes/factual inaccuracies in the book ? Please specify the chapters and the page numbers.

....

....

....

....

(iv) What other books on the same subject have you found/heard better than the present book ? Please specify in terms of price and quality.

....

....

....

....

(v) Further suggestions and comments for the improvement of this book.

....

....

....

....

(PS : If need be please attach a separate sheet to write your views)

OTHER DETAILS

(i) Who recommended you this book ? (Please tick (✓) in the box near the option relevant to you.

☐ Teacher ☐ Friends ☐ Bookseller

....

....

....

....

(ii) Name of the recommending teacher, his designation and address:

....

....

....

....

(iii) Name and address of the bookseller you purchased the book from :

....

....

....

....

(iv) Name and address of your institution (Please mention the University or Board as the case may be)

....

....

....

....

(v) Your name and complete address :

....

....

....

....

(vi) Course you are enrolled in :

....

....

....

....

The best assessment will be awarded each month : The award will be in the form of our publications as decided by the Editorial Board.

Please mail the filled up coupon at your earliest to :

S.K. KATARIA & SONS®

4760-61/23, Ansari Road,

Daryaganj, New Delhi-110002

Phone: +91-11-23243489, 23269324

TeleFax : +91-11-23243489

email : katariabooks@yahoo.com